# Basic Types of Life

Roger W Sanders, Editor

# Basic Types of Life

Published in the United States by Blyth Institute Press in Tulsa, Oklahoma.

Library of Congress Control Number: 2023949454

ISBN: 978-1-944918-24-8

For author inquiries, please send email to info@blythinstitute.org.

For more information, please see www.blythinstitute.org.

Information about The Blyth Institute can be found at www.blythinstitute.org.

# Contents

# Preface

The idea that interspecific hybridization signifies a fundamental biological unit of affinity was recognized by Linnaeus after a lifetime of study (as discussed in Chapter 3 of this book). However, it was not formalized until 1941 in an obscure, personally published book by Frank Marsh (1941)[1]. Although he promoted the concept over the years in several publications, it lay dormant, largely ignored by most readers.

However, toward the end of his career he reiterated the concept once again in a more widely available publication, *Variation and Fixity in Nature*, written to appeal to a broader audience (Marsh 1976). This time his seminal idea caught the attention of a new generation of biologists (Wood & Murray 2003, p. 17) who sought to test and apply Marsh's ideas to real groups of organisms.

Scherer and Hilsberg (1982) were the first biologists to apply Marsh's concept, citing his 1976 book and adopting "Grundtyp," the German for "basic type," to the study of the evolution of the Anatidae (water fowl). Ten years later Siegfried Scherer convened a number of mostly German colleagues to produce similar studies of other groups of organisms. The proceedings were published as the monograph *Typen des Lebens* (Scherer 1993a) in which Marsh's hybridization criterion was explicitly formulated and extended as an empirical taxonomic criterion. Looking back on my own career development, I realized that I had been taught by high school and college biology texts that "different animal species cannot hybridize." *Typen des Lebens* marshalls evidence that interspecific crossability is not as rare as generally thought and demonstrates the usefulness of the basic type concept.

In 1990, Kurt Wise learned of *Typen des Lebens* and recognized the value of the scholarship and insights of basic type biology to a much wider English-speaking audience. Thus, in 2000, he began the effort to have *Typen* translated into English.

Wise enlisted Georg Huber as translator. The English draft went through a minimum of three editorial passes. Kurt Wise smoothed the translation into native-sounding English for the first pass. I took over the role of editor in 2006. My goal was to use my experience in reading scientific German to double-check questionable passages, especially species' common names[2] and botanical and other technical English

---

1    Clausen *et al.* (1945, pp. 64, 65) generated limited popularity for the term comparium, originally coined in 1929. Even though comparium shares similar criteria with Marsh's concept, it is considered to have a different significance, being the upper limit of a hierarchy of classes of reticulating evolution, above which all evolution is exclusively branching.

2    The convention used applies American common names to species established or cultivated in the United States or commonly referred to in American publications and applies British common names to species restricted to European interest.

terms that are not direct translations of the German words.

Because the introductory chapter (Scherer 1993b) of the German edition was published in English, that chapter is not reprinted here but is replaced by an update graciously provided by Sheena Tyler. Where possible, the original authors have updated their own chapters, and Sheena Tyler has appended additional updates in her introductory chapter. Note that two new chapters have been added and one other has been deleted (see chapter one for details).

I would like to thank Kurt Wise, Georg Huber, Stephanie Wood (technical aspects of production), Sheena Tyler, Siegfried Scherer (granting permission to copy and translate), and Reinhard Junker (technical aspects of transfer from Germany) for their assistance in making this project a reality.

## References

Clausen, J., D.D. Keck, and W.M. Hiesey. 1945. Experimental studies on the nature of species II. Plant evolution through amphiploidy and autoploidy with examples from the Madiinae. *Carnegie Insititution of Washington Publication No. 564.*

Marsh, F.L. 1941. *Fundamental Biology.* Self published, Lincoln, NE.

Marsh, F.L. 1976. *Variation and Fixity in Nature.* Pacific Press, Mountain View, CA.

Scherer, S. (ed.). 1993. *Typen des Lebens.* (Studium Integrale) Pascal Verlag, Berlin.

Scherer, S. and T. Hilsberg. 1982. Hybridisierung und Verwandtschaftsgrade innerhalb der Anatidae–eine systematische und evolutionstheoretische Betrachtung. *Journal für Ornithologie* 123: 357-380.

Roger W. Sanders
June 30, 2015

# 1. Basic Types of Life: Introduction

## Sheena Tyler

**Summary**

Do distinct forms exist in nature, or not? The issue of how to analyse and organise the vast diversity of life has been disputed by scientists for thousands of years, down to this day. This is particularly so regarding what defines a species, but higher taxonomic categories are also in contention. These conflicts can be resolved or informed by proposing a systematic category which employs a genetic criterion based on interspecific hybridization. This category, known as the basic type, is defined objectively: two organisms belong to the same basic type if (i) they are able to hybridize or (ii) they have hybridized with the same third organism.

An overview of basic types delineated by this method is described, indicating that this methodology can be applied to both animal and plant kingdoms. Emerging from these data, relationships can indeed be observed as 36 distinct types to date. Findings from further and untapped data suggest that prospectively there may be at least 109 such types. These basic types can be distinguished from their neighbours by a combination of biological continuities and discontinuities, which a truly viable taxonomic system should seek to achieve. The processes of speciation and appearance of characters within these basic types is discussed in relation to the proposed existence of a hidden potential for variation, residing in a genetically complex ancestral population. This category has the advantage of being open to experimental validation. Since the inception of this category in the previous edition of this publication, a growing body of data indeed continues to support the notion that this is a real category found in nature, based on the members within a basic type possessing genetic and developmental mechanisms which are compatible and fundamentally in common.

## 1.1. The Fundamental Problem Of Biology: The Search For Order In Nature

From the dawn of human civilization there has been a search for order in nature, and the forces that generate diversity. This is not just of academic interest. Deciphering body form is arguably the fundamental problem of biology and medicine (Levin 2012). Knowledge of how form is generated (morphogenesis) may be vital in the development of regenerative procedures to correct diseases of degeneration, aging and cancer, and promote wound healing (Vishwakarma *et al.* 2020; Pensotti 2023). Why is it that some forms, such as salamanders, have remarkable regenerative capacities (Gómez and Echeverri 2021)? What is it about these creatures that enable them to regenerate whole limbs, whilst in other species, such as the human, such a capacity is lacking (Aztekin and Storer 2022)? Moreover, to this day, there remain seemingly intractable disputes regarding the nature of the "species" and the processes leading to the generation of new species (Pavlinov 2023). Disputes commonly exist as to how to classify particular groups of animals and plants. If it cannot be classified with certainty, it might be difficult to identify, and this affects biodiversity (often based on species estimates) and species conservation (Agapow *et al.* 2004). If a species fails to be recognized, it is less easy to protect in environmental law (Boudreaux 2020), as also may be the key species it inter-relates with in food webs. In some cases, accurate identification of a species can literally be a matter of life and death. When an unknown pathogen invades a human population or host of agricultural /veterinary importance, the priority is to identify the pathogen. If it is new and unknown to science, what features does it most closely share with other known organisms? This can be vital to discover, in order to bring measures to combat the pathogen. At the outbreak of the Covid19 pandemic, the nature of the virus was unknown. Application of phylogeny and taxonomy methods resulted in the recognition of the virus as a sister to severe acute respiratory syndrome coronaviruses (SARS-CoVs) and enabled it to be designated as severe acute respiratory syndrome coronavirus 2 (SARS-CoV-2) (Gorbalenya *et al.* 2020). As the pandemic continued, genomic surveillance monitored the diversity of SARS-CoV-2 viruses circulating in the community, to facilitate anti-viral strategies that targeted the emerging variants (Islam *et al.* 2021).

Taxonomic knowledge can also be harnessed to find new compounds of pharmacological importance from plants. There can be a clustering of medically therapeutic bioactive compounds between close relatives of plant species (Tyler and Tyler, 2023). Taxonomic knowledge of these relationships can thus be applied in drug discovery to provide leads to even higher concentrations of bioactive compounds (Saslis-Lagoudakis *et al.* 2012; Gras *et al.* 2021).

**1.1.1. The Bauplan.** In this search for order, distinct biological forms can be recognized, which can be distinguished from neighbouring forms (Brady 1985; Gould 1993). The forms may be real, sharing a suite of common essences, or contrived, in which any delineation of groups is viewed to be artificial, with a continuous gradation of intermediate forms interconnecting them (Rieppel 1988).

However, the reality of distinct forms is widely acknowledged (Arthur 1997; Raff 2012) at the higher, bauplan[1] level. Members of a

---

1    German: Bau, type of construction; plan, pattern

bauplan or body plan have in common major structural features and layout of cells, generally observable to the non-professional observer. In addition, it signifies the essence of the architectural range and limits (Gould 1993) and its functional design whereby the components are structurally and functionally compatible (Brusca and Brusca 2003; Brusca et al. 2022). It is often identifiable with either the phylum taxon or the class: for instance, the molluscan or echinoderm bauplan.

**1.1.2. Taxonomic Levels–Problems And Debates.** However, the placing of organisms in lower taxonomic levels can be a matter of debate. For instance, within the echinoderm phylum, for over a hundred years to the present, schemes ranging from 6 to 25 classes have been proposed. This is compounded by problems in defining the categories themselves. For instance, the genus can be regarded as the most definite and basic unit of practical and morphological taxonomy (Simpson 1945), in which the taxonomy specialist develops a 'feeling' of relatedness, but this feeling cannot be strictly defined (Scherer 1993). The family level appears to be the most easily recognisable form at a glance (Mayr et al. 1953), even to lay people, and separated from other families by a decided gap (Mayr 1969). Sibley and Ahlquist (1990) proposed a quantitative method for defining genus and family ranks within birds, according to measures of homology between DNA sequences. However, they calibrated their numbers according to pre-existing taxonomic rankings, which therefore does not add any further objectivity (Scherer 1993).

**1.1.3. The Species Problem.** These controversies extend down the taxonomic levels to the species concept, which continues to be a matter of great dispute and debate (Rieppel 2013; Barker 2022). There are at least two dozen definitions of the term species (Mayden 1997; Ereshefsky 1998) framed according to the research goals. Some of these are indicated in Table 1.

Table 1. Some of the main species concepts and their definitions

| Species concept | Definition | Reference |
|---|---|---|
| morphological | the smallest group that is consistently distinct and distinguishable by ordinary means | Cronquist 1978 |
| biological (BSC) | groups of actually or potentially interbreeding populations reproductively isolated from other groups | Mayr 1940 |
| phylogenetic | the smallest entity that is monophyletic, bound by a unique ancestry | Nixon and Wheeler 1990 |
| genetic | group of organisms transmitting a hereditary character to descendants | Simpson 1943 |
| genotypic | group of genotypes remaining distinct in the face of potential or actual hybridization and gene flow | Coyne and Orr 2004; Mallet 2007 |

| ecological | group of organisms that share a distinct ecological niche or adaptive zone | van Valen 1976 |
| --- | --- | --- |

Traditionally, the morphological species has been of most practical value in the field situation for the demarcation of species. Occasionally some species can be difficult to distinguish in this way, when they appear to be morphologically very similar. Some authors favour a molecular basis for species identification, such as DNA barcode taxonomy (e.g. Hebert 2003) over a morphological one. However, this has been sharply challenged (e.g., Scotland et al. 2003) due to the resulting implausible phylogenies, which can remain even with extensive genomic sampling. Moreover, the uncertain relevance of molecular characters in morphogenetic terms casts doubt on the authority of this treatment, a point to which we will return later. Although Mayr's biological species definition is probably the dominant concept, it too is problematic (reviewed by Stanford 1995; Devitt 2008; Nathan and Cracraft, 2020), leading to an unwieldy number of species being merged due to hybridization. Moreover, it cannot be applied in palaeontology and asexual populations (the prominent type of reproduction worldwide, especially in plants and fungi (Hull 1988). According to Scherer (1993), the biological species concept (BSC) is not as objective as some of its proponents would suggest, with gene flow between "good species" being common (Barton and Hewitt 1985, 1989). For instance, European fire-bellied toads interbreed freely within a 1000 km zone, so according to the BSC could be considered to be just one species, although they differ in morphology, ecology and genetics (Szymura 1993). Reproductive isolation, on which the definition depends, may be induced by other genomic constituents, such as transposable elements (Kidwell and Petersen 1991; Abbott et al. 2013), or even microorganisms in the egg cytoplasm (Breeuwer and Werren 1990; Bordenstein et al. 2001), leading to incompatibility of eggs and sperm cells. In other words, species may not necessarily be separated by absence of gene flow, since there are other isolating mechanisms, in the absence of which the species might be united. Regarding genetically-based traits, the forces of mutation, recombination and random drift can lead to the disappearance of that trait in a future member of the species, indicating that a trait may not be essential in defining the species (Stanford 2010). Also, perplexingly, morphologically highly distinct individuals can appear to be virtually identical genetically (Ford and Gottlieb 1992). The ecological species concept can also be problematic. Groups of organisms (e.g. tree nesters) occupying the same niche, and thus considered the same ecological species, may not be considered so by other taxonomists: for instance, the groups may comprise several morphologically distinct species. Moreover, it can be difficult to objectively define the niche in question.

Chronological definitions such as the Phylogenetic Species Concepts focus on tracing historically lineages of organisms that share a unique and common ancestry. However, other taxa are also lineages, so the concept can be too inclusive (Stanford 1995).

Thus, various species concepts employ distinctive unifying characteristics: for instance, biological species are united by the processes of interbreeding, ecological species are unified by stabilizing selection, and phylogenetic species are unified by historical processes. However, according to Mishler and Donoghue (1982) none of these processes are unique to species but are also found in higher taxa, with no clear conceptual distinction between.

Amongst this variety of species concepts, there is still no consensus on which species concept is best (Odenbaugh 2022). Monists argue that the aim should be to identify the single correct species concept, which should be the prevailing one. In disagreement, pluralists advocate that a number of species definitions are equally legitimate, according to the question being asked. For instance, the population geneticist is interested in gene flow, whilst the morphologist focuses on character groups of outward appearances (phenotypes). Sometimes these are in accord. At other times they are in conflict, such as when morphologically distinct individuals have a virtually identical genetic constitution. The "species problem" is also fraught with philosophical questions (Pigliucci 2003) such as the struggle to find the essence for defining a species. De Queiroz (1999) asserts that all species definitions share the characteristic that they are population-level lineages, but according to Pigliucci (2003) this is not sufficient for being a species. Ereshefsky (1998) considers the distinctions between species and other taxa might be vague, casting doubt on whether the term 'species' refers to a real category in nature. Sites and Marshall (2004) conclude that all of the operational methods of delimiting species give results in conflict with each other, and that virtually all methods require researchers to make qualitative judgments.

## 1.2. Basic Types Of Life - Highlight Themes

### 1.2.1. The Basic Type Category.
A resolution to the above conflicts would be found if there was a taxonomic category exhibiting both morphological and genetic patterns of similarity, which could be empirically determined. Such a pattern may indeed exist (Scherer 1993), derived primarily from observations of interspecific hybridization.

There is a wealth of hybridization data at the interspecific level, notably in fish, birds, mammals and plants. Interspecific hybridization occurs widely across a taxonomically diverse range of fish species (Scribner et al. 2000). Worldwide at least 1 in 10 species of bird hybridizes with other bird species and sometimes even between members of other genera (Grant and Grant 1994). For instance, the mallard duck

(*Anas platyrynchos*) hybridizes with over 60 other species within the family Anatidae (ducks, geese and swans), including 2 species of geese (McCarthy 2006). Similarly, 25% of plant species are known to hybridize with at least one other species (Mallet 2005). This hybridization is also considered to provide a major mechanism (known as hybrid speciation) for generating new species, leading Mallet (2007) to conclude that it would be hard to find another mode of speciation so readily documented historically and so amenable to experimentation, contributing to adaptive radiations such as African cichlid fish and Darwin's finches.

Scherer (1993), following Marsh (1976) defined groups distinguished by interspecific hybridization as "basic types", defined as follows:

> *Two individuals are envisaged to belong to the same basic type if they hybridize with each other, or if they hybridize with a third organism in common.*

They stated that fertility of the hybrids is not employed as a criterion of relatedness, because merely minor changes (such as loss of genetic material), or non-genic factors, can invoke sterility. In contrast, if hybridization is possible, morphogenetic programs must be highly compatible and similar, warranting the inclusion into the same basic type (BT).

This was not intended as a solution to the species problem (which is likely to remain intractable), but to provide an objective means to define a taxonomic rank prospectively based on real morphogenetic divisions of nature (further discussed later). From such data, basic type status was ascribed to the animal and plant groups featured in the original German version of this book, published in 1993.

This new edition comprises translation of the above book chapters into the English language, along with revisions and updates. Additional BTs, which exemplify further the BT concept, are proposed in new Chapters. Kutzelnigg (Chapter 9) features the "living stones" (Ruschiae), succulent plants so-named due to their pebble-like camouflage appearance. Within this tribe there is a striking unity of morphology, molecular similarities and hybridization, with intergeneric hybrids interconnecting most of the groups, and with evidence of a particularly rapid and extensive radiation from a basic ancestral stock. Tyler (Chapter 14) gives a resume of 12 bird groups that provide additional candidates for BT status, (such as the flamingos, cranes, penguins, and the parrot family) along with promising data for up to 35 yet further possible BTs. Strikingly, in some of these bird families, many or most of the species are united by hybridization.

**1.2.2. Genetically Complex Ancestors.** A number of processes are featured which may be sufficient to generate the observed morphological divergence from the ancestral stocks without macroevolutionary mechanisms. For instance, Junker (Chapter 2) gives a very useful review of speciation mechanisms by which the organisms in existence today are descended from ancestral stocks. He presents the little-known hypothesis that the ancestral forms of present-day groups possessed a higher genetic potential, providing a wide range of variation and thus adaptability to all kinds of environmental conditions. This could be the source of the diversification within basic types of organisms. This idea stems from Mayr (1991) who states that the source of speciation "is not mutations but the genetic variability of the original population." This is reiterated by Grant and Grant (1994), who estimate that interspecific hybridization in Darwin's finches provides a more abundant source of adaptive genetic variation than mutation. Similarly, hybridization has provided the source of an ecologically important trait in ragwort-groundsel hybrids, rather than "de novo mutation" (Kim et al. 2008). Abbott et al. (2013) conclude from such evidence that hybridization leads to the "rapid acquisition of a genetic architecture that would be difficult to evolve by sequential accumulation of mutations."

Junker shows how in small pioneer populations, catastrophic selection allows new species to be established rapidly, within just a few generations. He cites an interesting example in the emergence of new plant species that are able to grow on polluted mining deposits, due to their ability to neutralise toxic metals. Such plants become reproductively isolated from the ancestral species. The tolerance genes are not newly evolved, but are present in a small percentage of individuals growing on normal grounds, in which this is not an advantageous trait.

Similarly, a morphologically complex ancestor is proposed to be the derivation for a basic type of mosses (Adler Chapter 4) linked by intergeneric hybridization data. The same is true for the spleenwort fern family (750 species) (Kutzelnigg, Chapter 5). In the Maloideae, Kutzelnigg (Chapter 8) proposes that all present-day forms of this group can be traced back to a common ancestral population, containing a polyvalent genetic amalgam of the characters found today. Different combinations of this ancestral gene pool had the capacity sufficient to generate the subsequent phenotypic variations, without the need for macroevolutionary processes. Indeed, Kutzelnigg shows that this scenario can be explained by plausible mechanisms such as polyploidy. Interestingly, the literature increasingly reiterates this, with estimates that all extant flowering plants have polyploidy in their ancestry (Jiao et al. 2011).

**1.2.3. Ideas And Errors Behind The Word "Species".** Landgren (Chapter 3), in a compelling landmark study, traces the history of the ideas behind the word "species". The ancient Greek meaning of "species" was more akin to a set of essential characteristics, "ideas" or "forms," which were fixed or immutable. However, the Biblical account in Genesis refers to the created category of the "min", similar to "form" or "type" but with no suggestion as to their fixity. The Roman philosopher Cicero then coined the word "species", retaining its original meaning of an immutable "idea" or form. Unfortunately, the Latin Vulgate version of the Bible translated the original Hebrew word "min" as "species". As a result of this, Landgren demonstrates that the Greek view of fixity of species became wrongly imposed on the understanding of Genesis. Darwin too was mistaken in thinking that the "static" fixity of species concept was biblical, when in reality it was derived from Greek views of science. Thus, for Darwin to reject the fixity of species meant to reject his own naïve and flawed understanding of the Biblical account of origins. These errors commonly persist to this day, providing a source of conflict based upon the same mistaken view that to believe in the Genesis account of origins is to admit to fixity of species, when in reality the Biblical texts do not suggest this.

Darwin had claimed too that all the eminent scientists of his day adhered to the notion that species were "fixed" and had not varied from their inception, but Landgren shows that the opposite was the case, notably with Linnaeus. Linnaeus, the greatest taxonomist of all time, revised his earlier view of species fixity to state that from the originally created species and genera new ones could be generated by hybridization.

## 1.3. The Evidence For Basic Types
### 1.3.1. Hybridization Data.
The following chapters provide good and often strong evidence for the existence of basic types. These are distinguished mainly from interspecific hybridization data, and in some cases with additional supporting criteria such as morphological data. Relationships between organisms resulting from hybridization can be visualized by constructing a Venn diagram or a matrix (for example, see Junker Chapter 6, Table 2). These various groups of animals and plants are summarized in Table 2. In these 36 groups, either the number of species hybridizing is high or most/all of the genera are united by hybridization. In the latter case, the actual number of hybridizing species may not necessarily be high, but the BT may be distinguished merely by intergeneric hybrids (see connectivity criteria later: Section V). These often unite morphologically diverse forms, which provides further evidence to confirm BT status. The BTs can also be denoted by

adding the prefix bt to the taxon or taxa whose members they unite (e.g. btAnatidae).

Table 2. Summary of basic types of animals and plants distinguished in this publication (including those in the original German edition) primarily by interspecific hybridization data, with taxonomic location and rank indicated.

A. Animalia

| Taxonomic location | Basic type (BT) | Taxonomic rank | Common name or descriptive feature |
|---|---|---|---|
| Vertebrata: Aves | Anatidae | f | Ducks, Geese, Swans |
| | Anhimidae | f | Screamers |
| | Anseranatidae | f | Magpie goose |
| | Buteoninae [in part] | gg | Sea eagles |
| | Aquila | g | True eagles |
| | Aegypius + Gyps | gg | OW vultures |
| | Accipiter + Buteo + Milvus | gg | Hawks, buzzards + kites |
| | Falconidae/Falco | f/g | Falcons |
| | Cathartidae | f | NW vultures |
| | Sagittaridae | f | Secretary bird |
| | Pandionidae | f | Osprey |
| | Megapodiidae | f | Brush turkeys + scrubfowl |
| | Carduelinae | Sub-f | Cardueline finches |
| | Estrildidae | f | Estrildid finches |
| | Ardeinae | sub-f | Herons, Egrets |
| | Gaviidae | f | Loons and Divers |
| | Gruidae | f | Cranes |
| | Haematopodidae | f | Oyster-catchers |
| | Paradisaeidae | f | Birds of Paradise |
| | Pelicanidae | f | Pelicans |
| | Phasianidae +Numididae +Cracidae | ff | Grouse, Pheasant, Chicken Guineafowl, Curassow |
| | Phoenicopteridae | f | Flamingos |
| | Podicipedidae | f | Grebes |
| | Psittacidae | f/ff | Parrots, Lories, Macaws |
| | Rheidae | f | Rheas |
| | Spheniscidae | f | Penguins |
| | Threskiornithidae | f | Spoonbills, Ibises |

| Vertebrata:<br>Mammalia | Canidae*[1] | f | Dogs, Wolves |
| | Cercopithecidae | f | Old World Monkeys |
| | Equidae | f | Horses, Donkey, Zebra |

B. Plantae:

| Taxonomic location | Basic type (BT) | Taxonomic rank | Common name or descriptive feature |
|---|---|---|---|
| Bryophyta: Musci | Funariaceae | f | Ground Mosses |
| Pteridophyta | Aspleniaceae | f | Spleenwort Ferns |
| Spermatophyta: Aizoaceae | Ruschiae | f | Living Stones/Stone Plants |
| Spermatophyta: Poaceae | Tritieae | t | Wheat, Barley and Rye Grasses |
| Spermatophyta: Rosaceae | Maloideae | sub-f | Apple, Pear, Mountain Ash Trees, Quince, Cotoneaster, Hawthorn |
| | Geeae | t | Avens and allies |

Abbreviations: f, family; ff, assemblage of several families united by hybridization; g, genus; gg, assemblage of genera; NW, new world; OW, old world; sub-f, sub-family; t, tribe. Notes. The table also denotes the nearest taxonomic rank to which the BT is equivalent.

Although the equivalent taxonomic rank can vary according to the treatment by the authors, it can be seen that the basic types often correspond to the family or subfamily taxonomic level in birds or mammals, and between the tribe and family level in plants. A brief resume of the main findings is as follows.

**1.3.1.1 Animals.** Amongst the birds, Scherer (Chapter 10) demonstrates that, extraordinarily, 81 % of all species of ducks, geese and swans (Anatidae) hybridize with one another. Yet there are no known hybrids between the Anatidae and members of any neighbouring taxa (a recurrent theme). Moreover, over 50 % of the hybrids are intergeneric with over 30 % inter-tribal, and 20 % of all known hybrids are fertile. Together this suggests that the Anatidae are genetically related but clearly distinguished from other avian taxa. Fehrer (Chapter 13) concludes that 2 basic types of finches (the cardueline and estrildid finches) can each be distinguished by hybridization data. Similarly, from hybridization data, there may be 3 basic types of upland game birds (corresponding to the pheasant-like, mound-building and curassow families) (Klemm Chapter 11) and a number of diurnal birds of prey basic types such as hawks and buzzards (Zimbelmann Chapter 12).

Within the Equidae (horses, donkeys, asses and zebras), Stein-Cadenbach (Chapter 15) demonstrates that all of the species are connected together by hybridization. Another candidate BT is the Cercopithecoidea

(Old World monkeys). Hartwig-Scherer (Chapter 16) reveals that eight out of the nine genera are connected through hybridization.

**1.3.1.2 Plants.** In the spleenwort fern family, hybridization links the various genera, but with no hybridization extending beyond the spleenwort forms (Kutzelnigg Chapter 5). Similarly, BT status is suggested in the wheat, barley and rye grasses (combining tribes Triticeae and Bromeae) in which 35 out of 38 genera are connected through hybridization (Junker Chapter 6). In the Geeae too (a tribe within the rose family, typified by their hooked burr fruits) extensive hybridization and a distinctive morphology distinguish them from other rose taxa (Junker Chapter 7). Another subfamily within the rose family is the Maloideae, comprising 950 species including the familiar apple and pear trees, in which all of the most common genera are united by hybridization, including ones taxonomically more distant to each other (Kutzelnigg Chapter 8).

**1.3.1.3 Hybridization Data: Selection Criteria.** The hybridization data has been obtained from numerous sources including breeders' and zoo records, museum specimens and field observations. Citations in which the cross was deemed unlikely or worthy of doubt have been discounted. Multiple records add weight to the authenticity of a prospective cross. It can be difficult to establish if the cross was indeed viable, if the embryo developed only incompletely. One solution to this dilemma is to only accept records stating that embryos have hatched, with fully developed, viable individuals. Another solution is to accept data in which the embryo has developed to the stage at which the combined morphogenetic machinery of both parents is operating successfully. However, the nature of such machinery remains unknown, as to how the organism converts its genetic constitution into its outward form, *i.e.* genotype into phenotype. If embryos did not hatch but were fully developed, this cross could be accepted if confirmed with other data in which the offspring were fully viable. In contrast, data describing that "eggs were laid but did not hatch" (as certain records indeed state) are discounted, due to the degree of development not being apparent.

**1.3.2 Atavisms And Hidden Potential For Variation.** Several chapters provide observations of atavism: an outward expression of an ancestral characteristic or form, which is not normally evident. In the Anatidae, for instance, a "bimaculated" face pattern occasionally appears in hybrids between mallards and the domestic duck, although it does not occur in either parent species, even in successive generations (Scherer Chapter 10). Atavisms are also evident in the finches (Fehrer Chapter 13), the Equidae (Stein-Cadenbach Chapter 15) and the Old World monkeys (Hartwig-Scherer Chapter 16).

This atavistic character, once expressed, may be similar to a third species found elsewhere within the basic type, as in certain hybrids of the birds of paradise (Tyler Chapter 14) and the Anatidae. For instance, the face pattern unexpectedly appearing in the above duck hybrid is also part of the normal appearance of several other species within the family, notably the Baikal teal.

A similar observation concerns the expression of a mosaic or patchwork distribution of characters. This can lead to species being hard to place (the so-called aberrant types). For instance, within btMaloideae the genus *Pseudocydonia* is similar to *Cydonia* but also to *Chaenomeles* and *Pyrus*. Thus, *Pseudocydonia* cannot be assigned to either of the other genera (Kutzelnigg Chapter 8). Even fossil forms can display this phenomenon. *Romainvilla*, the oldest duck-like fossil within the order anseriformes, shows similarities to members of anseriform tribes distinct from each other, such as the whistling ducks and swans (Scherer Chapter 10).

One interpretation for these observations is convergence (homoplasy), in which selective forces would lead to the independent, multiple origin of similar characters. An alternative interpretation is that the common ancestor possessed a potential for variation (plesiomorphy). In this view, the genetic potential for this variation was hidden in the ancestral polyvalent gene pool (with a very high degree of polyallelism), which becomes visible when different species hybridize. The emergence of species from such "complex" ancestors would scatter different genetic characters throughout the descendant species, which could in turn generate the mosaic network of morphological features described above. Further evidence for this is provided by "bottlenecks" (populations with minimal variability) which can still acquire new adaptations. This suggests that loss of genetic variability is less than expected, which could be explained by the presence of a high level of polyallelism (Junker Chapter 2). Moreover during hybridization it is becoming increasingly evident that large subsets of genes, including transcriptional regulators, are known to be reprogrammed (reviewed by Abbott *et al.* 2013). Regulatory pathways are restructured in a compensatory fashion within complex networks (Johnson and Porter 2000; Birchler and Veitia, 2010). Transposable elements are activated, which can have profound effects, resulting in structural polymorphisms (He and Dooner 2009), trigger variation in functional genes (Hollister *et al.* 2011) and modify recombination patterns (Ungerer *et al.* 2006). The restructuring of these various genetic mechanisms could be driving the generation of new forms occurring during rapid speciation (Tsiantis and Hay 2003), and provide the mechanisms for expression of an ancestral reservoir of genetic information (see also Part V.2).

**1.3.3. Heterobathmy.** A recurring theme is Heterobathmy: the simultaneous occurrence of original and advanced characteristics. Kutzelnigg demonstrates that within the Maloideae whether traits are considered primitive or advanced varies according to the opinions of authors. He concludes, "However one decides, numerous cases of heterobathmy always result. For instance, if one considers leaf pinnation to be an advanced trait, it is an astonishing fact that the genus *Osteomeles* (with pinnate leaves) possesses five carpels, which is universally regarded as an ancestral trait." However, existence of a polyvalent ancestral population solves the problem of heterobathmy, if traits which some authors assign as primitive, whilst others assign as advanced, (such as "leaves non-pinnate vs. pinnate") both existed in the ancestral population together, side by side. It can also explain the disjunct occurrence of rare characteristics, which would require a high frequency of independent origins of similar structures, which is unlikely.

**1.3.4. Supplementary Criteria.** There are insufficient hybridization data to discern BTs additional to the ones specified in Table 2 unequivocally. However, in conjunction with supplementary criteria to support the incomplete hybridization data, further prospective BTs can be distinguished. Criteria of use in distinguishing BTs include adult and developmental morphology, behaviour, and molecular biochemistry.

**1.3.4.1. Morphology.** Morphological criteria have been largely successful in delineating the majority of plant and animal groups at various taxonomic levels, and controversial groups cannot detract from this fact. For instance, various authors consider that many bird families can be easily discerned due to their unmistakable form (such as the flamingo, pelican, stork, owl and toucan families) and unique skeletal characters. For instance, the hornbills (Bucerotidae) are one of the most recognizable of all birds (Kemp 2001), with their deep, curved great bill surmounted by a unique internally reinforced hollow casque, and uniquely modified vertebrae. Similarly, the pelican form is unmistakable, with its huge, distensible skin pouch and long bill (Elliot 1992). Molecular studies have led to taxonomic revisions in conflict with morphological criteria. For instance, DNA-DNA hybridization data has led to the placement of toucans with barbets. However, according to Höfling (1998), toucans are toucans, rather than large barbets, in possessing a suite of characters unique to toucans. The current treatment of toucans as a separate family (Ramphastidae) is recognized on morphological grounds, according to Short and Horne (2002). Thus, for groups in which hybridization is absent or inadequate, it may still be possible to ascribe BT status to a group based on morphological criteria, in which members of a group are united by a range of characters unique to that group (synapomorphies).

**1.3.4.2. Developmental Biology.** Developmental biology can offer a number of useful additional criteria to provide evidence that the taxa sharing the salient pattern are related in some fundamental way. Examples include:

- sperm ultrastructure has been used, for instance, to differentiate between five families of archaeogastropod molluscs (Haliotidae, Fissurellidae, Trochidae, Turbinidae, and Phasianellidae) (Hodgson and Foster 1992); and between various families of bivalve molluscs (Drozdov *et al.* 2009);
- *egg surface structure is of taxonomic significance in a number of insect orders, e.g. in mayfly families (Koss and Edmunds 1974; Domınguez and Cuezzo 2002); families of dipterans (Margaritis and Mazzini 1998) including sandflies (Jariyapan et al. 2022), even to the degree of being of diagnostic value to identify forensically important blowflies (Calliphoridae) (Mendonça et al., 2008); and in Phylliidae (leaf insects) (Büscher et al. 2023). Egg structure can also provide taxonomically useful information in fish, such as in Bitterlings from genera within Cyprinidae (Choi et al. 2023);*
- *egg capsules of some mollusc groups enable a number of families to be distinguished (Robertson 1974): Strombidae (conchs) capsules consist of long, sand covered tubes of eggs compacted into masses; the Epitoniidae (wentletraps) have clusters of capsules connected by a single thread like a necklace; the capsules of Littorinidae (periwinkles) possess a flat, spherical shape overlaid by concentric tiers; and Naticidae (moon snails) capsules are collar shaped.*

**1.3.4.3. Molecular Biochemistry.** Molecular criteria can also be employed to distinguish BTs. For instance, *Anseranas semipalmata* (Australian magpie goose) is variably placed in the family Anatidae or a separate monotypic family for this single species, according to the author. There are no known hybrids to date. Data from haemoglobin sequencing, along with morphology and behavioural characteristics, clearly separate *Anseranas* from btAnatidae, suggesting that *Anseranas* should indeed be placed separately from btAnatidae (Scherer Chapter 10).

**1.3.4.4. Behaviour.** In the bowerbirds (Ptilonorhynchidae), it is their behaviour that defines them, whereby they build an elaborate stage or "bower", used for courtship and mating (Frith and Frith 2009). The grebes (Podicipedidae) exhibit unique courtship rituals including the rushing ceremony and weed ceremony (Llimona and del Hoyo 1992). In some groups both structural and behavioural characteristics are together distinct: for example, the molluscs within the Eratoidae (sea buttons/false cowries) and Lamellariidae (ear shells) embed flask shaped egg capsules within the body tissues of ascidian hosts (Robertson 1974). Analysis of animal sounds (bioacoustics) can also be of taxonomic value (explored in Section V: areas for future research).

## 1.4. Discussion Of The Basic Type Category
**1.4.1. Advantages.** Scherer (1993) suggested that the BT definition has a number of advantages. First, BTs provide a category whose members share a common morphogenetic pattern, which can have phylogenetic implications. That is to say, it follows logically that this common developmental pattern will have a common connection to historical ancestors also sharing this pattern.

Second, a wealth of interspecific hybridization data is already available, which has rarely been used in classification. Thirdly, if data are missing, the experimental approach of artificial insemination or artificial pollination can be applied, as, for instance, in the crane breeding study described in Chapter 14 ((Maksudov and Panchenko 2002). Fourth, it allows a way previously impossible of how a taxonomic category in one group (say, a BT within birds) can be equivalent to the same category in another group (say a BT within mammals or angiosperms) because each group is demarcated chiefly by the same criterion: their ability to hybridize. Fifth, it allows morphological criteria in defining genera to be retained. Thus the information storage and retrieval capacity of such a classification scheme (*i.e.* species, genus, BT) remains high. Sixth, application of this criterion also retains the binomial system of naming (*i.e.* genus and species), therefore avoiding major changes in nomenclature.

The underlying purpose of the BT category is to provide an objective method, *i.e.* outside the qualitative judgement of the observer, of discerning what may fundamentally unite a group. It is not intended to replace the binomial classification (of genus and species), but can speak into taxonomic issues under dispute, which can be resolved by ascertaining membership of a particular BT. Thus, genera and species are defined independently of this, albeit with application of human judgement in selection of relevant membership criteria.

**1.4.2. Problems.** If hybrids cannot be found or induced, this does not necessarily mean that the two individuals belong to different basic types. Some species are notoriously difficult to breed even within the same species, as is the case with the giant panda. With nocturnal, cryptic and forest-dwelling species, it can be difficult to discern hybrids in the field situation; and rare species, or ones separated from other species by geographical barriers, are relatively less likely to encounter other species with which they might hybridize. In these situations, supplementary criteria can be invoked to assess whether the individual is sufficiently similar to other members of a basic type distinguished by hybridization data, or indeed if these criteria together suggest a stand-alone allocation to a separate taxon. Sometimes interspecific hybridization is deemed a threat to the viability of the parent species, so for reasons of conservation one might not wish to artificially induce hybridization or breed hybrids in captivity unless the hybrids are maintained securely in captivity, which can be difficult to ensure, and a prospective threat to native wild parental stock might be hard to predict. However, it should be stressed that there is sufficient interspecific hybridization data already in existence, for testing the plausibility of the concept, to render this a minor or unnecessary strategy.

Scherer (1993) proposed a future additional, tertiary membership criterion:

> *Two individuals belong to the same BT if embryogenesis proceeds beyond the maternal phase, including subsequent co-ordinated expression of both maternal and paternal morphogenetic genes.*

However, this stage varies between basic types and is also elusive, especially since we have not yet identified the morphogenetic genes implicated in this process, nor their modes of action. However, we can identify morphological stages at which mature embryogenesis has been attained, *i.e.* there are morphological markers for such stages. So, when it is difficult to resolve if a specific breeding cross has been successful, an alternative tertiary membership criterion could be proposed (which has not been necessary to apply in this book):

> *Two individuals belong to the same basic type if mature embryogenesis has been attained, in which the morphological form can be clearly distinguishable.*

**1.4.3. The Basic Type And The Biological Species Concept—Similar But Different.** Both concepts are limited in scope when environmental or other non-genic factors prevent hybridization. However, although hybridization is a defining criterion for both the basic type and biological species concept, there are also fundamental differences between the concepts, with important implications. The biological species definition is aimed solely at delineating the boundary of the species: according to this definition, if two individuals of interbreeding natural populations hybridize, they must be members of the same species. Adoption of this definition has, as already stated, led to profound taxonomic consequences of undue merging of species previously distinguished clearly on morphological grounds. In contrast, the BT criterion is not applied in order to delineate a species, but to identify all individuals which must share a morphogenetic program which is fundamentally in common, which in turn is logically expected to generate a common basic form. It is indeed recognized that the laws of development do seem to canalize development into such common forms. Thus, its emphasis is rather in elucidating the extent of fundamental similarity, in general comprising a much larger grouping, often of several genera.

**1.4.4. Issues Of Objectivity.** Some taxonomic methods require a tree structure at the outset, which creates the challenge of deciding which organisms are primitive (*i.e.* ancestral) or derived. Thus, the basic assumption is that the tree structure underpins the data. In contrast, morphological criteria need not assume this, but from the pattern of similarities and differences emerge the relationships suggestive of a common ancestry.

Thus, systematics overlaid with theoretical presuppositions which change with time should be avoided. Rather, the observations made, say, 100 years ago, should be just as valid, even if the interpretations should change. An example from medical history illustrates this dilemma. Until the 1860's, deaths from post-operative infections were at prohibitively high levels (Pitt and Aubin 2012). The idea of spontaneous generation, that life originates from non-living matter, and that infections arise similarly led people astray. A different conceptual framework was proposed by Louis Pasteur. His belief in a Biblical model (of life begetting life) rather than an evolutionary process led him to experimentally refute the idea of spontaneous generation (Farley 1978). He demonstrated that a nutrient broth heated in a flask with a swan-shaped neck remained sterile and free from micro-organisms indefinitely. In contrast, if the neck was broken off, the broth became contaminated with microbes (Farley 1974). From this became established the germ theory of disease, whereby germs were recognized as the causative agents of disease. This inspired the antiseptic innovations of Lister (Pitt and Aubin, 2012), sparing millions from hospital gangrene and mortality, and leading to the revolution in infection control which shapes current surgical practice (Toledo-Pereyra 2009; Nakayama 2018). However, in spite of Pasteur's discoveries at that time, Darwin's chief advocate Huxley persisted in believing in spontaneous generation for philosophical reasons (according to Huxley's son [Huxley 1903]) rather than from a scientific evidence base.

Thus, for the health of science, observations should not be too closely tied to a theoretical understanding of how the phenomenon may have come into existence. Observations should not be overladen with theory, but kept free to allow the reconstructions of scenarios that explain the observations. Of relevance to this book, all tree systems of taxonomy presuppose that the organisms can be arranged in a treelike pattern, so systems are employed to decipher it (e.g. cladism). However, this presupposes that they are all linked to a common ancestor. Hence the tree-based permutational analysis has this embedded assumption, which dictates that the "best fit" tree must be found. Rather, the tree structure should not be presumed and embedded in the methodology, but emerge from the data.

The notion was advanced earlier that the genus level cannot be strictly defined, implying that selection of a particular set of morphological characters depends upon the judgement of the observer (and is hence not objective?). Since BT analysis employs data on hybridization between the genera, does this in turn render the BT as dependent on a subjectively determined set of categories? This does not appear to be the case. Although the extent of intergeneric hybridization does give an idea of the inclusiveness and extent of genera comprising a prospective BT, ultimately, the degree of interconnectivity between hybrids exists independently of their assignment to genera. In other words, whatever the taxonomic categories of convenience happen to be, the BT is demarcated by hybridization data which depends not on that categorization but on a common morphogenetic programme that makes it possible.

More importantly, until the 1940s, species were assigned to genera according to morphological characters, which in general were not in dispute. They were determined objectively (*i.e.* observed independently of the observer's judgment)—for instance, quantity of observed elements, geometrical form, relative size and temperospatial distribution (Vogt 2008). However, subsequently there has been a trend to select the morphological characters that determine genera according to tree-based analyses, which has, as already stated, cannot escape from subjective judgements regarding what characters may be basal/primitive/ancestral, and what are derived/recent. Thus it is useful to be aware of the background methodology applied to demarcate genera, in order to maximize true objectivity. Some taxonomists have made a plea for this very same point—in, for instance, the publication "on the independence of systematics" (Brady 1985), as have others (e.g. Patterson 1982). This is reiterated by Fehrer (Chapter 13) who highlights the uncertainty of the biological relevancy of taxonomy based on genetic similarities, in contrast to the BT method, which is based primarily on the relationship of the complete genomes of crossbreeding partners.

## 1.5. Areas For Future Research

### 1.5.1. Degrees Of Interconnectivity Forming Core-Groups—A Useful Tool.
Some groups may exhibit considerable interspecific hybridization, but have a low degree of inter-connectivity between them. Such groups should of course be discounted as basic type candidates, pending further data. Conversely, even if the total number of species hybridizing in a taxon is not great, but the inter-connectivity is high, this would be a candidate for basic type status. This criterion is important in assessing prospective basic type status. Scherer (1993)

reiterates this, maintaining that comparatively few hybrids are necessary in order to discern a basic type, with intergeneric hybrids relatively more important than interspecific hybrids. Moreover, not all theoretically possible intergeneric crosses need be actually achieved. For instance, in *bt*Cercopithecinae (9 genera) only 9 combinations were reported (from a possible 36) but these connect 8 of the 9 genera, which clearly demarcates Cercopithecinae. In order to delimit a BT, one would need to produce (n-1) intergeneric hybrids, in which n represents the number of genera. So, for a prospective BT comprising 3 genera, just 2 intergeneric hybrids would be necessary. Assessment of inter-connectivity could be a useful tool to further confirm or reject (pending further data) basic type status.

Moreover, a systematics based on core groups whose members are interconnected by hybridization could aid the resolution of problematic taxa in flux, if the hybridization data are unequivocal. The rationale for this is that these core groups are empirically determined, and indicate that its members comprise morphogenetic closeness.

For instance, the guineafowl have traditionally been considered a subfamily within the family Phasianidae, but studies based on DNA differences resulted in their treatment as a separate family, the Numididae, which is reflected in avian checklists (Dickinson and Remsen 2013; Clements *et al.* 2023). However, guineafowl can hybridise with *Gallus gallus* (the chicken), a member of the Phasianidae (Klemm Chapter 11). Such crosses imply a biological or morphogenetic affinity between Numididae and Phasianidae.

A case can be made for a more inclusive connectivity between such families if they are considered together as a whole, whereby a family may be connected to another indirectly via hybridization with members of a third family in common. Indeed, the principle of the basic type definition is suggestive of this:

> *Rephrased at a higher taxonomic level, it can be stated that a group of species, connected together by hybridization, are related to another group of species (inter-connected by hybridization within that group), if some of their respective members hybridize with members of a third group of species.*

**1.5.2. Microevolution Vs. Evolution Involving Phenotypic Plasticity.** The term microevolution is used to distinguish between small-scale changes and macroevolution (the latter referring to the evolution of new constructional types of body plans). However, the term microevolution is associated with a Darwinian-derived concept of incremental evolution of novelty, which can be extrapolated, leading to large-scale innovation.

The Neo-Darwinian selection mechanism involves a variant which has arisen from a random mutation, then following an environmental change becomes selected because of its fitness advantage—this is known as 'positive selection'. Hughes (2012) asserts that verified cases of positive Darwinian selection (identified by analysis of genetic sequence data) remain few. An alternative non-Darwinian model, the plasticity–relaxation–mutation (PRM) mechanism, involves ancestral phenotypic plasticity followed by specialization in one alternative environment (Hughes 2012). It is non-Darwinian in that it does not rely on the mechanism of positive selection. Rather, variants are produced by other processes which recombine the existing genetic information.

Hughes also cites studies providing evidence that radiations in the evolutionary past have been due to traits present in the ancestors before the radiation. For example, in toads (Anura: Bufonidae) a suite of characters correlated with range expansion present in the ancestors of the family prior to its expansion (Van Bocxlaer et al. 2010). Similarly, rapid expansion of the avian family Zosteropidae was attributed to traits present in the family's common ancestor (Moyle et al. 2009). Although the role of selection may be a matter of debate, evidence of ancestral phenotypic plasticity promoting adaptive radiations is well documented, such as in the rapid diversification of East African cichlid fish (Schneider and Meyer 2017).

The variation documented by the various authors in the following chapters is consistent with this interpretation. If this (PRM) is in action, the basic type, rather than the species, is of more fundamental importance: from the basic type genotype the diversity is generated. In this working framework, the basic type is envisaged as a higher-level category that contains the species. In contrast, Myer's definition (i.e. the BSC) implies that the species are transformed to produce a divergent genome that is on an evolutionary trajectory. But the PRM is not on a trajectory. Rather, it is part of a diversification of an information-rich genotype. This idea is proposed for further discussion.

**1.5.3. Developmental Pathways—Basic Types May Lead The Way.** Why is there such abundant interspecific (and even intergeneric) hybridization within the taxa described in this book, yet hybridization between such taxa and their more disparate groups is lacking? Indeed, it is these very discontinuities that are sought for in a truly viable taxonomic system, which should not simply focus on similarities, but discriminate between them (Stanford 1995). Discontinuities may arise from isolating barriers, which can be external (e.g. geographical, such as mountain ranges or oceans), or internal (e.g. immunological, or chromosomal differences between the parents). However, viable hybrid offspring can still result from parents with highly dissimilar chromosomes. For instance, Muntiacus muntjak (Indian muntjak) (2n = 7) hybridizes readily with M.

*reevesi* (Chinese muntjac) (2n = 46) (McCarthy 2006, p. 68).

The basic types united by hybridization may be indicators of a common morphogenetic machinery, which is not shared with disparate basic types. This is salient, as Thomas and Reif (1993) ask, what are the distinctive embryological pathways which canalise the types into their distinct forms, making only certain forms possible and viable, whilst forbidding others? Pigliucci reiterates this question and adds that hybrids between phylogenetically disparate groups are made impossible by divergence in their behaviours, genetic architectures and developmental systems (Pigliucci 2007).

Taxonomic reconstructions using molecular data continue to produce conflicting networks. It is crucial to ask what comparative gene sequences are accurate indicators of phylogenetic relationships? The best candidates might be morphogenetic genes, but the genes implicated in the development of form are poorly characterized (Tyler 2014), with a lack of evidence for how changes in genes, or the interaction of their products, can explain morphogenesis (Newman and Linde-Medina 2013). Thus, there is still a knowledge gap in our ability to map the genotype to the phenotype, which remains unbridged (Gjuvsland *et al.* 2013; Sultan *et al.* 2022). However, the basic types distinguished by hybridization may facilitate the demarcation of such genes, notably by virtue of their type-specific distribution and tempero-spatial features during development. Thus, basic type biology can make an important contribution to the quest for how form is generated, the basis of which remains so elusive (Schock and Perrimon 2002; Linde-Medina 2020). Indeed, there are many hints in published scientific literature that there is a developmental basis for pattern differences in distantly related taxa. For instance, there are phylogenetic peculiarities between the avian, mammalian and fish developmental limb fields (Sordino *et al.* 1995); and the leg field gene expression patterns of spiders differs from that of insects (Prpic *et al.* 2003). In bats, a number of genes display unique expression patterns in embryonic wings and feet, different from those of mouse fore- and hind-limbs (Wang *et al.*, 2014), and numerous bat-specific limb enhancers are involved in driving limb-specific gene expression (Eckalbar *et al.* 2016). This provides a paradigm for future research, led not by molecular similarities of unknown developmental (and phylogenetic) significance but based on empirically-determined basic types which in turn may help to demarcate the underlying developmental features generating them.

**1.5.4. Supplementary Criteria.** Particularly where hybridization data is lacking, incomplete or inconclusive, further research is needed to explore additional criteria that can be employed. One promising example is bioacoustics, involving the recording and analysis of animal sounds. Bioacoustics data has been utilized in bird taxonomy (Lanyon 1969; Payne 1986; Prawiradilaga *et al.* 2022). In fish families, within the Sciaenidae (croakers and drums) sounds typically consist of a series of rapid drumming pulses (Ramcharitar 2006). Within Cichlidae (e.g. Angelfish, Oscars, *Tilapia*), fish sounds consist of a series of regular pulses with regular inter-pulse intervals (Lobel *et al.* 2021). Further details of bioacoustics are featured in Tyler, Chapter 14.

**1.5.5. Hierarchies Of Types.** Are higher taxonomic levels (such as orders, classes and phyla) above the basic type merely contrived for the benefit of classifying, or do they reflect too a real pattern found in nature? Do higher levels relate to each other in a system of increasingly inclusive groups, *i.e.* a hierarchical pattern? Although some higher groups of organisms are hard to place, the notion that there are indeed deep divisions in nature can be discerned by evidence from morphology and developmental biology. For instance, molluscs can clearly be distinguished as a phylum based on features including a muscular foot and a sheet of skin which forms the mantle cavity and glands that typically produce a covering shell. Indeed, as hinted at earlier, it is common nowadays to refer to these and other features as a fundamental molluscan bauplan (Brusca and Brusca 2003). Within the molluscan bauplan, distinct classes of shell form can be distinguished, each of which can be traced back to a group of cells on the dorsal surface of the embryo called the shell field (Yang *et al.* 2020). So, for instance, the molluscan gastropod class bears a helical shell, and this highly characteristic morphology can be derived from the relatively accelerated growth of the anterior and lateral margins of the shell field. In contrast, the scaphopods (tusk shells), accelerated lateral and posterior margins of the field, followed by their fusion, leads to formation of the tube-like shell morphology by which this class is distinguished (Kniprath 1981).

The embryological cleavage patterns can also be distinctly different between various higher taxa. For example, early development in eutherian mammals has unique characteristics, including a rotational cleavage during the second cell division, coupled with subsequent asynchronous divisions and a flattening of cells and then formation of an outer set of cells, which contributes to the placenta, and an inner cell mass (ICM), which forms the embryo itself (Marikawa and Alarcón 2009). However, marsupial mammals, in which development is completed within a maternal pouch, possess a unilaminar blastocyst with no ICM (Selwood 1992).

Similar distinctions can be found in plants. The early cell divisions have had important significance in taxonomic and phylogenetic interpretations (Steeves and Sussex 1989). For instance, the plane and pattern of these cell divisions contrasts between various species of angiosperms (flowering plants); however in gymnosperms (e.g. the conifers) the zygote nuclei divide freely, without cell wall formation, and multiple embryos can arise from a single zygote; and the lower vascular plants, such as ferns, feature a bewildering array of embryonic types (Steeves and Sussex 1989). Moreover, distinct sets of genes are involved in development of differences in shoot and root architecture between different species (Di Ruocco *et al.* 2018; Knauer *et al.* 2019).

### 1.5.6. Analysis Of Further Hybridization Data.

If the BT concept is valid, one would expect to find evidence in nature for further BTs to be identified. This indeed appears to be the case.

There is a wealth of hybridization data, particularly in mammals, fish, birds, reptiles and plants, (Appendix 2 and 3). In plants, in addition to the BTs described in this book, a further 18 prospective BTs can be proposed (primarily based on unpublished analyses by Herfried Kutzelnigg). For instance, within the orchid family, there are 286 inter-generic hybrids in the tribe Cymbidieae, with 50 crosses interconnecting all 11 sub-tribes (Appendix 3). In the cactus sub-family Cactoideae, 110 inter-generic hybrids interconnect 8 of the 10 tribes. In animals, 20 further BTs can be recognised, including 11 prospective mammal BTs. For instance, in the Camelidae (camel family), hybridization links both *Camelus* and *Lama* (camel and llama) genera (Junker, 2000). In Felidae (the cat family), hybridization links the big cats (e.g. puma) and small cats (Crompton and Winkler, 2006). In the family Hylobatidae (gibbons), all 4 genera are interconnected by hybridization (Kutzelnigg, unpublished). There are 500 intergeneric hybrids in 50 families of fish (including 276 intergeneric hybrids within Cyprinidae (the carp family), in which all 12 subfamilies are directly or indirectly linked by crosses (Kutzelnigg, unpublished: Appendix 3). There are great opportunities for more investigators to tap into such data, from which there are promising prospects for the discovery of many more BTs.

### 1.6. Update To Previous Edition

More recent hybridization data relevant to the subsequent chapters of this book concerning Aves can be found in Appendix 1.

### 1.7. Concluding Remarks

The themes of this book can be traced back to Carolus Linnaeus who, in his later publications, departed from the idea that species do not change (Landgren Chapter 3; Mayr 1982). The emphasis of his system was the genus (rather than the species), as the fundamental units of life through which new species could be generated by hybridization. Linnaeus categorized many of these genera using objective criteria such as quantity of observed elements, geometrical form, relative size and temperospatial distribution (Vogt 2008). Interestingly, most of his genera are still valid but some have been elevated to the rank of family today (Mayr et al. 1953).

Moreover, the BTs recognised in this volume share distinct synapomorphies (characters common to members of a group and their ancestors/descendants), which clearly separate them from other (apparently) closely related groups. In other words, there is an undisputed gap between basic types of living organisms. Where investigated, these gaps also appear to be present in the earliest known fossil forms of these groups.

In the 1993 publication, a modest number of 14 BTs had been distinguished, from which Scherer cautioned that the number is too low to provide for a reliable basis for generalization, and thus the BT concept "serves as a preliminary working hypothesis." However, with the advent of considerable further interspecific hybridization data, the numbers of BTs for which there is good evidence has significantly increased to 36, with preliminary and other published data demarcating up to 109 BTs prospectively, so the concept is developing on a promising, growing empirical base.

## Acknowledgments

Thanks to David Tyler for helpful discussions and ideas regarding the manuscript, and to Roger Sanders for his patient and constructive editorial support.

# References

Abbott, R. and 36 others. 2013. Hybridization and speciation. *Journal of Evolutionary Biology* 26(2):229-246.

Agapow, P.M., Bininda-Emonds, O.R., Crandall, K.A., Gittleman, J.L., Mace, G.M., Marshall, J.C. and Purvis, A., 2004. The impact of species concept on biodiversity studies. *The quarterly review of biology*, 79(2), pp.161-179.

Arthur, W., 1997. *Animal Body Plans*. Cambridge University Press, UK.

Aztekin, C. and Storer, M.A., 2022. To regenerate or not to regenerate: Vertebrate model organisms of regeneration-competency and-incompetency. *Wound Repair and Regeneration*, 30(6), pp.623-635.

Barker, M.J., 2022. We Are Nearly Ready to Begin the Species Problem. In *Species Problems and Beyond* (pp. 3-38). CRC Press.

Barton, N.H. and Hewitt, G.M., 1985. Analysis of hybrid zones. *Annual review of Ecology and Systematics* 113-148.

Barton, N.H. and Hewitt, G.M., 1989. Adaptation, speciation and hybrid zones. *Nature* 341(6242):497-503.

Birchler, J.A. and Veitia, R.A., 2010. The gene balance hypothesis: implications for gene regulation, quantitative traits and evolution. *New Phytologist* 186(1):54-62.

Bordenstein, S.R., O'Hara, F.P., and Werren, J.H., 2001. Wolbachia-induced incompatibility precedes other hybrid incompatibilities in *Nasonia*. *Nature* 409(6821):707-710.

Boudreaux, P., 2020. Species... in Law. *Tex. Envtl*. LJ, 50, p.1.

Brady, R.H. 1985. On the independence of systematics. *Cladistics* 1:113 126.

Breeuwer J.A.J. and Werren, J.H., 1990. Microorganisms associated with chromosome destruction and reproductive isolation between two distinct insect species. *Nature* 346:558-560.

Brusca, R.C. and Brusca, G.J., 2003. *Invertebrates*. 2nd Edition. Sinauer, Sunderland, Massachusetts, USA.

Brusca, R.C, Giribet, G. and Moore, W., 2022. *Invertebrates*. 4th Edition. Sinauer Associates /Oxford University Press, Oxford, UK.

Büscher, T.H., Bank, S., Cumming, R.T., Gorb, S.N. and Bradler, S., 2023. Leaves that walk and eggs that stick: comparative functional morphology and evolution of the adhesive system of leaf insect eggs (Phasmatodea: Phylliidae). *BMC Ecology and Evolution*, 23(1), pp.1-22.

Choi, S.J., Yun, S.W. and Park, J.Y., 2023. Comparative Morphology of the Zona Radiata in Oocytes of Korean Bitterlings from the Genera Rhodeus, and Acheilognathus (Cyprinidae). *Journal of Ichthyology*, 63, 781–787.

Clements, J. F., P. C. Rasmussen, T. S. Schulenberg, M. J. Iliff, T. A. Fredericks, J. A. Gerbracht, D. Lepage, A. Spencer, S. M. Billerman, B. L. Sullivan, and. Wood, C.L.,. 2023. The eBird/Clements checklist of Birds of the World: v2023. Downloaded from https://www.birds.cornell.edu/clementschecklist/download/

Coyne, J.A. and Orr, H.A., 2004. *Speciation*. Sinauer Associates, Sunderland, MA, USA.

Crompton, N.E.A., 1993. A review of selected features of the family Canidae with reference to its fundamental taxonomic status. In: Scherer, S., ed. *Typen des Lebens*. Pascal Verlag, Berlin, pp. 217–224.

Cronquist, A., 1978. Once again, what is a species? In: Romberger, J.A. ed. *Biosystematics in Agriculture*. Allanheld Osmun, Montclair, NJ, USA, pp. 3–20.

De Queiroz, K., 1999. The general lineage concept of species and the defining properties of the species category. In: Wilson, R.A., ed. *Species: New Interdisciplinary Essays*. MIT Press, Cambridge, MA, USA, pp 49–89.

Devitt, M., 2008. Biological Realisms. In: Dyke, H., ed. *From Truth to Reality: New Essays in Logic and Metaphysics*. Routledge, London, UK.

Dickinson, E.C. and Remsen, J.V., 2013. *Howard and Moore Complete Checklist of the Birds of the World, 4th edition, Vol. 1*. Aves Press, Eastbourne, UK.

Di Ruocco, G., Di Mambro, R. and Dello Ioio, R., 2018. Building the differences: a case for the ground tissue patterning in plants. *Proceedings of the Royal Society B*, 285(1890), p.20181746.

Domınguez, E. and Cuezzo, M.G., 2002. Ephemeroptera egg chorion characters: A test of their importance in assessing phylogenetic relationships. *Journal of Morphology* 253:148-165.

Drozdov, A.L., Sharina, S.N. and Tyurin, S.A., 2009. Sperm ultrastructure in representatives of six bivalve families from Peter the Great Bay, Sea of Japan. Russian Journal of Marine Biology, 35, pp.236-241.

Eckalbar, W.L., Schlebusch, S.A., Mason, M.K., Gill, Z., Parker, A.V., Booker, B.M., Nishizaki, S., Muswamba-Nday, C., Terhune, E., Nevonen, K.A. and Makki, N., 2016. Transcriptomic and epigenomic characterization of the developing bat wing. *Nature genetics*, 48(5), pp.528-536.

Elliot, A., 1992. Family Pelicanidae (Pelicans). *Handbook of the Birds of the World* 1:290- 311.

Ereshefsky, M., 1998. Species Pluralism and Anti-Realism. *Philosophy of Science* 65:103-20.

Farley, J., 1974. *The spontaneous generation controversy from Descartes to Oparin*. John Hopkins Unversity Press, Baltimore, USA.

Farley, J., 1978. The social, political, and religious background to the work of Louis Pasteur. *Annual Review of Microbiology*, 32, pp.143-154.

Ford, V.S. and Gottlieb, L.D., 1992. Bicalyx is a natural homeotic floral variant. *Nature* 358:671-673.

Frith, C.B. and Frith, D.W., 2009. Family Ptilonorhynchidae (Bowerbirds). Species accounts. *Handbook of the Birds of the World* 14:393-403.

Gjuvsland, A.B., Vik, J.O., Beard, D.A., Hunter, P.J. and Omholt, S.W., 2013. Bridging the genotype–phenotype gap: what does it take?. *The Journal of physiology*, 591(8), pp.2055-2066.

Gómez, C.M.A. and Echeverri, K., 2021. Salamanders: The molecular basis of tissue regeneration and its relevance to human disease. Current topics in developmental biology, 145, pp.235-275.

Gorbalenya, A.E., Baker, S.C., Baric, R.S., de Groot, R.J., Drosten, C., Gulyaeva, A.A., Haagmans, B.L., Lauber, C., Leontovich, A.M., Neuman, B.W. and Penzar, D., 2020. Coronaviridae Study Group of the International Committee on Taxonomy of Viruses. The species severe acute respiratory syndrome-related coronavirus: classifying 2019-nCoV and naming it SARS-CoV-2. *Nat. Microbiol*, 5(4), pp.536-544.

Gould, S.J. 1993. The first unmasking of nature. *Natural History* 102:14-21.

Grant, P.R. and Grant, B.R., 1994. Phenotypic and genetic effects of hybridization in Darwin's finches. *Evolution* 48:297-316.

Gras, A., Hidalgo, O., D'ambrosio, U., Parada, M., Garnatje, T. and Valles, J., 2021. The role of botanical families in medicinal ethnobotany: a phylogenetic perspective. Plants, 10(1), p.163.

He, L.M. and Dooner, H.K., 2009. Haplotype structure strongly affects recombination in a maize genetic interval polymorphic for Helitron and retrotransposon insertions. *Proceedings of the National Academy of Sciences* USA 106:8410–8416.

Hebert, P.D., Cywinska, A., Ball, S.L. and DeWaard, J.R., 2003. Biological identifications through DNA barcodes. Proceedings of the Royal Society of London. Series B: Biological Sciences, 270(1512), pp.313-321.

Hodgson, A.N. and Foster, G.G., 1992. Structure of the sperm of some South African archaeogastropods (Mollusca) from the superfamilies *Haliotoidea, Fissurelloidea* and *Trochoidea*. *Marine Biology* 113:89 97.

Höfling, E. 1998. Comparative cranial anatomy of Ramphastidae and Capitonidae. *Ostrich* 69(3-4):389-390.

Hollister, J.D., Smith, L.M., Guo, Y.L., Ott, F., Weigel, D. and Gaut, B.S., 2011. Transposable elements and small RNAs contribute to gene expression divergence between Arabidopsis thaliana and Arabidopsis lyrata. Proceedings of the National Academy of Sciences, 108(6), pp.2322-2327.

Hughes, A.L., 2012. Evolution of adaptive phenotypic traits without positive Darwinian selection. *Heredity* 108(4):347-353.

Hull, D., 1988. *Science as a Process*. University of Chicago Press, Chicago, USA.

Huxley, L., ed. 1903. *Life and letters of Thomas Henry Huxley*. 3 Volumes. Macmillan, New York, USA.

Islam, A., Sayeed, M.A., Kalam, M.A., Ferdous, J., Rahman, M.K., Abedin, J., Islam, S., Shano, S., Saha, O., Shirin, T. and Hassan, M.M., 2021. Molecular epidemiology of SARS-CoV-2 in diverse environmental samples globally. *Microorganisms*, 9(8), p.1696.

Jariyapan, N., Tippawangkosol, P., Sor-Suwan, S., Mano, C., Yasanga, T., Somboon, P., Depaquit, J. and Siriyasatien, P., 2022. Significance of eggshell morphology as an additional tool to distinguish species of sand flies (Diptera: Psychodidae: Phlebotominae). Plos one, 17(2), p.e0263268.

Jiao, Y. and 16 others., 2011. Ancestral polyploidy in seed plants and angiosperms. *Nature* 473:97–100.

Johnson, N.A. and Porter, A.H., 2000. Rapid speciation via parallel, directional selection on regulatory genetic pathways. Journal of Theoretical Biology, 205(4), pp.527-542.

Kemp, A.C., 2001. Bucerotidae (Hornbills). *Handbook of the Birds of the World* 6:436-523.

Kidwell, M.G. and Peterson, K.R., 1991. Evolution of transposable elements in Drosophila. In: Warren, L. and H. Koprowski, eds. *New perspectives on evolution*. Plenum Press, New York, pp 139 – 154.

Knauer, S., Javelle, M., Li, L., Li, X., Ma, X., Wimalanathan, K., Kumari, S., Johnston, R., Leiboff, S., Meeley, R. and Schnable, P.S., 2019. A high-resolution gene expression atlas links dedicated meristem genes to key architectural traits. *Genome research*, 29(12), pp.1962-1973.

Kniprath, E., 1981. Ontogeny of the molluscan shell field, *Zoologica Scripta* 10:61 79.

Kim, M., Cui, M.L., Cubas, P., Gillies, A., Lee, K., Chapman, M.A., Abbott, R.J. and Coen, E., 2008. Regulatory genes control a key morphological and ecological trait transferred between species. *Science*, 322(5904), pp.1116-1119.

Koss R.W. and Edmunds, G.F., 1974. Ephemeroptera eggs and their contribution to phylogenetic studies of the order. *Zoological Journal of the Linnean Society* 55:267–349.

Lanyon, W.E., 1969. Vocal characters and avian systematics. In: Hinde, R.A. ed. B*ird vocalizations: their relation to current problems in biology and psychology.* Cambridge University Press, Cambridge, UK, pp. 291–310.

Levin, M., 2012. Morphogenetic fields in embryogenesis, regeneration, and cancer: Non-local control of complex patterning. Biosystems doi:10.1016/j.biosystems.2012.04.005

Llimona, F. and del Hoyo, J., 1992. Family Podicipedidae (Grebes). *Handbook of the Birds of the World* 1:174 –196.

Linde-Medina, M., 2020. On the problem of biological form. *Theory in Biosciences*, 139(3), pp.299-308.

Lobel, P.S., Garner, J.G., Kaatz, I.M. and Rice, A.N., 2021. Sonic cichlids. In *The behavior, ecology and evolution of cichlid fishes* (pp. 443-502). Springer, Dordrecht.

Maksudov, G.Y. and Panchenko, V.G., 2002. Obtaining an interspecific hybrid of cranes by artificial insemination with frozen–thawed semen. *Biology Bulletin of the Russian Academy of Sciences,* 29(3), pp.311-314.

Mallett, J., 2005. Hybridization as an invasion of the genome. *Trends in Ecology and Evolution* 20:229–237.

Mallet, J. 2007. Hybrid speciation. *Nature* 446(7133):279-83.

Margaritis, L.H. and Mazzini, M., 1998. Structure of the egg. In: FW Harrison, F.W. and M. Loche, eds. *Microscopic Anatomy of Invertebrates, Vol. 11C: Insecta,* Wiley-Liss, New York, pp. 995-1037.

Marikawa, Y. and Alarcón, V.B., 2009. Establishment of trophectoderm and inner cell mass lineages in the mouse embryo. *Molecular Reproduction and Development: Incorporating Gamete Research*, 76(11), pp.1019-1032.

Marsh, F., 1976. *Variation and fixity in nature*. Pacific Press Association, Mountain View, CA., USA.

Mayden, R.L., 1997. A hierarchy of species concepts: the denouement in the saga of the species problem. In: Claridge, M., H.A .Darwah, and M.R. Wilson, eds. *Species: The Units of Biodiversity*. Chapman and Hall, London, pp. 381–424.

Mayr, E., 1940. Speciation phenomena in birds. *American Naturalist* 74:249-278.

Mayr, E., 1969. *Principles of Systematic Zoology*. McGraw-Hill, New York, USA.

Mayr, E., 1982. *The Growth of Biological Thought: Diversity, Evolution and Inheritance*. Belknap Press, Cambridge, MA.

Mayr, E., 1991. *Eine neue Philosophie der Biologie*. Piper, Munich, Germany.

Mayr, E., Linsley, E.G. and Usinger, R.L., 1953. *Methods and Principles of Systematic Zoology*. McGraw-Hill, New York, USA.

McCallum, A. 2011. Birding by Ear, Visually Part 2: Syntax. Birding 43(3) web extra: https://www.aba.org/birding_archive_files/v43n5p45w1.html

McCarthy, E. 2006. *Handbook of avian hybrids of the world*. Oxford University Press, England.

Mendonça, P.M., dos Santos-Mallet, J.R., de Mello, R.P., Gomes, L. and de Carvalho Queiroz, M.M., 2008. Identification of fly eggs using scanning electron microscopy for forensic investigations. *Micron*, 39(7), pp.802-807.

Mishler, B. and Donoghue, M. 1982. Species Concepts: A Case for Pluralism. *Systematic Zoology* 31:491-503.

Moyle, R.G., Filardi, C.E., Smith, C.E. and Diamond, J., 2009. Explosive Pleistocene diversification and hemispheric expansion of a "great speciator". *Proceedings of the National Academy of Sciences*, 106(6), pp.1863-1868.

Nakayama, D.K., 2018. Antisepsis and asepsis and how they shaped modern surgery. *The American Surgeon*, 84(6), pp.766-771.

Nathan, M.J. and Cracraft, J., 2020. The nature of species in evolution. *The theory of evolution*, pp.102-122.

Newman, S.A. and Linde-Medina, M., 2013. Physical Determinants in the Emergence and Inheritance of Multicellular Form. *Biological Theory* 8:274–285.

Nixon, K.C. and Wheeler, Q.D., 1990. An amplification of the phylogenetic species concept. *Cladistics* 6:211–23.

Odenbaugh, J., 2022. What Should Species Be?: Taxonomic Inflation and the Ethics of Splitting and Lumping. In *Species Problems and Beyond* (pp. 91-104). CRC Press.

Patterson, C., 1982. Morphological characters and homology. In: Joysey, K.A., and A.E. Friday, eds. *Problems of phylogenetic Reconstruction*. Academic Press, New York, USA, pp.1 74.

Pavlinov, I.Y., 2023. The Species Problem: *A Conceptual History*. CRC Press.

Payne, R.B., 1986. Bird songs and avian systematics. *Current ornithology* 87-126.

Pensotti, A., Bertolaso, M. and Bizzarri, M., 2023. Is Cancer Reversible? Rethinking Carcinogenesis Models—A New Epistemological Tool. *Biomolecules*, 13(5), p.733.

Pigliucci, M., 2003. Species as family resemblance concepts: the (dis-) solution of the species problem? *BioEssays* 25:596–602.

Pigliucci, M., 2007. Finding the way in phenotypic space: the origin and maintenance of constraints on organismal form. *Annals of Botany*, 100(3), pp.433-438.

Pitt, D. and Aubin, J.M., 2012. Joseph Lister: father of modern surgery. *Canadian Journal of Surgery*, 55(5), E8-9.

Prawiradilaga, D.M., Baveja, P., Suparno, S., Ashari, H., Ng, N.S.R., Gwee, C.Y., Verbelen, P. and Rheindt, F.E., 2018. A colourful new species of *Myzomela* honeyeater from Rote Island in Eastern Indonesia. *Treubia*, 44, pp.77-100.

Prpic, N.M., Janssen, R., Wigand, B., Klingler, M. and Damen, W.G., 2003. Gene expression in spider appendages reveals reversal of exd/hth spatial specificity, altered leg gap gene dynamics, and suggests divergent distal morphogen signaling. *Developmental biology*, 264(1), pp.119-140.

Raff, R.A., 2012. The shape of life: genes, development, and the evolution of animal form. University of Chicago Press.

Ramcharitar, J., Gannon, D.P. and Popper, A.N., 2006. Bioacoustics of fishes of the family Sciaenidae (croakers and drums). *Transactions of the American Fisheries Society*, 135(5), pp.1409-1431.

Rieppel, O.C., 1988. *Fundamentals of comparative biology*. Birkhauser Verlag, Basel, Switzerland.

Rieppel, O.C., 2013. Biological Individuals and Natural Kinds. *Biological Theory* 7(2):162-169.

Robertson, R., 1974. Marine prosobranch gastropods: larval studies and systematics. *Thalassia Jugoslavia* 10:213 23.

Saslis-Lagoudakis, C.H., Savolainen, V., Williamson, E.M., Forest, F., Wagstaff, S.J., Baral, S.R., Watson, M.F., Pendry, C.A. and Hawkins, J.A., 2012. Phylogenies reveal predictive power of traditional medicine in bioprospecting. *Proceedings of the National Academy of Sciences*, 109(39), pp.15835-15840.

Scherer, S., 1993. Basic types of life. In: Scherer, S., ed. *Typen des Lebens*. Pascal Verlag, Berlin, Germany, pp.11-30.

Schock, F. and Perrimon, N., 2002. Molecular mechanisms of morphogenesis. *Cell and Developmental Biology* 18:463-493.

Scotland, R.W., Olmstead, R.G. and Bennett, J.R., 2003. Phylogeny reconstruction: the role of morphology. *Systematic biology*, 52(4), pp.539-548.

Scribner, K.T., Page, K.S. and Bartron, M.L., 2000. Hybridization in freshwater fishes: a review of case studies and cytonuclear methods of biological inference. *Reviews in Fish Biology and Fisheries* 10:293–323.

Selwood, L., 1992. Mechanisms underlying the development of pattern in marsupial embryos. *Current Topics in Developmental Biology* 27:175 233.

Short, L.L. and Horne, J.F.M., 2002. Family Capitonidae (Barbets). *Handbook of the Birds of the World* 7:140-219.

Sibley, C.G. and Ahlquist, J.E., 1990. *Phylogeny and classification of birds.* Yale University Press, New Haven, Connecticut, USA.

Simpson, G.G., 1943. Criteria for genera, species, and subspecies in zoology and paleontology. *Annals of New York Academy of Sciences* 44:145–178.

Simpson, G.G., 1945. The principles of classification and a classification of mammals. *Bulletin of the American Museum of Natural History* 85:1–350.

Sin, Y.C.K., Eaton, J.A., Hutchinson, R.O. and Rheindt, F.E., 2022. Re-assessing species limits in a morphologically cryptic Australasian kingfisher lineage (Coraciiformes: Halcyonidae) using bioacoustic data. *Biological Journal of the Linnean Society.*

Sites, J.W. and Marshall, J.C., 2004. Operational criteria for delimiting species. *Annual Review of Ecology, Evolution, and Systematics* 35:199–227.

Sordino, P., van der Hoeven, F. and Duboule, D., 1995. Hox gene expression in teleost fins and the origin of vertebrate digits. *Nature* 375:678–681.

Stanford, P.K., 1995. For Pluralism and Against Realism about Species. *Philosophy of Science* 62:70-91.

Stanford, P.K., 2010. Species. http://plato.stanford.edu/archives/spr2010/entries/species

Steeves, T.A. and Sussex, I.M., 1989. *Patterns in plant development.* Cambridge University Press, England.

Sultan, S.E., Moczek, A.P. and Walsh, D., 2022. Bridging the explanatory gaps: What can we learn from a biological agency perspective? *BioEssays*, 44(1), p.2100185.

Szymura, J.M., 1993. Analysis of hybrid zones with *Bombina*. *Hybrid zones and the evolutionary process*, 261-289.

Thomas, R.D.K. and Reif, W.E., 1993. The skeleton space: a finite set of organic designs. *Evolution* 47:341–360.

Toledo-Pereyra, L.H., 2009. Louis Pasteur surgical revolution. *Journal of Investigative Surgery*, 22(2), pp.82-87.

Tsiantis, M. and Hay, A., 2003. Comparative plant development: the time of the leaf? *Nature Reviews Genetics* 4:169–180.

Tyler, S. E., 2014. The Work Surfaces of Morphogenesis: The Role of the Morphogenetic Field. *Biological Theory* 9(2):194-208.

Tyler, S.E. and Tyler, L.D., 2023. Pathways to healing: plants with therapeutic potential for neurodegenerative diseases. *IBRO Neuroscience Reports.* June;14: 210–234.

Ungerer, M.C., Strakosh, S.C. and Zhen, Y., 2006. Genome expansion in three hybrid sunflower species is associated with retrotransposon proliferation. *Current. Biology* 16:R872–R873.

Van Bocxlaer, I., Loader, S.P., Roelants, K., Biju, S.D., Menegon, M. and Bossuyt, F., 2010. Gradual adaptation toward a range-expansion phenotype initiated the global radiation of toads. *Science* 327:679–682.

Van Valen, L., 1976. Ecological Species, Multi-Species and Oaks. *Taxon* 25:233-9.

Vishwakarma, M., Spatz, J.P. and Das, T., 2020. Mechanobiology of leader–follower dynamics in epithelial cell migration. *Current opinion in cell biology*, 66, pp.97-103.

Vogt, L., 2008. Learning from Linnaeus: towards developing the foundation for a general structure concept for morphology. *Zootaxa* 1950:123–152.

Wang, Z., Dai, M., Wang, Y., Cooper, K.L., Zhu, T., Dong, D., Zhang, J. and Zhang, S., 2014. Unique expression patterns of multiple key genes associated with the evolution of mammalian flight. *Proceedings of the Royal Society B: Biological Sciences*, 281(1783), p.20133133.

Yang, W., Huan, P. and Liu, B., 2020. Early shell field morphogenesis of a patellogastropod mollusk predominantly relies on cell movement and F-actin dynamics. *BMC developmental biology*, 20(1), pp.1-10

Zbinden, Z.D., Douglas, M.R., Chafin, T.K. and Douglas, M.E., 2023. A community genomics approach to natural hybridization. *Proceedings of the Royal Society B*, 290(1999), p.20230768.Van Bocxlaer, I., Loader, S.P., Roelants, K., Biju, S.D., Menegon, M. and Bossuyt, F., 2010. Gradual adaptation toward a range-expansion phenotype initiated the global radiation of toads. *Science* 327:679–682.

# 2. Processes of Speciation

REINHARD JUNKER

**Abstract**

A thorough analysis of the processes involved in the formation and establishment of a species are vital aspects of evolutionary research. Depending on the chosen definition of a species, different aspects of speciation are more or less important. Presently, the most important models of speciation include: allopatric, parapatric, and sympatric speciation, speciation via polyploidization and hybridization, and speciation via formation of agamic complexes. The controversy over whether a geographic (external) barrier to gene flow is absolutely necessary for speciation, as required by the allopatric model, has still not been settled. A special form of the allopatric model is speciation via small pioneer populations (peripatric speciation). This type of process permits rapid establishment of new species based on catastrophic selection. Furthermore, mechanisms for conservation of genome rearrangements and the establishment of hybrid species are available in this model. An up-till-now largely disregarded model of speciation is diversification promoted by reduction of a high variation potential of the ancestral form.

The analysis of relationships between closely related races demonstrates that no single speciation mechanism exists. Numerous processes can lead to the formation of hybrid incompatibility. Incompatibility can arise based on genetic, karyotypic, ecological or ethological factors. Often, hybridization, karyotypic and genetic divergence, morphology, ecology, and evolutionary history are not correlated with one another. Correlation amongst various plant groups between habit types and speciation modes are demonstrated.

## 2.1. Introduction

The analysis of speciation processes is a foundational task of causal evolutionary research. Taking into consideration an estimated number of 2 million organisms which, according to evolutionary theory, all descended from one "original" organism, we need to ask the question which mechanisms caused the speciation, *i.e.*, the permanent splitting of a species into two or more stable daughter species. This question is also relevant within the framework of the "basic type" concept (Scherer 1993b, this volume), because here, too, it is assumed that the number of original polyphyletic "types" was much smaller than the number of present-day biospecies, morphospecies or evolutionary species.

**Biospecies**: Members of a biological species interbreed in their natural environment, producing viable offspring, and are, as far as their reproduction is concerned, isolated from the members of other biospecies (Mayr, *e.g.*, Mayr 1991).

**Morphospecies**: The members of a morphospecies show the same essential characters (Sedlag & Weinert 1987) (similar to taxonomic species).

**Ecospecies**: "Species are groups of organisms that, over the course of generations, develop and maintain a common ecological niche, forming, wherever necessary, a reproductive community" (von Wahlert 1981).

**Evolutionary species**: Evolutionary species are groups of populations with a common evolutionary history in time, and they are separated from other such groups which have a different evolutionary history (Templeton 1991, Wiley 1978).

Table 1. A selection of different definitions of the term species.

## 2.2. Definitions of the Term Species

We cannot discuss prerequisites and processes of speciation without defining the term species. The meaning of speciation depends on the definition of species (see Table 1; different definitions of species can be combined [see Sudhaus 1986, p. 98]; see Scherer 1993b for further discussion on the species problem).

- If we define species as biospecies, we must determine how crossing barriers arise.
- If we distinguish species according to morphological differences, we must establish criteria for those differences.
- If we define species by ecological criteria, the establishment of ecological niches must be researched.
- If our basis is the evolutionary species, the question arises, how new independently evolving populations could form.

Practical experience has shown that criteria based on different definitions of species, as well as other criteria, often do not correlate. Hybridization, divergences in chromosome structure, morphology, ecology, and evolutionary history are in many cases independent of each other. The following examples reflect situations that are often encountered in field research. They represent the data base for speciation models.

Hybridization barriers can form without causing a significant change of outward characters. Examples for this are sibling species and many annual plants. In the case of the genus *Stephanomeria* (*Asteraceae*) it was actually possible to observe the beginning stages of the formation of an autogamous species, *S. malheurensis*, splitting from the common species, *S. exigua*. Apart from the size of their fruit, the two species are very similar but are separated almost completely due to their different reproductive systems, lacking compatibility and showing partial F$_1$ sterility, in spite of the fact that they both occur in the same areas (Ehrendorfer 1984, p. 254). Similar situations are described by Lewis and Roberts (1956) and Lewis and Raven (1958) in the North American genus *Clarkia* (comp. Lewis 1966, p. 169). In this case, too, the hybrids are almost completely sterile, in spite of morphological and ecological similarities of the parents (and must therefore be considered different biospecies, in spite of their similar outward appearance). The reason for this is multiple changes in the chromosomes (Figure 1). On the other hand, a particular population produces fertile hybrids with other populations (some of which are considered to belong to different species) that clearly differ vegetatively and reproducitely.

On the other hand, morphological differences can arise without causing hybridization barriers. An example of this is the subgenus *Lepidobalanus* of the genus *Quercus* (oak): "*Lepidobalanus* is ... an example of a woody plant group in which morphological and eco-physiological differentiation far exceeds the formation of reproductive barriers. No one has yet proposed to include, due to the lack of such barriers, all taxa of this subgenus in one single species. We are obviously facing a discrepancy between the concepts of the 'taxonomic' and the 'biological' species" (Ehrendorfer 1984, p. 239).

Similar situations exist not only in many other woody plants, but also in some perennial herbs. For example, in the orchid genus *Ophrys* (Twayblade), the lower lip of the blossom imitates the form, appearance and scent of various pollinating insects, promoting reproductive isolation. In spite of the great variety of lower lip characters and specialization to certain different pollinators, species of the genus *Ophrys* usually interbreed without any problems (Danesch and Danesch 1972).

Phylogenetic relationship does not necessarily correlate to genetic compatibility. Two species that have been separated evolutionary for a long period may still interbreed, while species only recently separated may not (Cracraft 1989).

An example for the first case is the two sycamore species *Platanus occidentalis* from North America and *P. orientalis* from the eastern Mediterranean region and adjacent southwest Asia. According to the evolutionary model, these two species have been separated for at least 20 million years (and, thus, are to be considered to be two different evolutionary species). However, they interbreed without any loss of fertility, proving to belong to the same biospecies (Ehrendorfer 1984, p. 239). Stebbins and Day (1967) found a similar situation in the disjunct plantain species *Plantago insularis* and *P. ovata,* which occur on different continents. In spite of a presumably long period of separation, these two species show closer resemblance than all other *Plantago* species and are partially fertile when interbred.

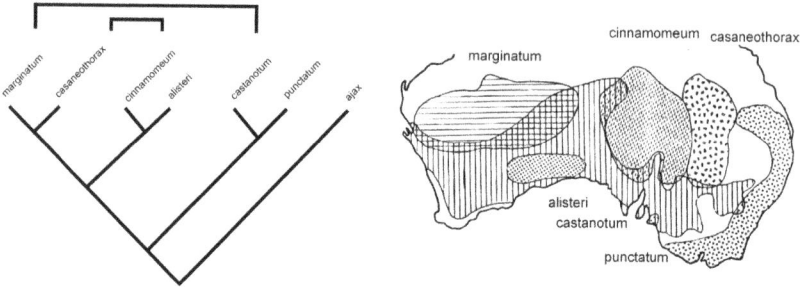

Figure 1. Distribution and hypothetical phylogenetic relationships of the Australian thrushes of the genus *Cinclosoma*, based on a comparison of 32 characters. The brackets indicate which species are known to interbreed (after Cracraft 1989, simplified).

Examples for the opposite case are species that were formed through polyploidization. Reproductive isolation develops rapidly in them, in spite of their close affinity to the original species. The same is true for species of hybrid origin, which arose without polyploidization (see below).

Cracraft (1989, p. 43) pointed out that, for the birds of the Australian genus *Daphoenositta,* the biospecies concept blurs the evolutionary history, which had been reconstructed by phylogenetic systematics (comparison of [exomorphic] characters). All five biospecies in that area are able to interbreed. Interbreeding between groups that were not sister groups was also observed among thrushes of the Australian genus *Cinclosoma*, whereas sister groups that formed for geographical and ecological reasons did not hybridize (Cracraft 1989, p. 37ff; see Figure1).

Cytogenetic processes leading to the formation of endogenous reproductive barriers are often very different from those related to adaptation to the environment (Ehrendorfer 1984, p. 248).

*What, Then, is Speciation?* Considering these different situations, we will use the term speciation to mean the splitting of one line into two daughter lines which are called "sister species" (according to Hennig) (Sudhaus 1984, p. 195). This includes the fact that species occurring in habitats that overlap at a later point in time are reproductively isolated and ecologically differentiated. *Speciation models must be able to explain the formation of isolation and coexistence mechanisms (divergence of niche characters).* This can include the explanation of the formation of morphological divergence.

"To explain speciation means primarily to explain the formation of independence and discontinuity between sympatric species caused by isolation mechanisms" (Sudhaus 1984, p. 195f).

We have to distinguish between speciation *sensu stricto* and those changes of the species which do not lead to the separation of populations.

More than a century after the establishment of evolutionary theory, the controversies around speciation processes are still continuing (comp.

Mayr 1991, p. 269f). Depending on the different speciation models a number of factors are taken into consideration which can play a role in speciation processes (as indicated in the examples mentioned above). The roles that these factors play can be very different in different cases and include karyotypic changes (reorganization of chromosomes), changes of the DNA-sequence, isozymal differentiation, morphological differences, or ecological requirements. However, none of these changes ever prove indispensable for speciation. There is no unified spectrum of requirements for speciation (Templeton 1981; comp. Lange 1992).

Data from a number of disciplines are necessary to evaluate speciation processes: taxonomic-systematic data, population size, migration behavior, patterns of geographic distribution, genetic differences, behavioral

Normal    Internal

Allopatric

Parapatric

Sympatric

Figure 2. Types of distribution of two populations (after Sudhaus 1984).

biology, mating biology, reproductive biology, ecological factors, and life histories have to be known in order to construct realistic speciation models. Extensive analyses are therefore indispensable.

Speciation processes are usually not observable. A sequence of events has to be designed hypothetically in order to interpret existing empirical data. It is the various relationships among and within populations that are observed.

The genus *Pinus* (pine) includes local breeds, geographically and ecologically differentiated subspecies and semispecies, as well as more or less isolated species (Ehrendorfer 1984, p. 236). These various relationships among *Pinus*-taxa are considered to be different degrees of speciation. Momentary snapshots are, thus, combined into a process.

## 2.3. Geographic Relationships

One of the basic questions concerning speciation is: What *geographic* relationship does an initially or newly forming species have to its parental population (Mayr 1991, p. 269)? Is a geographic separation necessary?

### 2.3.1. The Classical Model: Allopatric Speciation.
According to the classical allopatric model, speciation is only possible if there is at least a temporary geographic separation between daughter populations of the same species. The gene flow must be interrupted completely by factors *outside* the organisms (see Figure 2, above), so that a merging

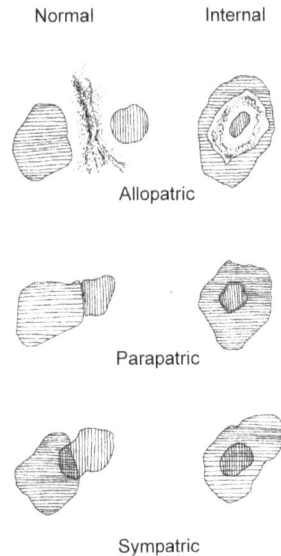

of the gene pool is prevented. The separation can occur in different phases of the splitting process (at the beginning, in the middle, or at the end). During the separation phase, either genetic, ecological and morphological differences are established, or existing differences in a polymorphic ancestral population are reinforced. These changes allow daughter populations to be isolated in terms of their reproduction and their niches when, at some later time, their geographic ranges overlap.

The differentiation during the phase of separation is ascribed to various genetic events (mutations, recombinations, reorganizations of the chromosomal structure), differing selective conditions and effects, or to genetic drift (coincidental fixation or elimination of alleles).

An important special case within the allopatric model of speciation is the "founder model" which we will discuss further below.

**2.3.2. Parapatric (Stasipatric) Speciation.** In the case of parapatric speciation, as in that of sympatric speciation, external factors separating populations (independent from organisms) do not play a vital role during the entire process of speciation (Zwölfer and Bush 1981). A complete interruption of the gene flow due to geography is considered unnecessary; the interruption of gene flow has to be caused by the organisms themselves, due to evolutionary events.

According to the parapatric model of speciation (White 1968; Sudhaus 1984, p. 204f; Zwölfer and Bush 1981, p. 215f; Figure 2, middle) new species are formed in marginal areas of polymorphic populations (therefore the term "parapatric"). This requires: an ecological gradient resulting in both a clinal or divergent selection and polymorphism; small populations; and low vagility (low mobility, small mating radius).

In the parapatric model, karyotypical changes (reorganization of the chromosomal structure due to translocations, fusions, inversions, *etc.*) are considered to be *pacemakers* of the speciation process, not the result thereof (as in the allopatric model). Karyotypical changes are supposed to *initiate* isolation and divergence.

However, the intial result is a selective disadvantage for the heterozygous state due to the reorganization of the chromosomal structure (reduced fertility of the descendants due to meiotic disturbances). In order for individuals with changed chromosomes to be able to survive and establish their own population, they need to overcome the selective disadvantages that are due to a heterozygous state. A homozygous state of the changed chromosomal structure has to be acquired as soon as possible, and its carriers have to be selectively superior (at least in their area).

However, due to the low vagility and the small population size required in this model, the chances for inbreeding and, thus, the chance to reach a homozygous state are relatively high. In the peripheral zone of the population, and in case of a high rate of inbreeding, the chromosomal changes that occur have a high chance of remaining. Then inbreeding furthers a homozygous state, whereas selection works against a heterozygous state due to the incompatibility of different karyotypes.

The parapatric model of speciation is based on the following observations:

1. A number of parapatrically distributed forms show differences in their karyotypes or different ecological requirements, causing a post-mating isolation (hybrids are usually subviable and sterile) (Mayr 1947, 1991, p. 280; Mayr and Vaurie 1948).

2. Character gradients exist universally (Endler 1977).

3. Another argument for parapatric (and sympatric, see below) speciation is that, in the history of species that supposedly were formed through this process of speciation, no geographic separation is plausible (this argument is used in the case of accumulations of species occurring in freshwater lakes, comp. Lange 1992, p. 106f).

4. Endler (1977) mentions statistical evaluations according to which subspecific differentiations have been described much more frequently in the case of parapatric than in disjunct distribution patterns—in contrast to the allopatric model.

It is debated whether speciation is possible without separation. Mayr (1991) argues that there are no examples of supposed para- or sympatric speciation where the existing data could not also be interpreted by the allopatric model. Critics like Mayr claim that they can make earlier separations of currently sympatric species plausible; they either interpret character gradients as secondary events following previous separations, or they deny that character gradients without separation can lead to speciation (reproductive isolation). Mayr (1991, p. 286) takes into consideration that the limits of parapatrically occurring species often do not correlate with physiographical limits or limits due to vegetation, a convincing proof for the secondary development of parapatry. For if parapatry were a primary condition, the geographical boundary would be expected to coincide with the boundary of life-limiting conditions. Futuyma and Mayer (1980), based on detailed investigations in natural environments, consider the pacemaker role of karyotypic changes for parapatric speciation to be unlikely.

**2.3.3. Sympatric Speciation.** In the case of sympatric speciation the daughter species arises in the same area as the original species due to a change of niches (a change of ecological needs) which simultaneously causes a pre-mating reproductive isolation (Bush 1975, p. 352ff; Tauber and Tauber 1989; Figure 2 below). The probability of a gene flow is supposed to depend only on the genetic makeup of the individuals involved (Kondrashov and Mina 1986, p. 202). Isolation mechanisms

Figure 3. Niche elements of the cherry fly *Rhagoletis*. 1 habitat of the fly, 2 food of the fly, 3 place of copulation, 4 place for laying the eggs, 5 habitat of the larvae, 6 food of the larvae, 7 place of forming the cocoon, 8 place of overwintering. The niche elements 3-6 are coupled to the host plant A by specific interactions. Shifting to host plant B means modifying several niche elements. (After Zwölfer & Bush 1984).

are supposed to occur within a population linked by interbreeding (Futuyma and Mayer 1980, p. 255). The model of sympatric speciation is heavily debated (Kondrashov and Mina 1986; Sudhaus 1986; Lange 1992; comp. Mayr 1991, p. 292ff).

Examples of sympatric speciation are usually parasites and stenoecious phytophagous invertebrates undergoing shifts in host species (Bush 1975, p. 352ff; divergent host specialization). Due to the strict limitation to a particular host organism (which, at the same time, has to be the place of copulation), the gene flow is channeled to a high degree (comp. Figure 3).

The change of niches can be initiated by adaptive processes or genetic drift. It results in a disruptive selection, which is regarded as the initiating force of diversification (comp. Kondrashov and Mina 1986). Only after diversification has started, a geographical separation gains increasing importance (Tauber and Tauber 1989, p. 311ff), for example, due to strict limitation to different host species.

Advocates of sympatric speciation believe it to be possible if the following prerequisites exist (Weninger and Kattmann 1981; comp. Maynard Smith 1966; Zwölfer and Bush 1981, p.220):
- the original species is polymorphic,
- it lives in a mosaic-shaped habitat,
- through mutation, genes arise which give its carrier a better chance to adapt to a certain type of habitat (e.g., change to a new host species),
- plus, genes that cause the organism to be bound to a certain type of habitat must be linked (through chromosomal mutations), which increases the fitness of the respective genotype,
- "assortative copulation" has to occur (copulation of individuals which are especially well adapted to a common habitat [e.g., host organism]).

According to this model a gradual linking of genes is taking place which causes selective advantages in certain areas (e.g., for specific host species), on one hand, and a preference for respective dependency to those areas (hosts), on the other hand.

For sympatric speciation, the structure of the niche is important. The elements of the niche have to be connected as closely as possible, for example, by the place of feeding being identical to the place of copulation and encounters with competitors and enemies (Figure 3). A close connection between the elements of the niche provides better channeling of the gene flow, and ecological parameters can be changed easier *en bloc*. Thus, a high degree of ecological specialization can be reached comparatively quickly, and a loss of the achieved differences tends to be prevented (Weninger and Kattmann 1981, p. 138f; Zwölfer and Bush 1981, p. 228; Tauber and Tauber 1989). One example would be the locusts of the species complex *Enchenopa binotata* (Tauber and Tauber 1989, p. 319). Six different species living on different sympatric host plants are known. Reproduction occurs within a limited time frame. The females lay their eggs inside the branches of the host plants where the eggs hibernate. In the spring, when the sap starts to flow through the branches again, the young hatch. The beginning point of the sap flow differs in the various species of host plants; therefore, the life cycles are asynchronous, which is also true of the mating times.

There are no distinct lines of separation between the modes of speciation described thus far (comp. Tauber and Tauber 1989, p. 311). High-degree specialization of parasites or phytophagous insects towards a host species, for example, resembles allopatry, even without geographical separation.

**2.3.4. Speciation by Polyploidization.** The formation of polyploids provides immediate genetic isolation from the original species, since breeding of individuals with differing degrees of polyploidy leads to considerable meiotic disturbances and therefore, sterility. Since probably three-quarters of all seed-bearing plants are polyploid, this mode of speciation plays an important role in the diversification of spermatophytes, frequently in connection with hybridization (allopolyploidy).

**2.3.5. Speciation by Hybridization.** New species can also be formed by hybridization between different biospecies (Grant 1966; Templeton 1981; Zwölfer and Bush 1981, p. 217). Hybrids usually are at a selective disadvantage. In disturbed environments, however, hybrids can have an advantage. Under certain circumstances it might suffice if at least some hybrids are viable and fertile with each other. Hybrids of different biospecies can exhibit genetic barriers with their parent species (Lewis 1966). In cooperation with speciation by small founder populations and subsequent hybridization, a splitting of species could happen rapidly. To this the observation can be added that hybridization of distantly related species can result in a disturbance of the genetic equilibrium and, under changed selective influences, can lead to a shift in that equilibrium.

Lewis (1966, p. 167ff) compares two cases. The species of sage *Salvia mellifera* and *S. apiana* form large accumulations of hybrids in their natural environment (in spite of considerable morphological differences). These hybrids are found in habitats that are severely disturbed by human activity. On the other hand, two species of columbine (*Aquilegia formosa* and *A. pubescens*), which show similar degrees of differentiation as the *Salvia* species and can be hybridized experimentally just as easily as the *Salvia* species, form only a very few hybrids in their natural environment. The reason could be a lack of suitable "hybrid habitats".

A similar example is cited by Templeton (1989, p. 18). In dense stands, red and black oaks rarely form hybrids because the seeds of the hybrids are unable to sprout in the dark and relatively cool forest floor. On the other hand, in open forests hybrids sprout much easier than their parent species.

**2.3.6. Formation of Agamic Complexes.** It should be mentioned that among plants and sporatically among animals the formation of agamic complexes plays a role in speciation processes. It occurs due to the loss of sexual reproduction (apomixis), for example, the formation of seeds without fertilization (agamospermy). "Thus, problems in the meiosis and fertility of hybrids are compensated by the fixation of heterotic or other advantageous recombinations. Often polyploidy is involved as well" (Ehrendorfer 1984, p. 252). Examples are the well known genus *Taraxacum* (dandelion) or the huge agamic complex of the subgenus *Eubatus* (blackberry) in the genus *Rubus* (Rosaceae).

## 2.4. Development of Isolation and Coexistence Mechanisms.

Explaining the development of isolation and coexistence mechanisms is an important goal of speciation models. What has been said about this subject will be summarized below. Table 2 provides an overview of the possibilities of interrupting the gene flow (Zwölfer and Bush 1981, p. 217; comp. Tauber and Tauber 1989, p. 311). According to the different models, speciation can be *initiated* by the formation of endogenous or exogenous barriers. Exogenous barriers cannot cause speciation, they can only provide a *prerequisite* for the formation of endogenous barriers (isolation factors) by preventing the mixing of variations. According to the allopatric model, exogenous barriers are an essential prerequisite (see above).

Isolation can occur by changes in ecological niches (pre-mating isolation factors, *e.g.*, differing periods of fertility, different ecological requirements), by changes in behavior (mating behavior, "non-overlapping communication systems" [*e.g.*, Spieth 1949]), or by reconstruction of the genome and other genetic changes (metagamic isolation mechanisms).

| Type of Barrier | Gene flow prevented by | Name |
|---|---|---|
| exclusively exogenous | geographical obstacles, great distances | allopatric speciation; peripatric speciation |
| exogenous and endogenous | clinal selection and divergent adaptation | parapatric speciation |
| | change of food and habitat niche; change of reproductive period | sympatric speciation |
| exclusively or predominantly endogenous | genetic incompatibility | sympatric speciation; transilience |

Table 2. Overview of the speciation types mentioned (modified after Zwölfer & Bush 1984).

To what degree isolation mechanisms can form directly through negative interactions with other species ("reinforcement"; see, e.g., Butlin 1987a, 1987b [disagrees]; Spencer et al. 1987 [agrees]), or whether they arise as a pleiotropic by-product of a divergent adaptive pressure, has not been clarified for sure. Thus, differing opinions exist as to whether isolation mechanisms have to form completely during a geographical separation, or whether incompletely formed isolation mechanisms can be completed when divergent populations overlap. Parallel to isolation mechanisms, coexistence mechanisms (settling in new niches; ecological separation) have to be established to avoid competition (Sudhaus 1984, p. 193ff). Unlimited competition due to niche equivalence will eventually lead to one of the competing species being pushed out of its habitat.

## 2.5. Genetics of Speciation

Several things have already been said about the genetics of speciation. The introductory examples showed that genetic changes of daughter species can vary considerably—from small point mutations (accompanied, e.g., by a change of behavior) to extensive chromosomal reconstructions (which can influence the formation of a species to a large degree) (Junker 1993a, this volume). Even though structural changes in chromosomes often cause genetic isolation, speciation is also possible without significant chromosomal changes (Mayr 1991, p. 379). Genes of enzymes usually do not play an important role in speciation as those of closely related species often do not show any differences (Ayala 1982; Patton and Smith 1989, p. 300; Mayr 1991, pp. 273, 378). However, sometimes different species can be distinguished more by enzymatic divergences than by morphological characters (Sperlich 1984). Regulatory genes may play a role in speciation, but this is presently rather speculative (Mayr 1991, p. 379). In most cases, the genetic basis of speciation is completely unclear.

## 2.6. Speed of Speciation

There is much evidence the speed of speciation is indirectly proportional to the size of the separated populations (Bush 1975, p. 346; Bush *et al.* 1977; Lange 1992; disputed by: Barton and Charlesworth 1984). The gene pool of a small founder population or a small residual population can most easily experience profound and permanent changes.

Mayr (1991, p. 280) mentions the example of the kingfisher species *Tanysiptera galathea*. On the mainland of New Guinea, it does not show significant divergences in spite of its distribution over an area of 1000 km (600 miles) and several different climatic zones, whereas all populations occurring on the islands near the coast of New Guinea differ enough to have been described as separate species. Each island was probably initially settled by a single founding pair.

Mayr (1991, p. 279) uses the term "peripatric speciation", Templeton (1980; 1981) uses the term "transilience", Carson argues for a "founder-flush model" (Carson and Templeton 1984). According to Mayr's model, (rapid) speciation occurs according to the following pattern (Mayr 1954; comp. Mayr 1991, pp. 281, 375ff):

- genetic variability decreases considerably;
- homozygosity increases considerably; thus, for the first time many alleles are being exposed to selection (due to a different ecological situation the selective conditions can differ from those of the original population);
- the coherence of the genotype due to disturbances of allelic and epistatic equilibria (cohesion; linkage of genes) is loosened (comp. Levin 1970);
- the status of detrimental heterozygosity (due to changes in chromosomal structure) can be rapidly overcome by reaching a homozygous state (intense inbreeding);
- there is a maximal chance for occupying new niches.

Newer results, however, question whether a "bottleneck" results in a significant loss of genetic variability (Carson 1990; comp. Carson and Templeton 1984). Moreover it is not easily explainable just why populations with minimal variability can acquire new adaptations and genetic equilibria rapidly and effectively (Carson and Templeton 1984, p. 119ff). These objections are being considered in the "transilience" model by Templeton (which can be applied to sympatric speciation as well) and in the "founder-flush" model by Carson. "Transilience" is a sudden shift in a genetic complex which influences its fitness, caused by a sudden disturbance in the genetic milieu (Templeton 1980, p. 1013; Carson 1982; Carson and Templeton 1984, p. 117). Thus, the founder population is exposed to an intense selective pressure which can only

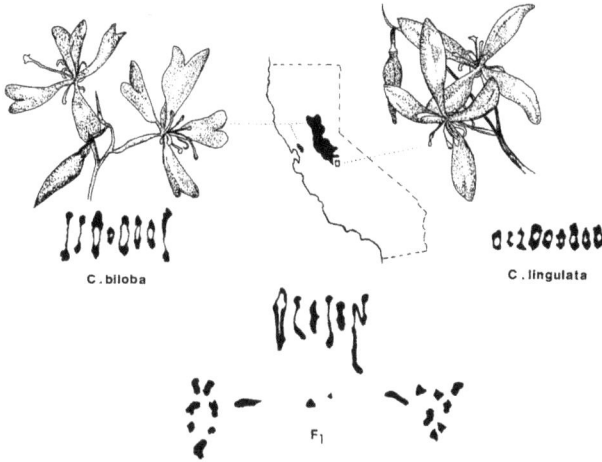

Figure 4. Peripatric origin of *Clarkia lingulata* (2n = 18) from *C. biloba* (2n = 16), which inhabits a much larger area of California. Included are floral structures and karyotypes. The F₁ hybrid is more or less sterile because meiosis is strongly disrupted. (After Lewis & Roberts 1956, modified).

be overcome by high variability in the founder population (in contrast to Mayr's model).

Carson's "founder-flush" theory is similar. According to this theory the founding event has to result in a "disorganization" of the polygenically balanced gene pool within a short period of time. Subsequently a reorganization and possible change of the gene pool takes place (Carson 1982). After the founding event, the population has to grow rapidly ("flush" phase), in order to preserve as much variability as possible. During a phase of temporarily diminished selective influences (in contrast to the transilience model) in the newly settled area, new recombinations can be established within that pool which would have otherwise been selected against. Later, selective influences increase; thus, a new "peak" can be reached (Carson 1982; Carson and Templeton 1984, p. 116). Carson (1982) calculates several hundred to a thousand generations for the entire process.

Lewis (1966) discusses a sudden speciation which could be caused by chromosomal reorganization in small marginal populations (parapatric or allopatric; Figure 4; there is a correlation between chromosomal reorganization and diversity of species [Bush et al. 1977; King 1987]). The fate of individuals with a changed chromosomal structure is similar to that of hybrids. Heterozygous descendants of hybrids usually show reduced fertility due to meiotic irregularities. If, however, such hybrids survive in appropriate (new) habitats (disturbed environment) and separate geographically, then there is a chance that the new chromosomal type occurs in a homozygous state. Thus, the meiotic problems are overcome.

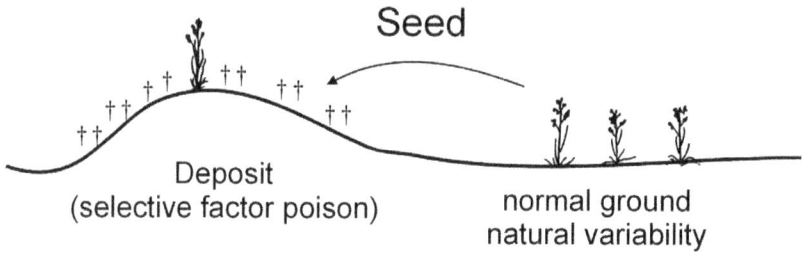

Figure 5. Colonization of mining deposits contaminated by heavy metals. Most of the seeds cannot germinate or those that do so usually soon die (†).

In this way, individuals with homozygous chromosome changes can establish a new population that is genetically isolated sufficiently from the previous population.

Besides these facts from population genetics, important support for peripatric speciation is the frequent observation, that in species groups, divergent populations and divergent taxa often occur in peripherally isolated locations (Mayr 1991, 280, p. 373; advocates of parapatric speciation also use this argument).

An interesting yet unanswered question in the context of peripatric speciation is whether chromosomal changes can possibly be caused by ecological disturbance or whether chromosomal changes arising coincidentally merely have a chance [for establishment] in disturbed situations. According to Mayr (1991, p. 377), it is not mutations but the genetic variability of the original population that is the source for peripatric speciation.

A similar possibility for rapid speciation occurs when, due to extensive environmental changes, a major part of the population is eliminated by selection ("catastrophic selection" [Lewis 1962]). Only those variants that can settle at the margin of the ecological range of the species in question are able to survive. The decimation of the population results in an increase in inbreeding and the establishment of homozygotes. Because of their rarity, these previously did not have a chance to spread within the population. However, here, too, the formation of isolation mechanisms is necessary so that the new population can successfully compete with the original species.

A very instructive example of this process can be found in plants growing on top of mining deposits produced at the end of the 19th century (Macnair 1987; Figure 5). The environmental conditions in this specific habitat are highly unfavorable. Besides the fact that the ground is contaminated by various heavy metals, the plants have to deal with dehydration and nutrient deficiencies. The plants that do grow on the contaminated ground probably possess the ability to bind heavy metals by means of specific proteins and thereby neutralize them. The ability to tolerate poison is also present in a small percentage of those individuals which grow on normal grounds. It is therefore not newly evolved (Table 3). The tolerance genes existed prior to colonizing the toxic deposits, but are a disadvantage under normal circumstances (and therefore present only in a small percentage of individuals). Besides the difficult living conditions that lead to catastrophic selection, there are other specific circumstances which promote speciation: there exists almost no competition on the mining deposits; at least initially, the small number of individuals results in a high degree of inbreeding; and hybrids with between normal and poison-tolerant plants are ecologically disadvantaged. It is also assumed that the tolerance genes are linked to traits which limit the ability to cross-breed ("hitch-hiking

| Species | a | b |
|---|---|---|
| Holcus lanatus | 0.16 | + |
| Agrostis capillaris | 0.13 | + |
| Festula ovina | 0.07 | - |
| Dactylis glomerata | 0.05 | + |
| Deschampsia jlexuosa | 0.03 | + |
| Anthoxanthum odoratum | 0.02 | - |
| Festuca rubra | 0.01 | + |
| Lolium perenne | 0.005 | - |
| Poa pratensis | 0.0 | - |
| Poa trivialis | 0.0 | - |
| Phleum pratense | 0.0 | - |
| Cynosurus cristatus | 0.0 | - |
| Alopecurus pratensis | 0.0 | - |
| Bromus mollis | 0.0 | - |
| Arrhenatherum elatius | 0.0 | - |

Table 3. Percentage of individuals tolerant of heavy metals in populations on uncontaminated grounds, in relation to the occurrence of spontaneous populations of tolerant species on contaminated grounds (from Macnair 1987). a: Percentage of tolerant individuals in normal populations on uncontaminated grounds; b: presence (+) or absence (-) of tolerant species on contaminated grounds.

effect"), resulting in the selection of genes which have nothing to do with tolerance (morphology; ability to cross-breed), as is the case in the genus Mimulus (monkey flower): "A gene giving postmating isolation has been discovered in M. guttatus, closely associated with the tolerance gene" (Macnair 1987, p. 357). Furthermore, a tendency to self-pollination, as well as a shift of blossoming phases, could be observed. All these factors promote reproductive isolation and, thus, the formation of new biospecies. In many cases the line to a new species is crossed. Thus, speciation may occur within just one or very few generations. Stebbins (1982, p. 33) estimates 10 to 20 generations for speciation to occur within Clarkia lingulata (Figure 4) and 100 to 500 generations for speciation in annual plants that are not self-pollinating.

The possibility of a sudden morphospeciation is argued by Ford and Gottlieb (1992), also in Clarkia. They proved that the homeotic mutant bicalyx, in which the petals were modified into sepals (forming the calyx), turned into a stable, self-fertilizing population with unlimited fertility (comp. Scherer 1993a, this volume).

Another observation that points in the same direction as the example of the plants growing on mining deposit concerns flowering annuals (*i.e.*, plants that survive winter only as seeds). Annuals show an especially high rate of chromosomal aberrations and, therefore, little tendency towards hybridization among different biospecies (Ehrendorfer 1984, p. 257). This obviously has to do with the fact that annual plants are often pioneers; they experience bottlenecks and phases of intense inbreeding. This situation can thus be compared to peripatric speciation and obviously tends to a conservation of chromosomal changes.

Since annuals cannot survive unfavorable living conditions by vegetative phases, sexual reproduction is absolutely necessary for survival. There exists, therefore, a selective pressure towards high fertility and, thus, relative homozygosity. As a result, the genetic material is hardly mixed. This situation promotes the fixation of aberrant chromosomal types, but there is no significant morphological or ecological (or any other) divergence connected to this—one reason for the rarity of interspecific or intergeneric hybrids within plants having this life history. In this case, changes of the genome, as well as the formation of reproductive barriers, may precede morphological differentiation (which is exactly the opposite in long-living woody plants and many perennial herbs; see above; comp. Stebbins 1982, p. 26ff).

Finally, in phytophagous and parasitic invertebrates, rapid speciation is made possible also by a host shift. Choice of, and dependency on, the host seem to be influenced by the genes of only a few loci (Zwölfer and Bush 1981, p. 222).

Spurway (quoted in Bush 1975, p. 357) draws the following pointed conclusion from these considerations: "Being an Adam and Eve gives a monster a chance to hope."

## 2.7. Diversification Through Differential Restriction of the Potential of Variation

A little-discussed hypothesis to partially explain speciation begins with the assumption that the ancestral forms of present-day groups possessed a rather high spectrum of variation. Over the course of time, this spectrum was more or less reduced by defective mutations or founder events. By narrowing the original range of variation, specialized biological species are formed (comp. Lönnig 1988, p. 586). This can be seen, for example, in the fact that the descendants will be able to colonize only a few specialized habitats (stenoecious), whereas their ancestors had a wide spectrum of possible habitats. This hypothesis is particularly of interest for the basic type concept (Scherer 1993a, comp. Scherer and Hilsberg 1982). An example of this is the genus *Geum* mentioned elsewhere in this volume (Junker 1993b, this volume), as well as plants which occur only in the Alps (Figure 6).

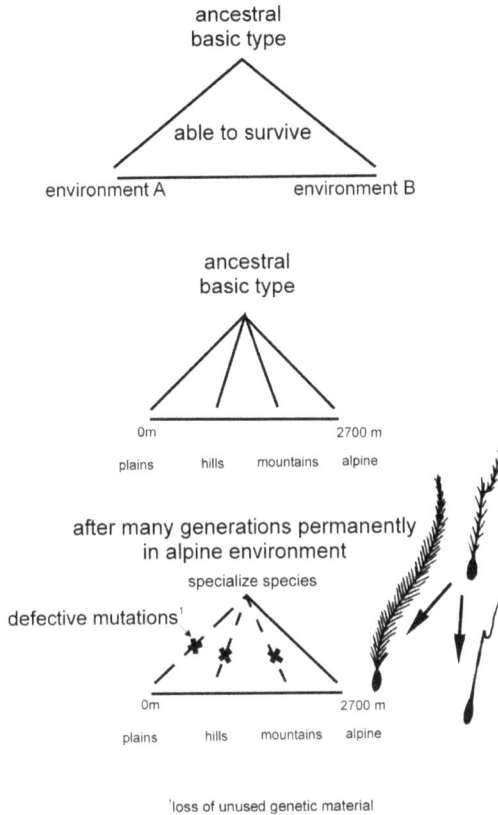

Figure 6. Variation potential of basic types (above), exemplified by the altitude-related distribution (middle) and narrowing of the potential in obligatory alpine plants (below). An example (right) is the avens (*Geum*). Above: The fruit of *Geum rivale* (showing a pinnate style, adapted to spread by animals, wind, and water; occurs in altitudes from 300m to over 2000 m); below left the fruit of *Geum montanum* (showing a pinnate style with even thicker hairs; spread mainly by wind; occurs only in alpine meadows); right the fruit of *Geum urbanum* (showing a harpoon-formed style; occurs in lower altitudes, near forests, in gardens, etc.; spread by animals). (Comp. Junker 1993b.)

This hypothesis basically claims that a transformation from species of wide habitat tolerance to species of narrow habitat tolerance occurred (and is still occuring). It is possible, however, that wide tolerance is retained. Thus, a high potential of variation, which was reduced in different ways over several generations, can be considered a source of the diversification within basic types.

The occurrence of parallel variations can provide the empirical basis for this model. That is, the changes of the soma (variations that were released, for example, by environmental causes) become genetically fixed ("copied"), *i.e.*, variations become genetically stable. Stubbe (1966), in his study of the herb *Antirrhinum*, points out that "all environmentally caused variations of the phenotype could also be caused by mutations."

Plus, numerous expressed traits caused by mutations can also arise as non-inherited modifications (phenocopies). These observations can be interpreted within the framework of the mode of speciation suggested here. Due to their higher genetic potential, the original species possessed a wide potential of variation and adaptability to all kinds of environmental conditions. It was narrowed by a subsequent accumulation of slightly disadvantageous mutations (narrowing of the range of variation). Due to the mutational degradation of the variation potential, the variations become "hereditary" and, thus, cease to be variations. Although it looks like it, there is no inheritance of acquired characters (comp. Lönnig 1988, pp. 473, 586). In many species only those areas of the adaptive potential that were necessary for the respective environmental conditions have been retained. The diversified narrowing of the spectrum of variation can result in morphological divergence; it may also promote reproductive isolation. Ho (1984, p. 267ff) writes: "It turns out that parallels between artificially induced modifications and naturally occurring variations are extremely common... these same structural modifications can be artificially induced in related organisms by simulating the appropriate environmental conditions in the laboratory."

## 2.8. Conclusions

The discussion about speciation modes is active. In spite of extensive data, the statement of Futuyma and Mayer (1980, p. 269) still holds true: the genetic and selective factors that cause reproductive isolation are still largely unknown. Especially the differences of opinion regarding the plausibility and meaning of the parapatric and sympatric mode of speciation persist to the present. One main reason for this is the difficulty in designing an *experimentum crucis* that could lead to a decision for or against one or the other concept. Numerous data can be interpreted by various speciation models. "Neither sympatric nor allopatric hypotheses are easily subjected to falsifying tests in the analysis of natural situations" (Kondrashov and Mina 1986). An example of this is the difficulty in distinguishing between primarily overlapping areas and secondarily overlapping areas (Endler 1977, p. 178). The plausibility of speciation models depends largely upon whether the conditions for specific case studies can be made plausible.

# References

Ayala, F. 1982. Gradualism versus punctualism in speciation: Reproductive isolation, morphology, genetics. In: Barigozzi, C., ed. *Mechanisms of Speciation*. Alan R. Liss, New York, pp. 51-66.

Barton, N.H. and B. Charlesworth. 1984. Genetic revolutions, founder effects, and speciation. *Annual Review of Ecology and Systematics*. 15:133-164.

Bush, G.L. 1975. Modes of animal speciation. *Annual Review of Ecology and Systematics*. 6:339-364.

Bush, G.L. 1984. Stasipatric speciation and rapid evolution in animals. In: Atchley, W.R. and D. Woodruff, eds. *Evolution and Speciation*. Cambridge University Press, Cambridge, England, pp. 201-218.

Bush, G.L, S.M. Case, A.C. Wilson, and J.L. Patton. 1977. Rapid speciation and chromosomal evolution in mammals. *Proceedings of the National Academy of Sciences* 74:3942-3946.

Butlin, R.K. 1987a. Speciation by reinforcement. *Trends in Ecology and Evolution* 2:8-13.

Butlin, R.K. 1987b. Species, speciation, and reinforcement. *American Naturalist* 130:461-464.

Carson, H.L. 1982. Speciation as a major reorganization of polygenic balances. In: Barigozzi, C., ed. *Mechanisms of Speciation*. Alan R. Liss, New York, pp. 411-433.

Carson, H.L. 1990. Increased genetic variance after a population bottleneck. *Trends in Ecology and Evolution* 5:228-230.

Carson, H.L and A.R. Templeton. 1984. Genetic revolutions in relation to speciation phenomena: The founding of new populations. *Annual Review of Ecology and Systematics* 15:97-131.

Cracraft, J. 1989. Speciation and its ontology: The empirical consequences of alternative species concepts for understanding patterns and processes of differentiation. In: Otte, D. and J.A. Endler, eds. *Speciation and its Consequences*. Sinauer Associates, Sunderland, MA, pp. 28-59.

Danesch, O and E. Danesch. 1972. *Orchideen Europas: Ophrys-Hybriden*. Hallwag, Bern, Germany.

Ehrendorfer, F. 1984. Artbegriff und Artbildung aus botanischer Sicht. *Zeitschrift für zoologische Systematik und Evolutionsforschung* 22:234-263.

Endler, J.A. 1977. *Geographic Variation, Speciation, and Clines*. Princeton University Press, Princeton, NJ.

Ford, V.S. and L.D. Gottlieb. 1992. *Bicalyx* is a natural homeotic floral mutant. *Nature* 358:671-673.

Futuyma, D.C. and G.C. Mayer. 1980. Non-allopatric speciation in animals. *Systematic Zoology* 29:254-271.

Grant, V. 1966. The origin of a new species of *Gilia* in an hybridization experiment. *Genetics* 54:1189-1199.

Ho, M.W. 1984. Environment and heredity in development and evolution.

In: Ho, M.W. and P.T. Saunders, eds. *Beyond Neo-Darwinism*. Academic Press, London, pp. 267-289.

Junker, R. 1993a. Der Grundtyp der Weizenartigen (Triticeae). In: Scherer, S., ed. *Typen des Lebens*, Pascal, Berlin, pp. 75-93 .

Junker, R. 1993b. Die Gattungen *Geum* (Nelkenwurz), *Coluria* und *Waldsteinia* (Rosaceae, Tribus Geeae). In: Scherer, S., ed. *Typen des Lebens*, Pascal, Berlin, pp. 95-111.

King, M. 1987. Chromosomal rearrangements, speciation and the theoretical approach. *Heredity* 59:1-6.

Kondrashov, A.S. and M.V. Mina. 1986. Sympatric speciation: When is it possible? *Biological Journal of Linnean Society* 27:201-223.

Lange, E. 1992. Artbildung auf den Hawaii-Inseln und in den großen ostafrikanischen Seen. *Biologisches Zentralblatt* 111:92-113.

Levin, D.A. 1970. Developmental instability and evolution in peripheral isolates. *American Naturalist* 104:343-353.

Lewis, H. 1962. Catastrophic selection as a factor in speciation. *Evolution* 16:257-271.

Lewis, H. 1966. Speciation in flowering plants. *Science* 152:167-172.

Lewis, H and P.H. Raven. 1958. Rapid evolution in *Clarkia*. *Evolution* 12:319-336.

Lewis, H. and M.R. Roberts. 1956. The origin of *Clarkia lingulata*. *Evolution* 10:126-138.

Lönnig, W.E. 1988. *Artbegriff, Evolution und Schöpfung*. Selbstverlag, Colonge, Germany.

MacNair, M.R. 1987. Heavy metal tolerance in plants: a model evolutionary system. *Trends in Ecology & Evolution* 2(12):354-358.

Maynard Smith, J. 1966. Sympatric speciation. *American Naturalist* 100:637-650.

Mayr, E. 1947. Ecological factors in speciation. *Evolution* 1:263-288.

Mayr, E. 1954. Change of genetic environment and evolution. In: Huxley, J., A. Hardy and E.B. Ford, eds. *Evolution as a Process*. Allen and Unwin, London, pp. 157-180.

Mayr, E. 1991. *Eine neue Philosophie der Biologie*. Piper, Munich, Germany.

Mayr, E and C. Vaurie. 1948. Evolution in the family Dicruridae. *Evolution* 2:238-265.

Patton, J.L. and M.F. Smith. 1989. Population structure and the genetic and morphologic divergence among pocket gopher species (genus *Thomomys*). In: Otte, D. and J.A. Endler, eds. *Speciation and its Consequences*. Sinauer Associates, Sunderland, MA, pp. 284-304.

Scherer, S. 1993a. Sprunghafte Bildung einer morphologisch klar unterschiedenen Art? *Spektrum der Wissenschaft* August:16-19.

Scherer, S. 1993b. Basic types of life. In: Scherer, S., ed. *Typen des Lebens*, Pascal, Berlin, pp. 11-30.

Scherer, S and T. Hilsberg. 1982. Hybridisierung und Verwandtschaftsgrade innerhalb der Anatidae—eine systematische und evolutionstheoretische Betrachtung. *Journal für Ornithologie* 123:357-380.

Sedlag, U. and E. Weinert. 1987. *Biogeographie, Artbildung, Evolution*

*(Wörterbücher der Biologie)*. Gustav Fischer, Jena, Germany.

Spencer, H.G., D.M. Lambert, and B.H. McArdle. 1987. Reinforcement, species, and speciation: A reply to Butlin. *American Naturalist* 130:958-962.

Sperlich, D. 1984. Populationsgenetische Aspekte der Artbildung. *Zeitschrift für zoologische Systematik und Evolutionsforschung* 22:169-183.

Spieth, H.T. 1949. Sexual behavior and isolation in *Drosophila*. II: The interspecific mating behavior of species of the *willistoni* group. *Evolution* 3:67-81.

Stebbins, G.L. 1982. Plant speciation. In: Barigozzi ,C., ed. *Mechanisms of Speciation*. Alan R. Liss, New York, pp. 21-39.

Stebbins, G.L and A. Day. 1967. Cytogenetic evidence for long continued stability in the genus *Plantago*. *Evolution* 21:409-428.

Stubbe, H. 1966. *Genetik und Zytologie von* Antirrhinum *L. sect.* Antirrhinum. Gustav Fischer, Jena, Germany.

Sudhaus, W. 1984. Artbegriff und Artbildung in zoologischer Sicht. *Zeitschrift für zoologische Systematik und Evolutionsforschung* 22:183-211.

Sudhaus, W. 1986. Wege der Artbildung. In: Laskowski, W., ed. *Evolution: Vortragsreihe der Gesellschaft Naturforschender Freunde zu Berlin, 1983/84*. Duncker and Humblot, Berlin, pp. 91-120.

Tauber, C.A. and M.J. Tauber. 1989. Sympatric speciation in insects: Perception and perspective. In: Otte, D. and J.A. Endler, eds. *Speciation and its Consequences*. Sinauer Associates, Sunderland, MA, pp. 307-344.

Templeton, A.R. 1980. The theory of speciation via the founder principle. *Genetics* 94:1011-1038.

Templeton, A.R. 1981. Mechanisms of speciation—a population genetic approach. *Annual Review of Ecology and Systematics* 12:23-48.

Templeton, A.R. 1989. The meaning of species and speciation: A genetic perspective. In: Otte, D. and J.A. Endler, eds. *Speciation and its Consequences*. Sinauer Associates, Sunderland, MA, pp. 3-27.

von Wahlert, G. 1981. Evolution als Geschichte des Ökosystems "Biosphäre". In: Kattmann U., G. von Wahlert, and J. Weninger, eds. *Evolutionsbiologie*. Aulis, Cologne, Germany, pp. 23-70.

Weninger, J and U. Kattmann. 1981. Genkoppelung und Selektion als Faktoren allmählicher Artaufspaltung. In: Kattmann U., G. von Wahlert, and J. Weninger, eds. *Evolutionsbiologie*. Aulis, Cologne, Germany, pp. 119-155.

White, M.J.D. 1968. Models of Speciation. *Science* 159:1065-1070.

Wiley, E.O. 1978. The evolutionary species concept reconsidered. *Systematic Zoology* 27:17-26.

Zwölfer, H and G.L. Bush. 1984. Sympatrische und parapatrische Artbildung. *Zeitschrift für zoologische Systematik und Evolutionsforschung* 22:211-233.

# 3. On the Origin of "Species:" Ideological Roots of the Species Concept

PER LANDGREN

**Abstract**

In a letter from Charles Lyell to the Swedish paleontologist Sven Lovén dated December 15, 1867, the English geologist expressed his thankfulness at having been informed about the species concept of Linnaeus. Lyell had been told that the great Swedish botanist later in life had changed his view of the fixity of species and integrated further experience within his basic concept of creation to accommodate his opinion about the created original kinds. Incidentally, Lyell also mentioned in his letter that his friend Charles Darwin was quite surprised, when he, in turn, was told that Linnaeus had changed his mind.

To that very date Darwin had believed that it was fair to generalize that according to Linnaeus the species were simply immutable, fixed and distinct entities. This kind of misrepresentation has followed the image of Linnaeus and of many of his contemporaries for more than two hundred years. Linnaeus himself expressed clearly his doubts on the immutability of the species. It can be read in all the editions of his *Philosophia botanica*.

One of the purposes in this chapter is to investigate this misrepresentation of Linnaeus and the astonishment of Darwin. Obviously Darwin had previously been unaware of Linnaeus' revised thinking, in spite of the fact that he had Linnaeus' *Philosophia botanica* at hand in his own library. The chief objective in this chapter, however, is to draw attention to the ideological roots of the species concept and to some fundamental ideas of Linnaeus for the benefit of modern taxonomy.

Later in their lives Linnaeus, as well as Darwin, believed in common descent of certain species. In the context of a theistically oriented *Weltanschauung*, Linnaeus changed taxonomical level for the created kinds in the Genesis account. By this move he stressed even more the importance of genera. Speciation occurred and it was always caused by hybridization. Darwin, for his part, rejected the concept of theistic creation and, entertaining a materialistic *Weltanschauung*, insisted that all modification in the evolution of life occurred by natural forces, *i.e.*, in a non-Aristotelian and therefore non-Linnaean sense of the word "natural" and in sharp contrast to supernatural. The young Darwin's creationist view of species was in his own time old-fashioned and heavily influenced by Greek thought.

Not very surprisingly, Darwin himself quite early in his life turned from his extreme form of idealistic creationism, which the greatest among the taxonomists, Carolus Linnaeus, had abandoned as early as 1742 when it seemingly clashed with empirical data. The views of the mature Linnaeus took Darwin by surprise in 1867. Probably Darwin had always looked upon the Swedish botanist as a stalwart upholder of the fixity of species and never thought of the possibility of a new reasonable synthesis within the framework of theistic creation.

## 5.1. Introduction

### 5.1.1. Linnaeus and Darwin.

In a letter (Lyell, 1867) from Charles Lyell to the Swedish professor of paleontology Sven Lovén, the English geologist expressed his thankfulness at having been informed about the species concept of Linnaeus. During a visit to the Lyells in London 1867, Lovén had been asked about Linnaeus' view on the nature and origin of species. Lovén had said something about the early opinions of his great countryman but also that Linnaeus later "speculated" and even called some species of plants "temporis filia."[1] Back home in Stockholm, however, Lovén prepared a more detailed answer. He consulted the Uppsala botanist Elias Freis, who, according to Lovén, was more familiar with the botanical works and the spirit of Linnaeus than any man then living.

In his answer[2] Lovén outlined three different Linnaean views on species. The immutability stance of the early Linnaeus was represented by the famous quotation from *Fundamenta botanica* (Linnaeus 1736, chap. VI §157): "We count as many species as there were created forms in the beginning."[3] Lovén added the short comment that it was a "petitio principii, purely theoretical, and an accommodation to the mosaic narrative, not unusual at the time of Linnaeus." In perceiving a second view Lovén stressed the Latin word "hodienum", *i.e.,* "today," in the same paragraph from *Philosophia botanica* (Linnaeus 1751a §157)[4] where it says in a quotation from *Classes plantarum* (Linnaeus 1738) that: "there are as many species as there are different forms or structures today." This "today" made the dogma practical for Linnaeus in his systematic works, Lovén explained. That was, according to Lovén, the slight difference from the first view.[5] The third position was, however, quite different. As

1    *i.e.* "daughter of time'.
2    See Ms. Lovén S. 17. A draft of a letter to Charles Lyell. My attention to the letter from Lyell and to the draft of the letter from Lovén was drawn by Ulf Danielsson (1965, p. 187). So far the original letter has not been traced. The letters of Lyell are unfortunately scattered among several international repositories.
3    "Species tot numeramus, quot diversae formae in principio sunt creatae."
4    "Ergo species tot sunt quot diversae formae sive structurae hodienum occurrunt."
5    Perhaps it is worth commenting that what Lovén perceived to be a second view of Linnaeus was rather a conclusion drawn by his reasoning, which started from the original thesis in *Fundamenta botanica* (Linnaeus 1736).

stated by Lovén, Linnaeus began quite early to "speculate widely on the biology of species" and "more than once expressed the view that a great number of species has originated" since creation.[6] To prove his point, Lovén cited and commented on a florilegium of corroborating passages from Linnaeus' works.

Lyell was happy to receive the letter just in time for the printing of his second volume of the 10th edition of his *Principles of Geology*. He could use several references, write some new passages and give Lovén and Fries due acknowledgment in a note. (see Lyell 1872, p. 325)[7] In passing, Lyell mentioned in his reply to Lovén that his "friend Charles Darwin was a good deal surprised," when he, in turn, was told that Linnaeus had changed his mind.

Accordingly, late in 1867 Charles Darwin was told that Linnaeus had changed his mind on the species question. To that very date Darwin had believed that it was fair to generalize that according to Linnaeus the species were simply immutable, fixed and distinct entities. This kind of misrepresentation has followed the image of Linnaeus and of many of his contemporaries for more than two hundred years. In textbooks, as well as in learned treatises, the views of Linnaeus have traditionally been represented as fixedly and rigidly as the very species concept itself. The truth of this issue, however, has not been hidden among dusty manuscripts. Linnaeus himself expressed clearly his doubts on the immutability of the species. It can be read in all the editions of his *Philosophia botanica* (Linneaus 1751aff). He even elaborated his overall picture of the creation and launched a new hypothesis, which harmonized better with available data and his own experience. Nevertheless, Linnaeus has been stereotyped and misrepresented, and too often – outside the sphere of serious Linnaean research (e.g., scholars like C. E. B. Bremekamp, G. Broberg, G. Eriksson, T. Fries, J. L. Larson)– he, *mirabile dictu*, still is.

**5.1.2. Objectives.** One of the purposes in this chapter is to investigate this misrepresentation of Linnaeus and the astonishment of Darwin. Obviously Darwin had previously been unaware of Linnaeus' revised thinking, in spite of the fact that he had Linnaeus' *Philosophia*

---

6    Lovén hesitated in choosing the words "view', "belief', "opinion" or "conviction" and quoted from Linnaeus (1749) where, in the dissertation "De Peloria" on pp. 70-71, Linnaeus *as praeses*, obviously for the first time in print, elaborated his new thoughts.

7    "Two eminent Swedish naturalists, Professors Fries and Lovén, have kindly pointed out to me these and many other passages in which Linnaeus shows that he had freely speculated on the variability and transmutation of species."

*botanica* at hand in his own library.[8] The chief objective, however, is to draw attention to the ideological roots of the species concept and to some fundamental ideas of Linnaeus for the benefit of the modern discourse on origins and taxonomy.

## 5.2. Methodology in history of science research

**5.2.1. Basic Principles.** Much has already been written on the species concept, but surprisingly little which explains Darwin's curious astonishment (see, e.g., Ghiselin 1969; Hull 1973; 1988; Kottler 1978; Gillespie 1979, 1988; Kohn 1985; Ereshefsky 1992). All too often in contemporary literature the species concept before the time of Darwin is associated with two general characteristics, where the first is considered to cause the second: firstly, literal interpretation of the biblical account of creation; and secondly, immutability (see, e.g., Gillespie 1979, pp. 20-21; Mayr 1982, p. 255, 1988, p. 313; Bowler 1989, p. 359).[9] This description, however, shows clearly that the interest of the Darwin industry has not yet been sufficiently focused on understanding in what specific ideological context Darwin wrote down his views on the species question, nor what opinion Darwin was reacting against. Furthermore, the questions why and how he wanted to accomplish his demarcation against his adversaries have not been given sufficient attention.

Of course, as Mary P. Winsor says in a book review, the pre-*Origin* days require "an imaginative power greater than most of us can muster" (Windsor 1985, p. 252). And to mobilize that, as Winsor maintains,"means to surrender not only our knowledge, but our materialist worldview" (Winsor 1985, p. 252). With these few words, Winsor has touched on some basic problems in the methodology of intellectual history. Historians of ideas were for a long time occupied with tracing "forerunners" or "precursors" to the "right" view (e.g., Glass et al.1959). These "right" views seemed, curiously often, to be the dominating ones in their own time. Consequently, the task to understand the background and context was neglected, and the opinions of the more or less "uninteresting" opponents were too easily dismissed. The heroes were men like Galileo, Newton, Lyell, Darwin and others. They were stripped of their worldview and, time after time, honored and crowned as the saints of science without a stain on their scientific records. These interpretations now belong to history and are looked upon as

---

8    "The Darwin Library - Down has both the *Systema naturae,* Ed. 13a, Cura J. F. Gmelin (bound in 10 vols.) (Linnaeus 1789-96) and the *Systema vegetabilium,* Ed. 15a ... (Linnaeus 1797)... The Darwin Library - CUL [Cambridge University Library] has Linnaeus's *Philosophia botanica* (Linnaeus 1783b), lightly annotated." (Burkhardt & Smith 1985, p. 260).

9    This view is often connected with the assumption that the biblical world picture is static. (*Cf.* Wuketits 1988, p. 18ff; Mayr 1992)

ideologically motivated.[10] With the demise of the positivist consensus in the form of logical empiricism, our images of the heroes have developed more human proportions.

The previous view of the Enlightenment and of the role of science as the sole torch of Truth is simply not sustainable any longer. The late positivist school of the philosophy of science was not really interested in the history of science,[11] while historians of science have played an important role in the dethronement of this school and in tracing the interactions between science and society at large. When we deal with science in any culture and any time, we have – at least to some extent – simultaneously to deal with the larger context: worldviews, ideologies, religion, politics, economics etc. How else could we get a firm grip on understanding the speculative nature of Galileo's ideas, the alchemy of Newton, the dogmatism of Lyell, the "Lamarckism" of Darwin or all the other apparent inconsistencies in the past and even in our own time?

**5.2.2. Rationality.** A primary methodological principle is that an historian of science ought to presuppose that there was a rationale for a view once held.

Often the evidence available at the time led to a perfectly reasonable position, which, however, can be hard to assess if it does not square with the historian's own views. According to the Dutch historian of science, Reijer Hooykaas (1982), an historian ought to look at the science of the past with a "phenomenological method" as if he were "its contemporary critical observer." He must, consequently, at least try to liberate himself from the burden of the common ideas of his own time. In his research he has to look for the contextual rationality of once accepted views, even if they no longer seem rational according to his own frame of reference.[12]

**5.2.3. Different Worldviews.** A second principle is to stress the crucial importance of distinguishing the different worldviews represented by the object of interpretation and by the interpreter himself.[13] The "materialist worldview" of Winsor and others has not been the dominating one in Western culture very long, but it has already caused

---

10   The literature which has contributed and still contributes in this Copernican revolution in the Philosophy of Science has been quite extensive since Kuhn (1962). Even though the issue on the arena of philosophy of science is not yet fully settled—and perhaps never will be—a post-positivist consensus among the defenders of scientific realism and the social constructivists has, according to Richard Boyd, emerged. See (Boyd et al. 1991, p. xi).

11   This assertion has been commented upon by many including Hodge and Cantor (1990).

12   Concerning the term and concept "Darwinism," James Moore (1991, p. 358) even opines that questions like "After all, who got it right? Who correctly interpreted what Darwin said? Who understood what Darwin really meant? Who has fair claim to represent authentic Darwinism?" are " unhistorical, and thus uninteresting."

13   Neal C. Gillespie (1979) argues in the same way, when he uses Foucault's "episteme," which is almost another word for epistemology, contrasting what he call "positivism" against "creationism."

too many scholars to misunderstand and even neglect the historian's
task of understanding and transmitting the historical background, the
rationale and the coherence of ideas held at a different time and age.
This chapter proposes that this is the case with some pre-*Origin* views
on species.[14] Therefore, it is vital to trace the concept "species" and its
different definitions in order to uncover ideological roots of the species
concept at the time of Charles Darwin. This implies trying to grasp
not only different historical and intellectual contexts but also different
worldviews.[15]

**5.2.4. History of "idea."** Concepts like "species" can not sufficiently
be treated as solitary ideas in human history. Like all human artifacts,
concepts have been formed by the minds of men. And as the minds
of men differ, so do concepts. Nevertheless, historians, having been
instructed by the intense dialogue on epistemology, on theory of science
and on methodology of history of science, need to write the history of
ideas all over again—not in the old "forerunner-fashion," even though
we still have much to learn from earlier generations of scholars.

This chapter, perhaps, will help us come a little bit closer to
answering the question why Linnaeus and Darwin in their different
social, ideological and scientific contexts reacted so differently to
nature. Both were men of science, wanting to contribute to the project of
enlightening mankind. They still have immensely prestigious positions
in our society. Already in their own time, they had become symbols of
widely different interpretations of nature. To write on the question of
species is, therefore, a challenge to write history and the history of ideas
at the same time. But, the history of "species" must, to a considerable
extent, be the history of "idea" itself. The simple reason for that, is that
the Latin word "species" is an often used translation of the Greek word
"idea (ἰδέα)." Therefore, this chapter could have been called: An outline
to the history of "idea."

### 5.3. Historical background

**5.3.1. Darwin and the Fixity of Species.** Traditionally it has been
said that the young Charles Darwin (1809-1882) during his studies in
Edinburgh and Cambridge was taught that species were eternal and
immutable and that they did not interbreed. God's immutability, it was
thought, secured the fixity of the species. Nevertheless, in the previous
century there had been an interest among scientists in hybridization.

---

14   In the history of geology, there has been a comparatively recent reevaluation of
     geological science before Lyell. R. Hooykaas triggered the reinterpretation by his
     seminal work (Hooykaas 1959). After Hooykaas, scholars like Susan F. Cannon with
     various male pseudonyms, M. J. S. Rudwick, Nicolaas A. Rupke, *et al.* have followed
     up this project.
15   The concept "worldview" could, perhaps, in certain contexts, be replaced by, e.g.,
     Kuhn's "paradigm" or Focault's "episteme."

Gunnar Broberg (1975, pp. 44-47) calls attention to the fact that even the old textbooks in natural philosophy talked about "animalia spuria," which were explained by unnatural and forbidden contacts. But these hybrids were the exceptions that confirmed God's original order of creation.

Although the immutability position was the officially received and accepted view in Great Britain, there were of course different views on the scope and limit of species. The majority of the naturalists believed that species always produced offspring according to its own kind, but assessing the border between species on the one hand and subspecies and varieties on the other was a tricky business and always open to criticism. The problem was to define the limits between the systematic categories.

There were also centers of free-thinking. Some universities like London and Edinburgh were known for their intellectual freedom, and some free-thinkers were well known for their unorthodox views. Quite early Darwin himself seems to have harbored heretical notions on the subject. There was the influence from his famous grandfather Erasmus Darwin and his work *Zoonomia*, which Charles had read as a young boy. In his autobiography, however, he asserted that he had read the work but that it did not produce any effect on him.[16] More important were his two years (1826-1828) of medical studies in Edinburgh,[17] "a haven for wealthy Dissenters" (Desmond and Moore 1991, p. 22), where he got to know the "passionate Francophile" (Desmond and Moore 1991, p. 34) and Lamarckian evolutionist Robert Grant.[18] Desmond and Moore (1991; Secord 1991, p. 36) assert that what he learned from Grant "was to shape his own initial approach to evolution ten years later." This view

---

16    "without producing any effect on me." (Darwin and Darwin 1892, p. 13).

17    Adrian Desmond (1989) has shown something of the complicated background of Darwin by tapping medical journals and analyzing the social and medical context of especially London. Darwin's friend and personal tutor Dr. Grant moved from Edinburgh to England for the London University chair of comparative anatomy, which he held from 1827 to 1874.

18    Secord (1991) convincingly argues that Darwin's old professor of Natural History at Edinburgh Robert Jameson was a Lamarckian and, with the words of Desmond & Moore (1991, p. 42) "one wonders what Charles made of his closing lectures on the 'Origin of the Species of Animals.' " An often overlooked circumstance is that Darwin, as most of his contemporaries, believed in the inheritance of acquired characters. This can clearly be seen e.g., in his *Origin* and in his various drafts to the same work. In the end of the second volume of *The Variation of Plants and Animals under Domestication* (Darwin, 1868), Charles also launched his all embracing hypothesis about heredity, called Pangenesis. This hypothesis was very important for Darwin, because it provided with a desperately needed cause to variations among plants and animals. The mechanism proposed in the hypothesis, with germs reproducing characters in each part of the body and shaping the following generation according to the present status at the reproduction, was definitely Lamarckian, not, however, in the sense that it was true to Lamarck, but in the modern sense that it was in accordance with the common belief in the heredity of acquired characters.

seemed plausible even to Darwin later in his autobiography. (Darwin
and Darwin 1892, p. 13)

**5.3.2. Linnaeus in England.** The botanists John Ray (1628-1705)
and Carolus Linnaeus (1707-1778) had dominated the field of taxonomy,
and their influential works on systematics had impressed the view of
the permanence of the created species on almost all English natural
scientists. (Stafleu 1971, esp. chap 7, pp. 199-240; see also e.g., Green
1914 and Gage 1938). The young Linnaeus was keen to organize the
three kingdoms of nature according to the same principles.[19] Among
his admiring contemporaries one often said: *Deus creavit, Linnaeus
disposuit.*[20]

Soon Linnaeus had gotten a firm footing in England. Contributing
to this was his own visit to England 1736 and his early contacts with
botanists such as Dillenius in Oxford. English friends even asked
Linnaeus to send Swedish disciples to England. The choice fell upon
Daniel Solander and Jonas Dryander, who worked as botanists, museum
curators and librarians to the mighty and wealthy president of the Royal
Society, Sir Joseph Banks. Above all, as an indication of the success of
the Linnaean taxonomy in England we have to take into account the final
abode of the Linnaean Collections and the foundation of the prestigious
Linnaean Society.

Natural Theology molded the minds of the intelligentsia at Oxbridge.
In *Historia plantarum* (1686-1704) Ray had asserted the constancy of
species, and in *Systema naturae* (1735) Linnaeus had declared that "if we
behold the works of God, it is more than sufficiently obvious to all, that
every animate being comes into existence out of a parent's egg, and that
all eggs produce offspring like its parents, wherefore no new species can
arise."[21] Furthermore he asserted in his *Fundamenta botanica* (1736 VI
§157) that "we count as many species as there were created forms in the
beginning."[22] In addition to the various editions of *Fundamenta botanica*
(Linnaeus 1736ff), this famous dictum was also restated and commented
in all the editions of *Philosophia botanica* (Linnaeus 1751aff). Linnaeus
meant that the essence of a plant or an animal enabled it to be classified
according to species and higher orders. However, species and genera
were clearly a result of God's creation. Variation was a product of culture;
classes and orders were products of God's creation and art. These rules of
Linnaeus were supported by empirical observation. Species, according
to Linnaeus, are "constantissimae," because their generation is "vera

---

19   Concerning, *e.g.*, the taxonomy of minerals see Hooykaas, (1952,1982) and Laudan
      (1989, p. 221-233, 1987)
20   *i.e.*, God created, Linnaeus arranged.
21   "Si opera Dei intueamur, omnibus satis superque patet, viventia singula ex ovo
      propagari, omneque ovum producere sobolem parenti simillimam. Hinc nullae
      species novae hodienum producuntur."
22   See note 3.

continuatio." In *Classes plantarum* (1738) he argued that the created species were forms, which according to the inherent laws of generation produced more but always like themselves,[23] but also the genera were a product of creation. In *Philosophia botanica* and in his other writings he asserted that genera are "naturalia" *i.e.,* an order of creation.[24] Most of the plants prove it, he wrote simply. Gardening, Linnaeus argued, proves that varieties are caused by cultivation and the natural orders even teach that most of the orders and classes are natural.[25]

**5.3.3. Natural Classification.** During his whole life Linnaeus searched for a natural classification. He strived for a "methodus naturalis", but confessed that a key for the plant kingdom, for instance, could not be given before all plants were related to their orders (see the chapter "Fragmenta methodi naturalis," principle 7, in Linnaeus 1738, pp. 485-486). In this context it is important to bear in mind that in using the Latin word "naturalis" or its various derivations Linnaeus normally intends the meaning "created as such" or "specially created."[26] During his whole life, Linnaeus wanted to identify the kinds of plants and animals God created in the beginning. Belief in God was fundamental to him, and divine intervention did not contradict his efforts to implement a system of nature.

The concept of genus was more fundamental for Linnaeus than the concept of species. Both were natural and, therefore, created by God's direct intervention. But in establishing order, the genera were prior in significance. In his introduction to *Genera plantarum*, Linnaeus (1737) quoted Andreas Caesalpinus, the father of systematic botany, who, following Linnaeus, first "cum publico" presented a system of genera according to what he believed were their natural affinities. Caesalpinus stated that if the genera are in disorder, everything will be in disorder.[27] The true demarcation of genera was, according to Linnaeus, the very foundation, the *punctum fixum* and even the *canon fundamentalis* for

---

23    "... quae formae, secundum generationis inditas leges, produxere plures, at sibi semper similes."
24    In reference to other writings of Linnaeus one example will suffice. In the dissertation *Vires plantarum* from 1747 with the respondent F. Hasselquist the text says in the beginning of § VIII: "Videmus igitur dari, fere in universum, genera naturalia, ab ipso Creatore ordinata" The text is quoted from Linnaeus *(1749).*
25    "NATURAE opus semper est *species* (157) & *Genus* (159), CULTURAE saepius *varietas* (158), NATURAE & ARTIS *Classis* (160) ac *Ordo* (161). *Species* constantissimae sunt, cum earum generatio est vera continuatio. *Genera* esse naturalia evincunt plurimae plantae: *Aconita, Nigellae, Bignoniae, Ranunculi, Mesembryanthema, Zygophylla, Gerania, Oxalides. Varietates* culturae opus esse, docet *Horticultura,* quae easdem saepius & producit, & reducit. *Classes & Ordines* plerasque naturales esse, docent ordines naturales § *77."* (Linnaeus 1736, chap. VI §162)
26    See e.g., Linnaeus (1751 §159), where it says quoting from *Systema naturae* on genera that "Genus omne est naturale, in primordio tale creatum..." (cf. Stafleu 1971, pp. 67, 125).
27    "Confusis generibus omnia confundi necesse est." (Linnaeus 1737, *Ratio operis* § 6).

all taxonomy. And if the botanists do not observe it, the splendid edifice will in the first tempest immediately collapse, Linnaeus feared.[28]

This combination of both species and genera as specially created made it easier for Linnaeus later in life to modify his views on species. In *Philosophia botanica* Linnaeus (1751a) seemingly without any hesitation called his own aphorism on fixity of species into question in the very paragraph where he developed the dogma. That was never the case with genera.[29] Accordingly for Linnaeus botany rests on fixed genera.

For Darwin natural classification was to ascertain actual relationships. These consisted in descent from a common stock. For him the word "nature" and its derivations had nothing to do with creation in the original sense of the word. In his *Origin of Species,* he contended that phrases like "plan of the Creator" added " nothing ... to our knowledge" (Darwin 1859, p. 413). In private letters he could be even more outspoken. To a Mr. Waterhouse he wrote simply that it was "empty high-sounding sentences" (Burkhardt and Smith1987, p. 375f); and to his close friend, the botanist Joseph Dalton Hooker, he wrote in March 1863: "I have long regretted that I truckled to public opinion, and used the Pentateuchal term of creation, by which I really meant 'appeared' by some wholly unknown process" (Darwin 1887, vol. 2, pp. 202-203).

Yet Darwin himself had once been happy to include the Creator in his theories, and he had believed in the immutability of species. That was, according to himself, the "common view", which he supplanted with his theory. In his autobiography Darwin nevertheless claimed that he "never happened to come across a single one [*i.e.,* naturalist] who seemed to doubt about the permanence of species" (Darwin and Darwin 1892, p. 45). And in his *Origin* he asserted: "Why, it may be asked, have all the most eminent living naturalists and geologists rejected this view of the mutability of species" (Darwin1859, p. 480)? In his introduction to the same work he wrote that "the view which most naturalists entertain, and which I formerly entertained—namely, that each species has been independently created—is erroneous" (Darwin 1859, p. 6).

The immutability concept, according to Darwin himself, would have led him to extreme conclusions. When he worked with his voluminous classification of barnacles his "most frequent source of doubt was whether others would not think this or that was a God-created Barnacle & surely deserved a name" (Burkhardt & Smith1990, p. 194). By then

---

28  "... quem si non observant aediles, illico ruit prima oborta tempestate splendissimum quamvis aedificium." (Linnaeus 1737, *Ratio operis* § 8)

29  "NOVAS SPECIES dari in vegetabilibus negat generatio continuata, propagatio, observationes quotidianae, Cotyledons. Dubium movere *Marchant. act. paris.* 1719; Ego in *Peloria* 1744; *Gmelinus* in *orat. inaugur.* 1749. vide *Amoenit. acad.* 71." (Linnaeus 1751)

he had become convinced that the species were mutable, but he should have known that he was far from alone in that respect.[30]

**5.3.4. Plato and Aristotle.** Tracing back the formation of the Linnaean species concept, we have to turn to the classical Greek and Roman world. Linnaeus worked as a natural scientist in the Aristotelian tradition. When Aristotle classified animals or when Theophrastus, the great disciple of Aristotle, classified plants, they both used the concepts genus (Gk. "genos", γένος) and species (Gk. "eidos", εἶδος) (Hall 1991, p. 111ff; Lennox 1980, p. 321-346).[31] The use of the two concepts had a clear technical distinction in the field of logic and later there developed a similar hierarchic distinction in biology.[32] Behind the Aristotelian concept of "eidos" was hidden the metaphysics of his master, Plato.[33] In Plato's worldview the concept of "idea" (Gk. e.g., ἰδέα and εἶδος) had an essential importance.[34] There was a world of ideas separated from the world of senses. This world of ideas was for Plato the real world, and it consisted of perfect ideas of everything from qualities like "beauty" to living beings such as "man" and "cat."

The question is, what kind of relation these ideas had to the world people normally refer to as "reality." Traditionally, Plato has been interpreted as having claimed that one gained knowledge of true reality by philosophical contemplation of the eternal ideas, while Aristotle

---

30   See below in the section "Lost credibility".

31   *Cf.* Gotthelf and Lennox (1987) and P. Pellegrin (1986). In a preliminary Ibycus-scanning through the corpus of Aristotle, it was noticed that the nominative of the words ἰδέα, εἶδος, μορφή, □γένος, and σχῆμα occurred respectively 226, 561, 192, 869 and 618 times.

32   Commenting on the technical discussion on the classification of Aristotle, Hall (1991, p. 112) concludes: "However, despite the lack of special vocabulary there are passages in Aristotle which suggests the construction of a hierarchy of progressively narrower genera, roughly similar to that used by modern zoologists." Bäumer (1991, p. 228) says that it was Caspar Bauhin (1519-1603), who first made the completely accomplished hierarchic classification of species and genera: ("Hier gelangen Caspar Bauhin erste Fortschritte, da er durch die Einführung einer binären Nomenklatur, die auf einer erstmals konsequent durchgeführten Unterscheidung zwischen Genus und Spezies beruhte, erste Anfänge einer natürlichen Systematik vorlegte, die von Andreas Cesalpin ausgearbeitet wurden.")

33   See *e.g.,* Bäumer (1991, p. 42) "Aristoteles übertrug nämlich den Platonischen Ideenbegriff auf die biologische Art und bezeichnete mit ihm das Gleichbleibende im biologischen Naturgeschehen. Die Arten sind für Aristoteles ewig und unveränderlich, Ziel eines jeden Organismus ist es, wie bereits angedeutet, sich selbst zu reproduzieren, um auf diese Weise durch den Weiterbestand seiner Art an der Ewigkeit teilzuhaben." *Cf.* Lennox (1980), Greene and Egerton (1983), and Gotthelf and Lennox (1987).

34   The original Greek meaning of the words ἰδέα and εἶδος is nearly the same. The words stem from the same root and both mean "that which is seen". The basic English rendering is "form", and as stated by Liddell & Scott (1977), the words are also in the writings of Plato and Aristotle translated according to the different contexts, *e.g.,* in logic – "class" or "kind" and in philosophy – "ideal form" and "archetype" (ἰδέα) or formal cause and essence (εἶδος).

thought that one acquired this knowledge by sense experience of the palpable instances of the different ideas in this world. This well-known difference between Plato and his disciple Aristotle can easily be seen in the subsequent history of the natural sciences, in which there have often been controversies between scientists of the two traditions. The Platonists have stressed and preferred philosophical contemplation to empirical studies and looked with condescending disdain at the foolishness and naïveté of those Aristotelian colleagues who have worked with hands and eyes. The conception of "idea" or "form" (*i.e.*, Gk. μορφή) was therefore common to the two great philosophers, although they stressed different ways of arriving at knowledge of the ideas.

The purpose here is not to go into the sophisticated debate on the meaning of these terms in the writings of the great Greek philosophers.[35] Rather, the interest is focused on the understanding and use of these terms in the generations which were to follow. Three strands of understanding in the history of the Latin West will be mentioned here.

**5.3.5. Roman Philosophy.** Although there is some originality in the philosophical writings of the Romans, their greatest contribution was to hand down the Greek philosophical thought to the Latin-speaking world well into the Renaissance and even beyond. One very important figure in this respect was Cicero, the Roman politician and author. Approximately three hundred years after Greek philosophy reached its apex, Cicero made a tremendous contribution through his philosophical writings. Without any regard for modern copyright laws, he wrote influential treatises that depended heavily upon the Greek heritage. In choosing Latin words for the Greek "eidos," "idea" or "morphe" he used "species"[36] and sometimes "forma."[37] The very terms and the ideas behind them became a cornerstone in the foundation of the species concept of Caesalpinus, Ray, Linnaeus and other early botanists.

**5.3.6. The Vulgate.** The Vulgate, Jerome's Latin translation of the Bible, has had an immense impact on the language and thought of Western culture. The Vulgate is influenced by Cicero's terms and uses the

---

35    For example, Scott Atran (1987, p. 244ff) denies that the traditionally referred ingredient to Aristotelian essentialism of the eternal fixity of the forms "appropriately appl[ies] to Aristotle." On the other hand David Hull (1988, p. 82) states that "according to Aristotle and generations of Aristotelians, all species are as eternal and immutable as are the physical elements. No one today expects lead to evolve into gold; Aristotle had no higher expectations of fish evolving into frogs. To be sure, an organism might change its species, just as a sample of lead might be transmuted into a sample of gold, but the species themselves remain unchanged in the process. Species are at least potentially eternal."

36    See *e.g.*, *Tusculanae disputationes* I. XXIV. 58 and *Academica* I.IX.

37    See *e.g.*, *Orator 10, De natura deorum* II.145.

translation "species" or "genus" of the Hebrew "min,"[38] which roughly can be translated with "sort," "kind" or "type."[39] The Vulgate's account of Genesis reiterates phrases like 'secundum speciem suam,'[40] and in the history of theology and biblical interpretation the word "species" came to mean the unchanged form or idea of a created being.[41] My conclusion after studying the subject so far is that the Platonic and Aristotelian view of the immutability of species imposed itself upon the understanding of Genesis. This conflation of the Hebrew Scriptures and Greek philosophy

---

38    The technical expression that is used in the OT is, in transcribed form, *lĕmînô,* "according to its kind." It consists of the preposition *lĕ,* "according to," the noun *mîn,* "kind," and some form of the suffixed possessive pronoun in the third person, as e.g., here, *-ô,* "its." The noun *mîn* is always in singular, but it is among Old Testament scholars understood as a collective singular. The translation is not controversial. The expression occurs 31 times in the OT. Genesis 1:11, 12(2), 21(2), 24(2), 25(3), 6:20(3), 7:14(4), Leviticus 11:14, 15, 16, 19, 22(4), 29, Deuteronomy 14:13, 14, 15, 18 and Ezekiel 47:10. For the Hebrew technical data in this note I'm thankful to the Old Testament scholar Åke Viberg.

39    The most common English and American Bible translations use "kind," e.g., King James Version, Revised Version, American Standard Version, New English Bible, New American Standard Bible. The Revised English Bible has "kind of seed" and The Jerusalem Bible "species." The New International Version and the English Old Testament scholar Gordon J. Wenham (1987) uses "type." Martin Luther used "nach seiner Art." Carl Friedrich Keil writes about "Art" (1878) and so does also Franz Delitzsch (1887). In explaining the Hebrew word *mîn* Delitzsch says that it is "genau dem griech. εἶδος, dem lat. species entsprechend" (1887, p. 58). Claus Westermann writes: "Zusammen mit den Definitionen kann das Wort nichts anderes bedeuten als der moderne Begriff, wie er u. a. auch in der Naturwissenschaft verwendet wird: Art oder Gattung" (1974, p. 174-175). On the other hand Walter Kaiser (1980, p. 503-4) writes: "Some have argued that when God created *mîn,* he thereby fixed the "species." This is a gratuitous assumption because a link between the word *mîn* with the biologists descriptive term species cannot be substantiated..." Kaiser concludes the article on *mîn* in the following way: "God created the basic forms of life called *mîn* which can be classified according to modern biologists and zoologists as sometimes species, sometimes genus, sometimes family or order. This gives no support to the classical evolutionist view which requires developments across kingdom, phyla, and classes."

40    *I.e.,* "according to its species." In Genesis 1 of the Vulgate we find, as correspondingly with the Greek words in Aristotle's biological writings, the terms "species" and "genus" used interchangeably: "secundum speciem suam" (v. 12), "secundum species suas" (v. 24) "iuxta species suas" (v. 25), "in species suas" (v. 21), "iuxta genus suum" (v. 12), "secundum genus suum" (v. 21) and "in genere suo" (v. 25). In the Septuagint (LXX) all these phrases are translated from the Hebrew with κατά γένος.

41    In *De Genesi ad litteram* (III.12) Augustine with his neoplatonist cast of theology used the word "genus" throughout and explains the concept in the following quote: "Hoc est ergo secundum genus, ubi et seminum vis et similitudo intelligitur succedentium decedentibus, quia nihil eorum ita creatum est, ut semel existeret, vel permansurum vel nullo succedente decessurum." Thomas Aquinas integrated the Aristotelian essentialism in his massive and elaborated systematic theology. Aquinas wrote in his *Summa Theologiae* I.98.1: "Quia igitur in rebus corruptibilibus nihil est perpetuum et semper manens nisi species, bonum speciei est de principali intentione naturae, ad cuius conservationem naturalis generatio ordinatur."

had far-reaching consequences for the development of science and theology.

**5.3.7. Scala Naturae.** Another classical notion of vast influence, essentially suggested by Aristotle, was the Great Chain of Being, the *scala naturae* (Lovejoy 1936) This concept had two closely related components, the idea of constancy and the idea of plenitude. Later in the Christian world these ideas contained at least two distinct beliefs: first, that God had created all things that possibly could have been created and, second, that God had created these perfect and immutable forms or species in a hierarchic order from the simplest physical organism to the most complex spiritual being. The chain was so complete that all things from angels to formless matter composed a ladder of increasing complexity and minute differences. And furthermore, it had all existed simultaneously since creation. *Natura non facit saltus, i.e.,* nature makes no leaps, was the famous dictum for the perfection of the Creator's chain. The structure of the whole universe had the highest degree of perfection. Even Charles Darwin, as most of the biologists before him, took this dictum to heart, but later with an important twist. Using it for slow accumulative change he reiterated the phrase to his friends and followers including the comparative anatomist Thomas Henry Huxley, who among others from time to time leaned towards saltationism. With Darwin, however, triumphed the temporalization of the Great Chain of Being. According to him, the chain had not been created all at once, but had successively evolved in time without any supernatural intervention. Together with the ideas of immutability, constancy and plenitude, the Great Chain of Being had formed a view of nature that had made classification possible. With Darwin and his followers the chain became a tree with no borders between the forms to transgress and, in fact, no species to transmute.[42]

**5.3.8. Lost Credibility**. The pre-*Origin* stage of the species question was in a fluctuating state. The old Aristotelian position, which can duly be called a scientific hypothesis, was in its extreme form a dead end for research. Zoology and botany were no longer sciences characterized by collecting, describing and naming. More and more scholars, including dominating men such as Linnaeus, Buffon, Georges Cuvier and even contemporaries like Richard Owen, worked with the same problem as Darwin. Some of them wanted to be free to speculate not only on historical questions but on questions of origin. A growing number of scientists also found that their experiences of reality could not be harmonized with a rigid immutability stance. J. H. F Kohlbrugge published an animated article (Kohlbrugge 1915), in which he asserted that 87 of Darwin's contemporary scientists believed in the variability of species to such an

---

42    That was the critique of Darwin by Agassiz (1869, p. 89-90, note 1) and Hopkins (1861-2)

extent that the author accused Darwin of being unaware of the positions of his colleagues and incapable of mastering foreign languages.[43] Others, seemingly, did partake in a gentlemen's agreement not to discuss the issue in public (Hull 1973, p. 215). After all, since Newton it was felt to be more a question of philosophy than of science.

The standard application of the old scientific Aristotelian view, which the young Linnaeus as well as the young Darwin had endorsed and which had been married to the biblical interpretation of the creation account long before, lost its intellectual and scientific credibility little by little (see Gillespie 1979, pp. 19-40). Of course there were many able defenders of a more sophisticated application of the old immutability stance like the Cambridge educated entomologist Thomas Vernon Wollaston, who three years before the *Origin* was published had written the work *On the Variation of Species* (Wollaston 1856). Darwin had seen Wollaston as a potential convert, but was badly mistaken. In a review of the *Origin* in 1860 in *Annals and Magazine of Natural History,* Wollaston thought that Darwin had "pressed his theory too far". Varieties were "at times mistaken by naturalists for true species ..." and that was "surely no argument against the genuineness of the latter: it merely shows," Wollaston continued, "the imperfection of our limited judgment, and that the best observers are liable to err, and either not to catch the true characters of a species intuitively (which, in point of fact, they could scarcely be expected to do), or else to assign at times undue importance to differences which they may afterwards detect not to be in reality specific" (Hull 1973, pp. 131-132). For Wollaston and many others, there were "fixed specific bounds" (Desmond and Moore 1991, p. 434) and a legitimate variation within them.

To the Cambridge mathematician William Hopkins, there was an important difference between natural and artificial species, which Darwin had not grasped. In *Frazer's Magazine* Hopkins (1861-2) reviewed Darwin's opus. A natural species was, Hopkins maintained, a "group of organic beings which can only have been derived by descent from beings similar to themselves," and artificial species were "equally distinguished by particular characters, but ... such that ... they may not have been derived from each other or from some original stock" (Hull 1973, p. 241). Darwin had, according to Hopkins, mixed these different groups of species in systematic literature, when he tried to "break down the line of demarcation between *varieties* and *species,* founded on the perfect fertility of the former when crossed with each other, and the sterility or very imperfect fertility of the latter when similarly crossed" (Hull1973, p. 255). This was Darwin's serious mistake, according to

---

43   The accusation about Darwin's incapability to master any foreign language was supported with a quotation from Darwin himself. See Darwin (1887, vol. 1, p. 29) where Darwin says: "During my whole life I have been singularly incapable of mastering any language."

Hopkins. In doing "this he [Darwin] confounds artificial with natural species" (Hull 1973).

Darwin and his sympathizers meant they perceived a bad smell of theology from views like Hopkins' and Wollaston's. For the latter category of scientists as for the early Linnaeus, a natural species had by definition an independent and separate origin. The idea of one origin for all classes of animals or plants was tantamount to a flat denial of natural species altogether. Darwin and others, on the contrary, could not accept special creation as a natural explanation. That was plainly to give up what natural science was all about. Darwin's antagonists were accused of using theological presuppositions. Natural science should only reckon with the world of phenomena. A natural scientist should never invoke anything but known forces, processes and materials. The rejection of the supernatural on questions of origins, however, required, as Ernst Mayr cogently has observed, "an erosion of the worldview prevailing in the western world prior to the adaption of evolutionary thinking" (Mayr 1982, p. 310). That slow erosion undermined imperceptibly the credibility of the belief in creation by divine fiat.

**5.3.9. After the Voyage.** When Darwin and the British ornithologist John Gould in March 1837 studied the finches that had been collected at the Galapagos islands, Darwin was taught by Gould that what he had thought were uninteresting varieties were actually different species (Sulloway 1982a, pp. 21–22, and 1982b, pp. 359–362). Darwin thought that, according to the view of the fixity of species, God should have created the different species of finches at each island of the archipelago. It was the logical implausibility of such a conclusion that lead Darwin in the direction of the French and English free thinkers.[44]

In this situation Darwin did not doubt Gould's classification. After all, Gould was his tutor on birds. Nor did Darwin try to adjust the Linnaean concept of species. Instead he started more seriously to write down his thoughts and speculations on transmutation in a series of notebooks (Barrett et al.1987). These notebooks make it obvious that transmutation was a starting point for his speculations. What Darwin did was to try old and new ideas. According to Desmond and Moore, "fossil monkeys," which were looked upon as immensely old, had, "to everyone's surprise," been discovered and publicly announced in spring 1837 (Desmond and Moore 1991, p. 222). This news gave credibility to Lamarck's view and incited Darwin in the same year to unhesitatingly speculate on the most delicate subject of man's common ancestry with apes (Barrett et. al. 1987, p. B169 etc.).

---

44   We have good reasons to believe that Darwin was well acquainted with the views of e.g., his grandfather Erasmus Darwin, Jean Baptiste Lamarck, Robert E. Grant and Robert Chambers. See Desmond (1989), and Schwartz (1990).

**5.3.10. Darwin and Criticism.** In his autobiography Darwin once refers to his first notebook on transmutation, which he started in July 1837: "I worked on true Baconian principles, and without any theory collected facts on a wholesale scale" (Darwin and Darwin 1892, p.42). That was the image of science of the Enlightenment in a nutshell. After the publication of *The Origin of Species,* friends like Prof. Sedgwick criticized him severely for just the opposite. "You have deserted ... the true method of induction ... Many of your wide conclusions are based upon assumptions which can neither be proved nor disproved, why then express them in the language and arrangement of philosophical induction?" asked Sedgwick in a letter of November 1859 (Darwin and Darwin 1892, p. 229). According to George Grinnell, the "extent to which he was willing to push one model, and after its collapse, to entertain new models suggests that he was philosophically inclined to transmutation theories for reasons that transcended the empirical data with which he originally worked" (Grinnell 1974, p. 273).

The old positivistic picture of Darwin's scientific method and work with the evolutionary theory does not seem to fit the historical data now available. On the contrary the data reveal an early shift of worldview and epistemology, obviously partly triggered by a rigid application of the immutability position. Darwin's later followers and biographers have all too often tried to uphold an artificial distinction between his theory on the one hand and his observations on the other. But there was no such distinction. His observations were dependent on theory, and his theory was dependent on his *Weltanschauung.*

**5.3.11. Fixity and Fluidity.** In describing Charles Darwin's intellectual development concerning the species concept one must conclude that at least officially he started with fixity and ended up with fluidity. From an Aristotelian and early Linnaean static view, he advanced a plastic one with ingredients like natural selection and pangenesis. This was serious business in Victorian England. In a letter to Hooker in January 11, 1844, Darwin revealed his views on the mutability of species and confided that it was "like confessing a murder" (Burkhardt and Smith 1988, p. 2).

In Darwin's own understanding of his former view, the static species concept was biblical. What he did not know was that the roots of his former view were Aristotelian and, in that respect, a part of the Greek heritage of science. He was also unable or perhaps even unwilling to see that the immutability position did not have to be applied as strictly as some taxonomists did. One could say, without entering the debate on when modern science had its beginning, that Darwin switched from one scientific paradigm to another, none of which had anything to do with the Mosaic creation account. The roots were not biblical for the simple reason that it is not hermeneutically defensible to force the content of the early Linnaean concept of "species" into the old Hebrew word "min."

## 5.4. Linnaeus and his new synthesis

**5.4.1. Linnaeus' Change.** This was roughly the historical *Sitz im Leben* when Charles Lyell expressed his thankfulness to the Swedish professor Sven Lovén for having been informed about the species concept of Linnaeus. The Swedish botanist had changed his view on species, and not until 1867 was Charles Darwin informed about this change of mind of Linnaeus.

During a botanical excursion in 1742, a student from Uppsala had discovered a strange flower, later called *Peloria*, which looked like a hybrid stemming from *Linaria*. When Linnaeus later got the report, he could not fit it into his natural system. His belief in the fixity of species was challenged and Linnaeus soon felt forced to abandon his earlier belief. In the dissertation *De Peloria* (1744) Linnaeus and his disciple Daniel Rudbeg made the discovery public and discussed the species problem. In the later edition of his dissertations, *Amoenitates academicae* (1749), Linnaeus even added an extensive passage on hybridization in *De Peloria* which did not occur in the original work. There he wrote that even if it at first sight looked like a paradox for him, new species, and even genera, seem to arise from crossing by different species in the vegetable kingdom.[45] Linnaeus obviously meant that by interspecific hybridization new species could come into existence, and by intergeneric hybridization even new genera could originate.

**5.4.2. New Proposals.** Linnaeus abandoned the fixity of species. He could have cast the net more widely and made the definition of "species" more flexible, but that would have caused great confusion in terminology and classification. After all he had only a decade before laid the very foundation for taxonomy through his great systematic works. The genus level in his system was, however, as stated above, also seen as natural, *i.e.*, original at the time of creation. That, obviously, made it natural for Linnaeus to launch a new proposal. In his *Fundamentum fructificationis* from 1762 and in his sixth edition of *Genera plantarum* (1764), he proposed a hypothesis which was more adapted to nature and to his own and others' experience.

In the preface to the sixth edition of *Genera plantarum*, the *Ratio operis*, Linnaeus made some important changes. In the introductory note on "Species," he omitted the word *semper (i.e.,* always) where it says that these forms produce more like themselves. [46] In the note on "Genera,"

---

45    "Novas species immo et genera ex copula diversarum specierum in Regno vegetabili oriri primo intuitu paradoxon videtur." (Linnaeus 1749, p. 70-71)

46    Linnaeus changed "... produxere plures, at sibi semper similes, ut Species nunc nobis non sint plures, quam quae fuere ab initio." from the first edition to the less dogmatic "... producunt plures, sibi similes, quam quae fuere." (Linnaeus 1764, p. 2)

he passed over the word 'species' where it said in earlier editions that all genera and species are natural. And where this phrase *Genera & species esse naturalia omnia* from *Classes plantarum* occurs a second time on page 5, Linnaeus inserted a new passage which states: "Let us suppose that the thrice exalted God only created one species from a genus. Let us also suppose that these first species (in the beginning or in time) were fertilized by species from other genera; then it would follow that many species originate, and that the structure of the flower would be like the mother and structure of the herbage like the father."[47] This was a remarkable but wholly understandable turn.

In his *Fundamentum fructificationis* Linnaeus (1762) was even more explicit. There and in a supplement to his sixth edition of *Genera plantarum*[48] he stated that God created as many individuals as there were orders, and that God later mixed these to form genera. Nature in its turn mixed these genera to form species and fate mixed these species to beget varieties. For Bremekamp, a Dutch botanist, it seemed reasonable to interpret the new Linnaean system as if God in the beginning created 50 to 60 plants and around 7000 animals (Bremekamp 1953, p. 245ff).[49]

Broberg suggests that Linnaeus' shift of position was due to his daily difficulties of separating species from varieties (Broberg, 1975, p. 84) Soon after the discovery of the Peloria, Linnaeus rather freely proposed cross-breedings between different species. In 1751 he published *Plantae hybridae* (Linnaeus 1751b) in which he suggested several hybrids. He also proposed that dog, wolf and fox belonged to the same created genus and that they were products of hybridization. The contemporary French biologist Buffon objected and argued for a wider definition of the concept, in which their interbreeding argued their inclusion in the same species (Buffon 1755, p. 210).[50]

The different views here by Linnaeus and Buffon show plainly that sometimes the problem was more semantic than empirical. The early

---

47    "Supponamus D. T. O. ab initio creasse unicam tantum speciem e quovis genere. Supponamus etjam has primas Species dein (vel in primordio, vel in tempore) ab aliorum generum speciebus foecundatas; sequeretur inde quod plures orirentur Species, dum hae Floris structura evaderent quodammodo Matri similes, at Herbae structura Patri. confer. *Amoen. acad.* 6. p. 279." (Linnaeus 1764, p. 5)

48    Unfortunately I have not yet been able to find and read this supplement. See Bremekamp (1953), Larson (1971, p. 107f), and G. Broberg (1975, p. 85). Bremekamp (1953, p. 243) quotes Linnaeus: " 1. Creator T. O. in primordio vestitit Vegetabile *Medullare* principiis constitutivis diversi Corticalis, unde tot difformia individua, quot Ordines Naturales prognata." "2. *Classicas* has (1) plantas Omnipotens miscuit inter se, unde tot *Genera* ordinum, quot inde plantae." "3. *Genericas* has (2) miscuit Natura, unde tot *Species* congeneres, quot hodie existunt." "4. *Species* has (3) miscuit Casus, unde totidem, quot passim occurrunt *Varietates.*"

49    Mayr (1982, p. 259) did not understand the ideological and contextual rationale in this view and called it only "a curious belief."

50    This information about Buffon comes from Broberg (1975, p. 47; See also Lovejoy 1959).

Linnaeus had a strict view on the permanence of species. Many of his followers, however, were even more strict in their application, pushing the species concept further down in the hierarchy into the group of varieties or subspecies. Ever since Linnaeus there have been innumerable quarrels between taxonomists where to draw the line between a variety and a species. And the more tightly the boundaries of species were set, the more absurd the species concept became. When Linnaeus later kept the terminology but switched level concerning the original created order of species, men like Buffon used the word species as Linnaeus later used "genus" and "ordo."

**5.4.3. Speciation by Hybridization.** The Linnaean view that new species could be generated by hybridization were later supported by several great scientists.[51] The difficulty at the time of distinguishing between observations and imaginary hybrids gave further support to Linnaeus' new position. It did not, however, reach a broad popular audience, and it did not really alter his formal taxonomic publications. When Linnaeus was confronted by empirical observation he changed his view of the fixity of species and integrated further experience within his concept of creation to accommodate his opinion about the created original kinds. Darwin, on the other hand, changed his view rapidly and totally from a belief in fixed species to one in "fluctuating forms" (Darwin 1859, p. 52). It is quite evident from his *Origin* that he wrote against the position that God had created every individual species (Darwin 1859, pp. 55,59, 133, 152, 155; see also Gillespie 1979). In Darwin's own words, that was "the ordinary view". This statement of Darwin, however, did not reflect the thought of his more advanced contemporary colleagues, nor did it reflect the thoughts of the mature Linnaeus about speciation after God's original creation by means of interspecific and intergeneric hybridization. It is quite certain, however, that Darwin wrote against what he himself claimed to be his own former opinion. In that case, one could even say that the *Origin* and, in fact, most of his writings was a long argument against himself as a young scientist.

With regard to the species concept itself, Darwin perhaps used it in the *Origin* for strategic reasons in spite of his disbelief in species. From his viewpoint it could be an eye-opener for at least some of his colleagues.[52]

---

51  Including Linnaeus, Kohlbrugge (1915, p. 98) mentions Gmelin, Koelreuter, Adanson, Ackermann, Henschel, Knight, Sageret and Puvis, who all lived late in the 18th or early in the 19th century. As far as Koelreuter is concerned, Ernst Mayr has kindly informed me that Kohlbrugge was wrong.
52  See Sulloway (1979) and Beatty (1985). According to Ernst Mayr (1992a), Darwin later fell back on a more typological treatment of the species.

## 5.5. Conclusion

On Christmas Eve in 1856 Darwin wrote to Hooker, telling him how much he missed him in the midst of his work. Darwin had "just been comparing definitions of species" and he felt that it was "really laughable" to read about all the different meanings of species among the naturalists. Even to try to define species was for himself to try to define the "undefinable" (Burkhardt and Smith 1990, p. 309). For Darwin it was no longer meaningful to define species. In the Introduction to his *Origin* Darwin wrote: "I am fully convinced that species are not immutable; but that those belonging to what are called the same genera are lineal descendants of some other and generally extinct species, in the same manner as the acknowledged varieties of any one species are the descendants of that species" (Darwin 1859, p. 6).

According to Lyell, Darwin was, as pointed out above, "a good deal surprised" when he was told of the "hypothetical and theoretical notions of Linnaeus." "This," continued Lyell, "prove how little his [Linnaeus] mind was really fettered and how freely—if he had lived to our time—he would have welcomed the new ideas, which are gradually becoming general (Lyell 1867)." In the eleventh edition of his *Principles*, Lyell expressed a similar astonishment and, quoting from *Amoenitates academicae*, wrote that Linnaeus "even throws out the idea that the day may come when botanists may hold that all the species of the same genus may have sprung from the same mother (Lyell 1872, p. 325)."[53] Later the American historian of botany, Edward L. Greene, even asserted that Linnaeus was an evolutionist but that he did not dare to make his views public (Greene 1909, p. 21ff as cited in  Greene 1959, p. 355 note 13).

What was then the difference between Linnaeus and Darwin? Obviously, according to the quotations above, they both believed in common descent of certain species. Greene's view goes much beyond Linnaeus' plain statements. In the context of a theistically oriented worldview, Linnaeus changed taxonomical level for the created kinds in the Genesis account. By this move he stressed even more the importance of genera and all speciation after the divine creation occurred by hybridization. Darwin, for his part, rejected the concept of theistic creation and, entertaining an alternative *Weltanschauung*, insisted that all modification in the evolution of life occurred by natural forces such as natural selection, *i.e.*, "natural forces" in a non-Linnaean sense of the word "natural" and in sharp contrast to supernatural.

Darwin's own scientific position on creation as a young natural scientist was not that well supported by field experience. Linnaeus, who lived in the previous century, had a much more sophisticated view on

---

53    Lyell inserted a footnote into this edition with the Latin quote of Linnaeus: " Tot species dici congeneres quot eadem matre sint progenitae."

the created kinds and on their historical development after creation. Despite the old fixity view of species being old-fashioned in his own time and heavily influenced by Greek thought, young Darwin held it as a creationist but later saw it as his foremost object of attack. Not very surprisingly, Darwin himself quite early in his life turned from that extreme form of idealistic creationism, which the greatest among the taxonomists, Carolus Linnaeus, had adjusted as early as 1742 when it seemingly clashed with empirical data. The views of the mature Linnaeus took Darwin by surprise in 1867. Probably he had always looked upon the Swedish botanist as a stalwart upholder of the fixity of species and never thought of the possibility of a new reasonable synthesis within the framework of theistic creation.

## Acknowledgements
I would like to thank Oliver Barclay, Åke Bergvall, Judy and John Brenemann, Nigel Crompton, Alvar Ellegård, Gunnar Eriksson, Annika Landgren, Magnus Landgren, Sven-Eric Liedman, Bo Lindberg, Ernst Mayr, Siegfried Scherer, Franz Stuhlhofer, David Tyler, Åke Viberg, Nicolai Winther-Nielsen and the higher seminar at the Institution for history of ideas and learning at the University of Göteborg, for stimulating comments, corrections, and discussions. I would also like to thank the warden, Mr. Winter, of Tyndale House at Cambridge, the staff of the university libraries of Uppsala and Gothenburg, especially at the botanical institution in Gothenburg, and now also Dr. Roger Sanders for his kind and fine revision of my chapter to this English edition.

Corresponding address: Per Landgren, Grinnekärrsvägen 26, 436 56 Hovås, Sweden. E-mail: p.landgren@telia.com

# References

Agassiz, L. 1860. *Contributions to the Natural History of the United States of America: Volume 3.* Little, Brown and Company, Boston. [*non vidi,* cited by Beatty (1985)]

Atran, S. 1987. Origin of the species and genus concepts: An anthropological perspective. *Journal of the History of Biology* 20(2):195-279.

Barrett, P.H., P.J. Gautrey, S. Herbert, D. Hohn, and S. Smith. 1987. *Charles Darwin's Notebooks 1836-1844: Geology, Transmutation of Species, Metaphysical Inquiries.* Cornell University Press, Ithaca, NY.

Bäumer, Ä. 1991. *Geschichte der Biologie, Band 1: Biologie von der Antike bis zur Renaissance.* Peter Lang, Frankfurt, Germany.

Beatty, J. 1985. *Speaking of Species: Darwin's Strategy.* In: Kohn, D.Q., ed. *The Darwinian Heritage.* Princeton University Press, Princeton, NJ, pp. 265-281. [also pp. 227-245 *in* Ereshefsky (1992)]

Bowler, P. 1989. *Evolution: The History of an Idea.* University of California Press, Berkeley, CA.

Boyd, R., P. Gasper, and J.D. Trout, eds. 1991. *The Philosophy of Science.* MIT Press, Cambridge, MA.

Bremekamp, C.E.B. 1953. Linné's views on the hierarchy of the taxonomic groups. *Acta botanica Nederlandica* 2:242-253.

Broberg, G. 1975. *Homo sapiens L. Studier i Carl von Linnés naturuppfattning och människolära.* Almquist and Wiksell, Stockholm.

Buffon, G. L. LeC., C. de. 1755, *Histoire naturelle, générale et particulière, avec la description du Cabinet du Roi,* Volume 5. L'Imprimerie Royale, Paris.

Burkhardt, F. and S. Smith, eds. 1985. *The Correspondence of Charles Darwin, Volume 1, 1821-1836.* Cambridge University Press, Cambridge, England.

Burkhardt, F. and S. Smith eds. 1987. *The Correspondence of Charles Darwin, Volume 2, 1837-1843.* Cambridge University Press, Cambridge, England.

Burkhardt, F. and S. Smith, eds. 1988. *The Correspondence of Charles Darwin, Volume 3, 1844-1846.* Cambridge University Press, Cambridge, England.

Burkhardt, F. and S. Smith, eds. 1990. *The Correspondence of Charles Darwin, Volume 6, 1856-1857.* Cambridge University Press, Cambridge, England.

Danielsson, U. 1965. Darwinismens inträngande i Sverige: I. *Lychnos* 1963-64: 157-210.

Darwin, C. 1859. *On the Origin of Species.* John Murray, London.

Darwin, C. 1868. *The Variation of Plants and Animals under Domestication.* John Murray, London.

Darwin, F. 1887. *Life and Letters of Charles Darwin.* 3 volumes. John Murray, London.

Darwin, C. and F. Darwin, ed. 1892. *The Autobiography of Charles Darwin and Selected Letters.* D. Appleton Publishers, New York.

Delitzsch, F. 1887. *Neuer Commentar über die Genesis.* Dörffling und Franke, Leipzig, Germany.

Desmond, A. 1989. *The Politics of Evolution.* University of Chicago Press, Chicago.

Desmond, A. and J. Moore. 1991. *Darwin.* Michael Joseph, London.

Ereshefsky, M. 1992. *The Units of Evolution: Essays on the Nature of Species.* MIT Press, Cambridge, MA.

Gage, A.T. 1938. *A History of the Linnean Society of London.* The Linnean Society, London.

Ghiselin, M.T. 1969. *The Triumph of the Darwinian Method.* Dover Publications, New York.

Gillespie, N.C. 1979. *Charles Darwin and the Problem of Creation.* University of Chicago Press, Chicago.

Glass, B., O.Temkin, and W.L. Strauss, Jr., eds. 1959. *Forerunners of Darwin: 1745-1859.* Johns Hopkins University Press, Baltimore, MD.

Gotthelf, A. and J.G. Lennox, eds. 1987. *Philosophical Issues in Aristotle's Biology.* Cambridge University Press, Cambridge, England.

Green, J.R. 1914. *A History of Botany in the United Kingdom from the Earliest Times to the End of the 19th Century.* J.M. Dent and Sons, London.

Greene, E.L. 1909. Linnaeus as an evolutionist. *Proceedings, Washington Academy of Sciences* 11:17-26. [*non vidi,* cited in Greene 1959, 355n13]

Greene, E.L. and F.N. Egerton, ed. 1983. *Landmarks of Botanical History.* 2 volumes. Stanford University Press, Stanford, CA.

Greene, J.C. 1959. *The Death of Adam: Evolution and its Impact on Western Thought.* Iowa State University Press, Ames, IA.

Grinnell, G. 1974. The rise and fall of Darwin's first theory of transmutation. *Journal of the History of Biology* 7:273.

Hall, J.J. 1991. The classification of birds: In Aristotle and early modern naturalists (I). *History of Science* 29:111-152.

Hodge, M. J. S. and G. N. Cantor. 1990. The development of philosophy of science since 1900. In: Colby, R.C., G.N. Cantor, J.R.R. Christie, and M.J. Hodge, eds. *Companion to the History of Modern Science.* Routledge, London, pp. 838-852.

Hooykaas, R. 1952. The species concept in eighteenth-century mineralogy. *Archives International d'Histoire des Sciences* 1:45-55.

Hooykaas, R. 1959. *A Historical-Critical Study of the Principle of Uniformity in Geology, Biology and Theology.* EJ Brill, Leiden, Germany.

Hooykaas, R.1982. Wissenschaftsgeschichte, eine Brücke zwischen Natur- und Geisteswissenschaften. *Berichte zur Wissenschaftsgeschichte* 5:162-170.

Hopkins, W. 1861-2. Physical theories of the phenomena of life. *Fraser's Magazine* 61:739-52; 62:74-90. [reprinted in Hull 1973, 229-275]

Hull, D.L., ed. 1973. *Darwin and His Critics: The Reception of Darwin's Theory of Evolution by the Scientific Community.* University of Chicago Press, Chicago.

Hull, D.L. 1988. *Science as Process: An Evolutionary Account of the Social and Conceptual Development of Science*. University of Chicago Press, Chicago.

Kaiser, W. 1980. mîn. In: Harris, R.L., G.L. Archer, Jr. and B.K. Waltke, eds. *Theological Wordbook of the Old Testament*. Moody, Chicago, pp. 503-504.

Keil, C.F. 1878. *Genesis und Exodus*. Brunnen, Gießen, Germany.

Kohn, D., ed. 1985. *The Darwinian Heritage*. Princeton University Press, Princeton, NJ.

Kohlbrugge, J.H.F. 1915. War Darwin ein originelles Genie? *Biologisches Zentralblatt* 35:93-111.

Kottler, M.J. 1978. Charles Darwin's biological species concept and theory of geographic speciation: The transmutation notebooks. *Annals of Science* 35:275-297.

Kuhn, T.S. 1962. *The Structure of Scientific Revolutions*. University of Chicago Press, Chicago.

Larson, J.L. 1971. *Reason and Experience*. University of California Press, Berkeley, CA.

Laudan, R. 1987. *From Mineralogy to Geology: The Foundation of a Science 1650-1830*. University of Chicago Press, Chicago.

Laudan, R. 1989. Individuals, species and the development of mineralogy and geology. In: Ruse, M., ed. *What the Philosophy of Biology Is: Essays Dedicated to David Hull*. Kluwer Academic, Dordrecht, Netherlands, pp. 221-233.

Lennox, J.G. 1980. Genera, Species and 'the more and the less' in Aristotle. *Journal of the History of Biology* 13:321-346.

Liddell, H.G. and R. Scott. 1977. *Greek-English Lexicon*. Clarendon Press, Oxford, England.

Linnaeus, C. 1735. *Systema naturae, sive Regna tria naturae, systematice proposita per classes, ordines, genera, et species*. Theodor Haak, Leiden, Netherlands.

Linnaeus, C. 1736. *Fundamenta botanica, quae majorum operum prodromi instar theoriam scientiae botanices per breves aphorismos tradunt*. Salomon Schouten, Amsterdam.

Linnaeus, C. 1737. *Genera plantarum, eorumque characteres naturales secundum numerum, figuram, situm & proportionem omnium fructificationis partium*. Conrad Wishoff & fil., Leiden, Netherlands.

Linnaeus, C. 1738. *Classes plantarum, seu systema plantarum omnia a fructificatione desumpta [...] secundum classes, ordines et nomina generica cum clave cuiusvis methodi et synonymis genericis*. Conrad Wishoff, Leiden, Netherlands.

Linnaeus, C. 1749. *Amoenitates academicae, seu dissertationes variae physicae, medicae, botanicae, antehac seorsim editae, nunc collectae et auctae cum tabulis aeneis. Accedit hypothesis nova de febrium intermittentium causa*. Cornelius Haak, Leiden, Netherlands.

Linnaeus, C. 1751a. *Philosophia botanica, in qua explicantur fundamenta botanica cum definitionibus partium, exemplis terminorum, observationibus rariorum, adjectis figuris aeneis*. G. Kiesewetter, Stockholm.

Linnaeus, C. 1751b. *Plantae hybridae*, Uppsala, Sweden.

Linnaeus, C. 1762. *Fundamentum fructificationis*. Uppsala, Sweden.

Linnaeus, C. 1764. *Genera plantarum*. Sixth Edition. Stockholm, Sweden.

Lovejoy, A.O. 1936. *The Great Chain of Being*. Harvard University Press, Cambridge, MA.

Lovejoy, A.O. 1959. Buffon and the problem of species. In: Glass, B., O.Temkin, and W.L. Strauss, Jr., eds. *Forerunners of Darwin: 1745-1859*. Johns Hopkins University Press, Baltimore, pp. 84-113.

Lyell. C. 1867, 15 December. [Letter to Sven Lovén] available in library of the Academy of Science, Stockholm, Sweden.

Lyell, C. 1872. *Principles of Geology*. Eleventh edition. John Murray, London.

Mayr, E. 1982. *The Growth of Biological Thought: Diversity, Evolution and Inheritance*. Belknap Press, Cambridge, MA.

Mayr, E. 1988. *Toward a New Philosophy of Biology: Observations of an Evolutionist*. Harvard University Press, Cambridge, MA.

Mayr, E. 1992a. Darwin's principle of divergence. *Journal of the History of Biology* 25(3):343-359.

Mayr, E. 1992b. The idea of teleology. *Journal of the History of Ideas* 51(1):117-135.

Moore, J. 1991. Deconstructing Darwinism: The politics of evolution in the 1860s. *Journal of the History of Biology* 24: 353-408.

Pellegrin, P. 1986. *Aristotle's Classification of Animals: Biology and the Conceptual Unity of the Aristotelian Corpus*, translated by A. Preus. University of California Press. Berkeley, CA.

Ray, J. 1686-1704. *Historia plantarum, etc.* London, England

Schwartz, J.S. 1990. Darwin, Wallace and Huxley, and vestiges of the natural history of creation. *Journal of the History of Biology* 23:127-153.

Secord, J.A. 1991. Edinburgh Lamarckians: Robert Jameson and Robert E. Grant. *Journal of the History of Biology* 24:1-18.

Stafleu, F.A. 1971. *Linnaeus and the Linnaeans: The Spreading of their Ideas in Sytematic Botany, 1735-1789*. International Association for Plant Taxonomy, Utrecht, Netherlands.

Sulloway, F. 1979. Geographic isolation in Darwin's thinking. *Studies in History of Biology* 3:23-65.

Sulloway, F.J. 1982a. Darwin and his finches: The evolution of a legend. *Journal of the History of Biology* 15:1-53.

Sulloway, F.J. 1982b. Darwin's conversion: The Beagle voyage and its aftermath. *Journal of the History of Biology* 15:325-396.

Wenham, G.J. 1987. *Word Biblical Commentary: Genesis 1-15*. Word Books, Dallas, TX.

Westermann, C. 1974. *Genesis*. Neukirchener Verlag, Neukirchen, Germany.

Winsor, M.P. 1985. Book Review: *The philosophical naturalists: Themes in early Nineteenth-Century British biology* by Philip R. Rehbock. *Isis*

76:252-253.

Wollaston, T.V. 1856. *On the Variation of Species: With Especial Reference to the Insecta, Followed by an Inquiry into the Nature of Genera.* J. Van Voorst. London.

Wuketits, F. M. 1988. *Evolutionstheorien: Historische Voraussetzungen, Positionen, Kritik.* Wissenschaftliche Buchgesellschaft, Darmstadt, Germany.

# 4. Formation of Characters and Hybridization in the Funariaceae (Bryophyta, Musci)

## Martin Adler

**Abstract**

The moss family Funariaceae is characterized by a wide spectrum of sporophyte characters. However, numerous intra- and interspecific hybrids indicate a close relationship between the species of this family, which can thus be assigned to the same basic type. The derivation from a morphologically complex ancestor is discussed.

### 4.1. Introduction

Unlike the situation in seed plants, where the interest of cultivators has always promoted experimentation, the experimental results on the hybridization of mosses are quite rare. Our knowledge of moss hybrids is based on reports of spontaneous hybridization, and often some doubt remains over the actual hybrid nature of the findings. With reference to the application of hybrid formation to the analysis of systematic relationships or basic types, it must be pointed out that the few spontaneous hybrids which have been found cover only a small part of the potential spectrum, so that there is still ample room for additional research in this area.

In view of this background, the works of the Berlin botanist Wettstein on crossbreeding experiments with Funariaceae, although already seventy years old, must be viewed as groundbreaking (Wettstein 1924a, 1924b, 1925). They form the basis for the present summary.

The Funariaceae are ground mosses and, as such, relatively easy to cultivate. In addition, although they are frequently monoecious, the formation of unisexual gametangium-bearing branches simplifies controlled hybridization. Thus, this family represents a good choice for crossbreeding experiments. The flora of central Europe, to which Wettstein restricted himself, provides a sufficient cross-section of the entire family (over 200 species) to allow for generalizations.

### 4.2. Family Characteristics and Division of Genera

The family Funariaceaee includes small, annual to biannual acrocarpous ground mosses with typically rosette-shaped gametophytes. The leaves have a central rib; the leaf cells are large and thin-walled, and usually contain relatively few chloroplasts. The sporophytes (see Figure 1) show a variety of shapes and form the basis of the current systematic division of the genera. The following five genera are found in Europe; four of them are relevant for the analysis of the formation of hybrids:

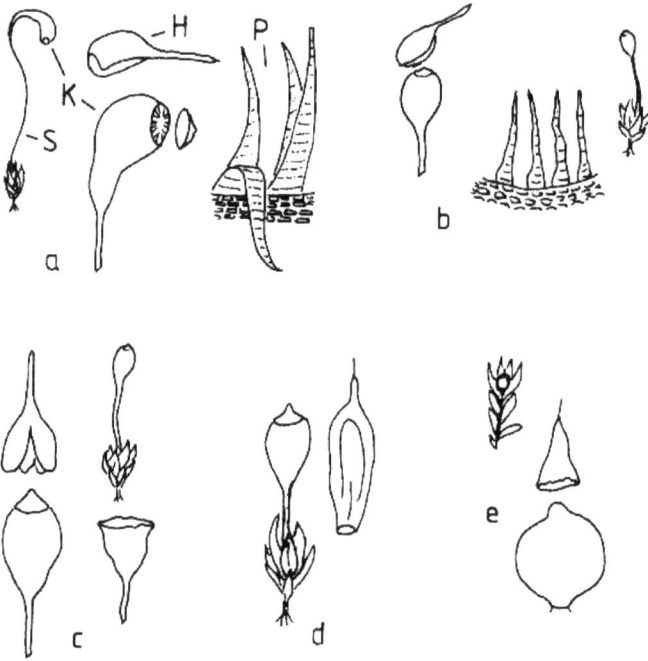

Figure. 1: Sporophyte characteristics in Funariaceae (schematic): *Funaria* (a), *Enthostodon* (b), *Physcomitrium* (c), *Pyramidula* (d), *Physcomitrella* (e). S = seta, K = capsule, H = calyptra, P = peristome.

*Funaria*: Seta long and curved, capsule dorsoventrally bent, and consequently bilaterally symmetric. Calyptra hood-shaped. Peristome well developed, in the shape of two rows of teeth. Capsule lid differentiated. Approximately 150 species, worldwide distribution.

*Enthostodon*: Seta long and straight, capsule upright and radially symmetric. The calyptra is hood-shaped. Peristome with one row of teeth, which in European species is occasionally incomplete or missing. Capsule lid present. Worldwide, approximately 100 species.

*Physcomitrium*: Usually with straight seta, capsule upright, broad to spherical. Calyptra small, cap-shaped and lobed. Peristome teeth are missing; a capsule lid exists. Worldwide, approximately 100 species.

*Pyramidula*: Small forms, very short seta, capsule upright and symmetrical. Calyptra large, inflated, four angled. Peristome teeth are missing. Capsule lid developed, however, according to Düll (1987), it does not open when the capsule reaches maturity. Monotypic genus in Europe and North America.

*Physcomitrella*: Small forms, seta practically nonexistent, capsule spherical, without lid and peristome (hence, cleistocarpous). Calyptra small and cone-shaped. Five species in Eurasia, North America, and Australia.

|  | male |  |  |  |  |  |  |  |
|---|---|---|---|---|---|---|---|---|
| female | F.m | F.h. | E.f. | P.e. | P.py | P.s. | P.t. | P.pa. |
| Funaria muehlenbergi |  | e |  |  |  |  |  |  |
| F. hygrometrica | e |  | ? |  |  |  |  |  |
| Enthostodon fascicularis |  | n |  |  |  |  |  |  |
| Ph. eurystomum |  | e+ |  |  | e |  |  | ne |
| Ph. pyriforme |  | ne |  | e |  |  |  |  |
| Ph. sphaericum |  |  |  |  |  |  |  |  |
| Ph. turbinatum |  |  |  |  |  |  |  |  |
| Physcomitrella patens |  | ne+ |  | ne |  | n | n |  |

Table 1. Crossbreeding matrix of the Funariaceae. n = spontaneous hybrid, e = experimentally induced hybrid, e+ = simple experimental hybridization. The experimental data are based on the works of Wettstein, who also makes mention of some spontaneous hybrids. With *Funaria microstoma*, no experimental hybridizations were successful. For *Physcomitrium turbinatum* × *Physcomitrella patens* see Andrew (1942).

## 4.3. Hybridization

The results of Wettstein's crossbreeding experiments are summarized in the crossbreeding matrix (Table 1), and are supplemented by some well-documented spontaneous hybrids.

The genera examined are interconnected through hybridizations, especially due to the high potential for hybridization of *Funaria hygrometrica* as a male partner. Crossbreeding within genera is normally possible. The hybrid sporophytes of the F$_1$ generation are mostly sterile and show characteristics of both parents.

## 4.4. Discussion

The division of the Funariaceae, as shown above, can be called classical-typological, and largely disregards phylogenetic considerations. It is very common in newer bryological standard works (e.g., Frahm and Frey 1983), but may not fulfill the requirements of more modern taxonomic methods, like Hennigian systematics (comp. Mishler 1986). Its strength lies in a clear presentation of the characters and character combinations within the family.

Within the Funariaceae, the close relationship of the genera is well demonstrated by the potential of crossbreeding. The forms that were examined are extensively interconnected through hybridization, and consequently belong to a basic type in the sense of Scherer (1993, this volume). If one presupposes a common ancestral form of the species belonging to the same basic type and if one considers the morphological variety of the Funariaceae, one can ask which morphotype would be closest to the postulated ancestral form.

A comparison of the individual genera points to the fact that the most important differences are based on different degrees of morphological complexity. Within the family, the genus *Funaria* shows the most extensive morphological differentiation of its sporophytes. With the long seta, the curved capsule with hood-shaped calyptra and the well developed doubled peristome, it best represents the concept of a "complete" deciduous moss. *Enthostodon* possesses a simpler peristome, and the capsule lacks the dorsoventral differentiation. The bilaterally symmetric calyptra on top of the radially symmetric capsule seems somewhat out of place. *Physcomitrium* and *Pyramidula* do not possess any peristome, and their setae are shorter. In *Physcomitrella*, neither a peristome nor a capsule lid exists, and a seta is not properly developed.

In spite of the theoretical possibility of other options, one should point out that the postulated ancestral form of the Funariaceaee is most likely a "complex" and "complete" form, like *Funaria*, whereas the other types can be regarded as reduced forms. The reduction – the direction of the evolution in this case – leads to different stages of reduced morphological differentiation. Indeed, all important deviations from the *Funaria*-type consist of morphological simplifications within the family; only the large, angular calyptra in *Pyramidula* may be regarded as an independent, new formation. However, it would not be right to interpret the simplified morphology as inferior, or to speak of a general degeneration. Rather, the reduction processes are closely connected with certain specializations, for example, the stunted growth and short life spans of *Physcomitrella* and *Pyramidula*. Such morphologically minimized—or maybe optimized—specialists usually possess a lower potential for development and adaptation than generalists, such as *Funaria hygrometrica*.

In phylogenetic discussions it is common practice to arrange the known forms, whether fossil or recent, in series that are meant to be a model of sequential evolutionary stages (the fossil horses are a popular example of such a series). The Funariaceaee could also be arranged in an "evolutionary series" (in this case a reduction series) with the sequence *Funaria, Enthostodon, Physcomitrium,* and *Physcomitrella. Enthostodon,* for example, looks like a stunted *Funaria.* Within *Enthostodon,* the peristome shows different degrees of reduction, which, at least theoretically, could have lead to *Physcomitrium.* On the other hand, the crossbreeding experiments indicate that the two morphological extremes, *Funaria hygrometrica* and *Physcomitrella patens,* hybridize very easily and are possibly close relatives. Morphologically intermediate forms like *Physcomitrium* do not offer any hints of being genealogically intermittent forms. This suggests a model of evolutionary radiation, which led from an ancestral form to the individual morphotypes without the necessity of morphological intermediate stages. In the Funariaceaee, further proof for this view is the central position of *Funaria hygrometrica* in the crossbreeding matrix, which, as presumably similar to the ancestral form, connects the various genera with each other. For this reason, the systematic relationships within the family might be depicted by a star-shaped rather than a sequential pattern. However, to offer further proof for this model, additional crossbreeding experiments among other species would be required. Molecular-biological data (for example sequence comparisons) would also be desirable, in order to be able to correlate genetic and morphological similarities.

# References

Andrews, A.L. 1942. Taxonomic notes, II: Another natural hybrid in the Funariaceaee. *Bryologist* 45:179-181.

Düll, R. 1987. *Exkursionstaschenbuch der Moose.* Second edition. IDH, Rheurdt, Germany.

Frahm, J.P. and W. Frey. 1983. *Moosflora.* First edition. Ulmer, Stuttgart.

Mishler, B.D. 1986. A Hennigian approach to bryophyte phylogeny. *Journal of Bryology* 14:71-81.

Scherer, S. 1993. Basic types of life. In: Scherer, S., ed. *Typen des Lebens,* Pascal, Berlin, pp. 11-30.

Wettstein, F. v. 1924a. Morphologie und Physiologie des Formwechsels der Moose auf genetischer Grundlage, I. *Zeitschrift für Inductive Abstammungs und Vererbungslehre* 23:1-236.

Wettstein, F. v. 1924b. Kreuzungsversuche mit multiploiden Moosrassen, II. *Biologisches Zentralblatt* 44:145-168.

Wettstein, F v. 1925. Genetische Untersuchungen an Moosen. *Bibliographia Genetica* 1:1-38.

# 5. The Spleenworts (Filicatae, Aspleniaceae) in the Basic Type Concept

HERFRIED KUTZELNIGG

**Abstract**

The spleenworts are a group of genera within the true ferns (order Polypodiales). Most authors consider them to be a separate family named Aspleniaceae. This group includes some 7-10 segregates or genera. With about 700 species, the genus *Asplenium s. str.* (spleenworts) shows the greatest variety of forms. Morphological and molecular data show the close affinity of all taxa. Therefore, most modern authors accept only one genus *Asplenium s.l.* or the two genera *Asplenium* and *Hymenasplenium*. Many cases of hybridization have been observed, both within *Asplenium* and between different genera (or subgenera) of the Aspleniaceae. Due to apomictic reproduction, some hybrids then became fixed species. No hybridizations beyond the framework of the Aspleniaceae are known. In addition, transitional forms have been found between different genera. These observations suggest that the Aspleniaceae can be regarded as a basic type.

## 5.1. Characteristics and Systematic Position of the Family

The spleenworts are small to moderate-sized ferns with mostly a short radial rhizome, only very few species having a long creeping rhizome (e.g., *Hymenasplenium, Loxoscaphe*). Their spore receptacles are arranged in stripe-shaped sori, following the course of the side veins on the underside of the leaves. The veil protecting the spore receptacles (the indusium) is attached to the side of the sori. In rare cases it is reduced. The petiole is often dark and sclerotic, typically with two C-shaped vascular bundles (= leaf traces) back to back joining to form one X-shaped bundle. The basic chromosome number is in most cases $x = 36$, and only in a few exceptions (*Hymenasplenium* incl. *Boniniella*) $x = 38$ or $39$ (Murakami & Moran 1993; Murakami 1995); *H. costarisorum* has $x = 36$.

Many of the older authors considered the spleenworts to be a subgroup of the family Polypodiaceae *s.l.* (the common or true-ferns), which include the majority of the commonly known ferns. Whereas there has been considerable disagreement in the past, they are today most often placed in their own family: Aspleniaceae Mettenius ex A. B. Frank, *Leunis Syn. Pflanzenkunde*, 2nd ed., 3:1465. 1877. Peculiarities of the anatomy of the dictyostele as well as the nature of the latticed scales are named as criteria for their separate systematic position.

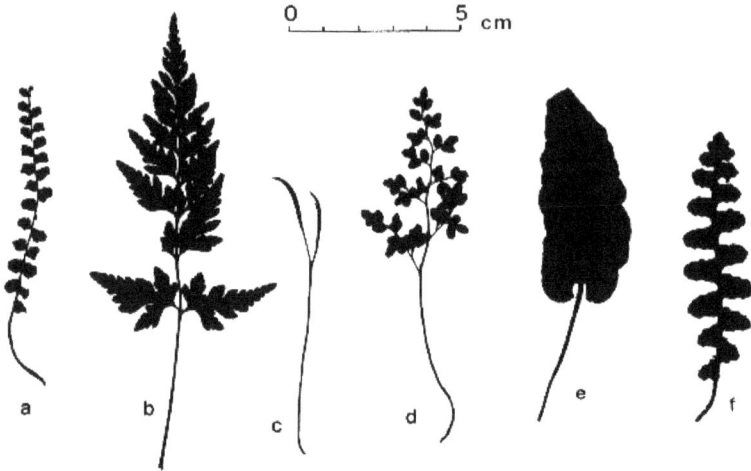

Figure 1. Examples of central European representatives of the spleenwort ferns (Aspleniaceae): silhouettes (photocopies) of leaves. a. maidenhair spleenwort (*Asplenium trichomanes*), b. black spleenwort (*Asplenium adiantum-nigrum*), c. forked spleenwort (*Asplenium septentrionale*), d. wall-rue (*Asplenium ruta-muraria*), e. hart's-tongue fern (*Phyllitis scolopendrium*), f. rustyback fern (*Ceterach officinarum*).

Reviews of the family are given by Reichstein (1984), Kramer and Viane (1990), Murakami *et al.* (1999), Schneider *et al.* (2004), and Smith *et al.* (2006).

The Athyriaceae (synonym Woodsiaceae; the lady fern family), with the genera *Athyrium* (lady ferns), *Cystopteris* (bladder ferns), *Woodsia* (cliff ferns), and *Matteuccia* (the ostrich ferns), are often considered to be their closest relatives.

Aspleniaceae are sometimes regarded as the only family of the order Aspleniales Pic. Serm. ex Reveal, but most modern authors place them within the large order Polypodiales Link, comprising as many as 15 families (Smith *et al.* 2006).

No successful hybrids are known to cross the limits of the Aspleniaceae as defined above.

## 5.2. Delimitation of genera

Due to the limited or inconsistent morphological and molecular differences among taxa, delimitation of the segregates is very difficult (Smith *et al.* 2006). The commonly known genera or sections are *Asplenium* (spleenworts), *Camptosorus* (walking ferns), *Ceterach* (spleen ferns), *Phyllitis* (hart's tongue ferns), and *Pleurosorus* (blanket ferns). The "List of genera in Aspleniaceae" compiled by the Royal Botanical Gardens, Kew (Brummitt 1992, www.rbgkew.org.uk), names 9 genera (*Antigramma, Asplenium, Camptosorus, Ceterach, Diellia, Diplora, Holodictyum, Pleurosorus, Schaffneria*), 2 intergeneric hybrids, and 28 generic synonyms, most of which are included in *Asplenium s.str.*, e.g., *Phyllitis* and *Hymenasplenium*.

For some time many authors accepted only the genus *Asplenium s.l.* However, recent molecular data in accordance with morphological and cytogenetic findings suggest that the tropical genus *Hymenasplenium* (incl. *Boniniella*) has to be separated from *Asplenium s.l.* (Murakami 1995; Schneider *et al.* 2004; Smith *et al.* 2006), so that the family now consists of two genera, sometimes enlarged by *Phyllitis* as a third genus.

## 5.3. Number of Species and Distribution

The Aspleniaceae include some 750 species, the majority of them (approximately 700 species) in *Asplenium s.str.*, while the remaining segregates have species numbers between 2 and 10. An exception is *Hymenasplenium* which is now thought to comprise 30(-60) species (Murakami 1995).

The distribution of Aspleniaceae is world-wide and extends from the tropics to the polar circles, most species living in the tropics.

Conspicuous differences between the species can be found in the appearance of the leaves, which are either undivided, lobed, or pinnately divided once to several times (Figure 1).

The morphological variability of *Asplenium* is demonstrated by the fact that several species have specific common names. For example, *Asplenium ruta-muraria* is called wall-rue; the epiphytic *Asplenium nidus* (a popular indoor plant) is known as bird's-nest fern.

In central Europe, *Ceterach officinarum* DC. (the rustyback fern) and *Phyllitis scolopendrium* L. (the hart's-tongue fern) occur together with a number of other species of the genus *Asplenium*. A detailed treatise can be found in Reichstein (1984). Another European genus is *Pleurosorus*, in which the indusium is reduced.

## 5.4. Hybridization within the Aspleniaceae

First of all it should be said that speciation through hybridization plus chromosome duplication (and thus through allopolyploidy) is very widespread within the ferns. Approximately half of the European ferns—as far as they have been examined—are considered to be allotetraploid (Kramer 1984). Among them are species as abundant as *Dryopteris filix-mas* (male fern) or *Dryopteris dilatata* (broad buckler thorn fern).

Mechanisms for chromosome duplication—which is rare in nature, but significant for the formation of new species—are tetraploidy in the leaves or apospory, the formation of restitution nuclei [in the sporangium]. Further growth then takes place apomictically, *i.e.*, through nonsexual reproduction.

Among the species of *Asplenium s.str.*, numerous primary hybrids have been observed in central Europe (Meyer 1960; Vida 1976; Reichstein

1981, 1982). *Asplenium adulterinum* (adulterated spleenwort), for example, represents a secondary hybrid and, as such, has become a separate species. As to its appearance, this hybrid lies between the parents *A. trichomanes* L. and *A. viride* L. and is allotetraploid. It can also be generated experimentally. It is remarkable that the associated primary hybrid, which is occasionally capable of germinating, produces spores through diploidization, and that in planting experiments grew into prothallia, from which fertile, tetraploid plants resulted.

The taxanomy of natural hybrids and taxa of hybrid origin (reticulate evolution) in North America has been pictured by Wagner (1954, 1963).

Numerous taxa have been reported in *Asplenium* world-wide, most of them alloploid hybrids behaving like species. As an example, the hybrid origin of at least seven taxa in subgenus *Ceterach* is pictured by van den Heede *et al.* (2003). Knobloch (1996) in the second version of his "Pteridophyte hybrids and their derivatives" lists some 220 hybrids within *Asplenium s.str.*, three within *Phyllitis*, two within *Ceterach* and two within *Hymenasplenium*. In the latter genus, at least three hybrids must be added (Murakami & Moran 1993).

Intergeneric (or intersubgeneric) hybrids, too, are not rare within the Aspleniaceae. The following combinations have been observed:

*Asplenium* × *Camptosorus* = ×*Asplenosorus* Wherry: Knobloch (1996) lists seven combinations of this type.

*Asplenium* × *Ceterach* = ×*Asplenoceterach* D. E. Meyer: Two combinations are known (*A. majoricum* × *C. officinarum*, and *A.* cf. *trichomanes* × *C. aureum*). A third supposed hybrid *A. ruta-muraria* × *C. officinarum* in the Kaiserstuhl uplands region (Figure 2a) turned out to be a monstrous form of *Ceterach officinarum* (Rasbach *et al.* 1989).

*Asplenium* × *Phyllitis* = ×*Asplenophyllitis* Alston: Knobloch (1996) lists seven combinations of this type, e.g., *A. trichomanes* × *P. scolopendrium* = ×*Asplenophyllitis confluens* (T. Moore & Lowe) Alston which grows spontaneously in England and Slovenia and can also be produced artificially (Figure 2b), and *A. adiantum-nigrum* × *P. scolopendrium* = ×*Asplenophyllitis jacksonii* Alston, which has thus far only been produced experimentally (Figure 2c).

*Asplenium* × *Pleurosorus*: Two hybrids are known, *A. trichomanes* × *P. pozoi* (Meyer 1963) and *A. petrarchae* × *P. hispanicus* (Lovis 1973)

*Camptosorus* × *Phyllitis*: The hybrid *C. rhizophyllus* × *P. scolopendrium* has been described by Wagner (1954, 1963).

*Ceterach* × *Phyllitis* = ×*Ceterophyllitis* Pic. Serm.: *C. officinarum* × *P. sagittata* = *Phyllitopsis hybrida* (Milde) Reichstein is an example of a stable generic hybrid that behaves as a species. The natural distribution area of the allotetraploid plant lies in Dalmatian Croatia. It is usually regarded as a species of the genus *Phyllitis* (*P. hybrida*). That it is indeed a hybrid was demonstrated by Vida (1976), who produced this plant experimentally by using a diploid cytotype of *Ceterach officinarum*.

Recently, even a tri-generic hybrid was experimentally synthesized: *Asplenium montanum* × *Camptosorus cf. rhizophyllus* × *Phyllitis scolopendrium* (Wagner & Hagenah 1989).

## 5.5. Conclusion

The results clearly demonstrate the connections among the genera of the Aspleniaceae based on hybridization. This fully agrees with the observation of transitional forms between the genera (see Reichstein 1984), the close affinities between the genera, and the monophyly of the clade shown in morphological and molecular studies (Murakami *et al.* 1999; Schneider *et al.* 2004; Smith *et al.* 2006).

The close relationships among

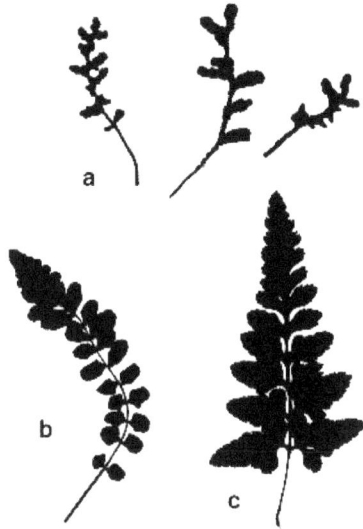

Figure 2. Intergeneric hybrids within the Aspleniaceae. a. *Asplenium ruta-muraria* × *Ceterach officinarum*, b. *Asplenium trichomanes* × *Phyllitis scolopendrium*, c. *Asplenium adiantum-nigrum* × *Phyllitis scolopendrium*. (After Reichstein 1984).

the genera, on one hand, and the absence of intermediate forms and hybrids with other families of the ferns, on the other hand, suggest that the Aspleniaceae represent a basic type.

In this context it is interesting that Linnaeus as early as 1753 summarized all representatives of this group known to him under the generic name *Asplenium*. He called *Ceterach officinarum* "Asplenium ceterach" and *Phyllitis scolopendrium* "Asplenium scolopendrium."

As mentioned above, the *Asplenium* clade was for a long time thought to be composed of about 7-10 genera, but in recent decades many authors have submerged them again in *Asplenium*. Since the work of Murakami (1995), section *Hymenasplenium* is mostly regarded as a second genus because it is the sister group of *Asplenium s.*

# References

Brummitt, R.K. 1992. *Vascular Plant Families and Genera*. Royal Botanical Gardens, Kew. (see also http://www.rbgkew.org.uk)

Knobloch, I.W. 1996. Pteridophyte hybrids and their derivatives. Dep. Botany and Plant Pathology, Michigan State University, East Lansing.

Kramer, K.U. ed. 1984. *Pteridophyta in G. Hegi, Illustrierte Flora von Mitteleuropa* 1/1.Third Edition. Parey, Berlin.

Linnaeus, C. 1753. *Species Plantarum*. Second Volume. Salvius, Stockholm.

Lovis J. D. 1973 A biosystematic approach to phylogenetic problems and its application to the Aspleniaceae. In: Jermy A.C., J. A. Crabe, and B. A. Thomas, eds. *The phylogeny and classification of the ferns*. Academic Press, London, pp. 211–228.

Meyer, D.E. 1960. Zur Cytotaxonomie der Asplenien Mitteleuropas. *Berichte der Bayerischen Botanischen Gesellschaft* 73:386-394.

Meyer, D.E. 1963. Über neue und seltene Asplenien Europas. 2. Mitteilung. *Berichte der Deutschen Botanischen Gesellschaft* 76:13-22.

Murakami, N. 1995. Systematics and Evolutionary Biology of the Fern Genus Hymenasplenium (Aspleniaceae). *Journal of Plant Research* 108:257-268.

Murakami, N. and R.C. Moran. 1993. Monograph of the Neotropical Species of *Asplenium* Sect. *Hymenasplenium* (Aspleniaceae). *Annals Missouri Botanical Garden* 8:1-38.

Murakami, N., S. Nogami, M. Watanabe, and K. Iwatsuki. 1999. Phylogeny of Aspleniaceae inferred from rbcL nucleotide sequences. *American Fern Journal* 89:232-243.

Rasbach, H., K. Rasbach, and R. Viane. 1989. A new look at the fern described as ×*Asplenoceterach badense* (Aspleniaceae). *Willdenowia* 18:483-496.

Reichstein, T. 1981. Hybrids in European Aspleniaceae (Pteridophyta). *Botanica Helvetica* 91:89-139.

Reichstein, T. 1982. Hybrids in European Aspleniaceae: Addenda et corrigenda. *Botanica Helvetica* 92:41-42.

Reichstein, T. 1984. Aspleniaceae. In: Kramer, K.U., ed. *Pteridophyta in G. Hegi Illustrierte Flora von Mitteleuropa* 1/1.Third Edition. Parey, Berlin, pp. 211-275.

Schneider, H., and 6 others. 2004. Chloroplast phylogeny of asplenioid ferns based on rbcL and trnL-F spacer sequences (Polypodiidae, Aspleniaceae) and its implications for the biogeography of these ferns. *Systematic Botany* 29: 260-274.

Smith, A.R., K.M. Pryer, E. Schuettpelz, P. Korall, H. Schneider, and P.G. Wolf. 2006. A classification for extant ferns. *Taxon* 55:705-731.

Van den Heede, C., R.L.L. Viane, and M.W. Chase. 2003. Phylogenetic analysis of *Asplenium* subgenus *Ceterach* (Pteridophyta: Aspleniaceae) based on plastid and nuclear ribosomal ITS DNA sequences. *American Journal of Botany* 90:481-495.

Vida, G. 1976. The role of polyploidy in evolution. *Evolutionary Biology*

*(Prague)* 9:267-294.

Wagner, W.H., Jr. 1954. Reticulate evolution in the Appalachian aspleniums. *Evolution* 8:103-118.

Wagner, W.H., Jr. 1963. A biosystematic survey of United States ferns - preliminary abstract. *American Fern Journal* 53:1-16.

Wagner, W.H., Jr. and E. Hagenah. 1989. A synthetic „trigeneric" hybrid, ×*Asplenosorus pinnatifidus* × *Phyllitis scolopendrium* var. *americanum*. *American Fern Journal* 79:1-6.

# 6. The Basic Type of the Wheat Grasses (Poaceae, Tribe Triticeae)

REINHARD JUNKER

**Abstract**

Because of extensive crossings and morphological and anatomical data, the Triticeae together with Bromeae can be clearly distinguished from other grasses of the family Poaceae. Successful crossings with species from other tribes are not known. Among the perennial Triticeae, hybrids exist of almost 2/3 of all possible genus pairs. The proportion is similarly high among the annual genera. In addition, there are several successful crossings of perennial and annual species. Almost every genus and subtribe is directly interrelated by hybridization.

Hybridization and numerical studies reveal a complex taxonomy that does not allow for a reconstruction of the phylogenies of the Triticeae within the present scope of knowledge. Numerous character inconsistencies within the basic type of the Triticeae + Bromeae suggest rapid speciation, whereby different characteristics of the probably complex parent form were spread either stochastically or by means of selection throughout the individual genera and species. Other processes like introgression, "pivotal-differential-evolution," and different rates of evolution of different characters probably have contributed to the complication of the relations of descent.

## 6.1. Introduction

Among the monocot flowering plants (Angiospermae: Monocotyledoneae), the family of the Poaceae (=Gramineae, the grasses), with approximately 4,000-10,000 species, can be clearly distinguished from other families. The Poaceae are regarded as a highly developed group, whose tribes (groups of genera) are, without exception, specialized in one way or the other, partially in combination with so-called "archaic" characteristics (Stebbins 1956b). The few known grass fossils are very similar to present-day genera (Stebbins 1972; comp. Crepet and Feldman 1991).

The systematics of the Poaceae is still in a state of flux. "The evolutionary 'tree' of most grass genera is not a simple branching affair, but a highly complex network" (Stebbins 1956b). This assessment is still valid. The division into subfamilies and tribes is based on anatomical, cytological, cytogenetic, chemical, ecological, physiological, and other criteria. Different groupings emerge, depending on the validity ascribed

Figure 1. Some European grasses of the tribe Triticeae; from the left: *Hordelymus europaeus, Hordeum murinum*, and *Agropyron caninum*, photos by the author. Right column, above: structure of the spike of the Triticeae; below: generalized spikelets dissected, showing glumes (two basal scales) and one vs. three florets. (after Porter 1967).

to various combinations of traits. Thus, 32 tribes are listed in the "Flora Europaea" (Melderis *et al.* 1980); Porter (1967) divides the Poaceae into 27 tribes, Hutchinson (1973) also into 27, Bor (1960) into 38, and Hegnauer (1963) into 26. (For the history of the taxonomy of the Triticeae see Barkworth 1992.)

Tribe Triticeae Dumortier (*Obs. Gram. Fl. Belg.* 82, 84, 91. 1824. as Hordeeae, Conert 1983) is traditionally separated from the other tribes of the Poaceae on the basis of morphological characteristics (Stebbins 1956a). The spikelets each comprise one or more florets and are located on opposite sides of a single symmetrical spike (Figure 1). When the seeds are ripe, the florets fall, leaving behind the glumes (basal pair of spikelet scales).

The Triticeae include important species of grain—wheat (*Triticum*), barley (*Hordeum*), and rye (*Secale*)—as well as several other edible and some ornamental grasses (McIntyre 1988). Approximately 325 species are counted within the Triticeae, among which approximately 250 are perennial with the others mostly annual (Dewey 1984).

The main areas of the distribution of the Triticeae are Eurasia and North America. The annual species are mostly found in the Near East, but the perennial species are concentrated in the prairies of Eastern Europe, Southern Russia, Western North America, and Argentina (West *et al.* 1988). Only a few perennial species of the Triticeae exist both in the Old and New Worlds. Although representatives of the Triticeae are found worldwide, many genera are limited to just one continent.

Due to the enormous economic importance of this tribe, extensive hybridization experiments have been conducted to improve the quality

*Sitopsis speltoides*
  *Aegilops speltoides*
  *Triticum speltoides*
  *Agropyron ligusticum*
  *Aegilops ligustica*
  *Aegilops aucheri*
  *Aegiiops macrura*
  *Sitopsis bicornis*
  *Triticum bicorne*
  *Aegilops bicornis*
  *Crithodium aegyptiacum*

*Sitopsis longissima*
  *Aegilops longissima*
  *Triticum longissimum*

*Sitopsis searsii*
  *Aegilops searsii*

*Sitopsis sharonensis*
  *Aegilops sharonensis*
  *Aegilops longissima ssp sharonensis*

*Orrhopygium caudatum*
  *Aegilops caudata*
  *Aegilops c. ssp dichasians*
  *Aegilops dichasians*
  *Aegilops cvlindrica*
  *Aegilops markgrafii*
  *Triticum caudatum*
  *Triticum dichasians*
  *Triticum markgrafii*

Table 1. A small excerpt from the "Conspectus der Triticeen" (Löve 1984), which illustrates the very irregular nomenclature of five species. The indented names are synonyms.

of various species of grain adding desirable traits by crossing them with other species (e.g., Cauderon 1979; Asay 1992; Merker 1992.

In recent times, results of anatomical studies and numerical analyses have become available. Thus, an extensive amount of data exists that can be used to test whether the basic type concept contributes to solving the challenging taxonomical problems within this group of plants.

## 6.2. Crossbreeding Within the Triticeae
### 6.2.1. The Genomic System of Classification. The Triticeae have long been known as a "crossbreeding-happy" group. Due to the vast amount of data and especially because of the very irregular nomenclature, it is, however, not easy to gain an overview of the currently known extent of hybridizations. In just the two genera *Hordeum* and *Critesion* for example, almost 300 hybrid pairs are known, and for the four genera *Triticum*, *Aegilops*, *Hordeum* and *Secale* (*sensu lato* in each case) over 1,000 species names were used until 1959 (Bowden 1959; comp. Table 1).

Depending on which criteria are used, certain species are assigned to different genera, certain races are given the status of a species, or certain

Agropyron (s.lat.)

Elytrigia (SSXXXX) ———⋀——— Agropyron (PP)
Pseudoroegneria (SS)⁄ ⎜ ⟍ Pascopyrum (SSHHJJNN)
Thinopyrum (JJ)      Elymus (SSHHJJ)

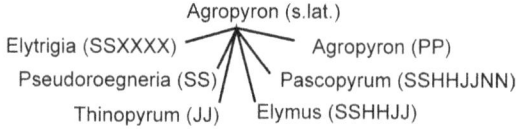

Figure 2. The "traditional" genus *Agropyron* is divided into six different genera, based on the genomic system of classification. The letters in parenthesis indicate the genome types.

species are considered to be only subspecies, *etc*. Thus, according to past terminology, over 100 species were included in the genus *Agropyron* on the basis of a single morphological criterion: "spikes with 1 spikelet per node". In contrast, Löve (1984) includes only 3 species in this genus by applying the "modern" criterion of the type of genome (see below). He divided the original genus into six different genera (Figure 2).

In order to give an overview of the extent of successful hybridizations within the Triticeae, a certain nomenclature must first be chosen as a basis.

Löve (1982, 1984, 1986) and Dewey (1984) proposed to divide the genera of the Triticeae according to the characteristics of their genome. An overview can be found in Table 2. The genome analysis is based on an examination of the behavior of the chromosomes during meiosis and on a comparison of the karyotypes. All Triticeae have the chromosome number x = 7. Thus, diploid types have 2n = 14 chromosomes, polyploid —of which there are many in this group—a multiple of seven. A simple set of chromosomes, and thus a monoploid element, is also called haplome. Allopolyploid plants therefore possess two or more different haplome pairs. The genome consists of one or several haplome pairs. The genus *Crithodium* for example possesses the haplome A (and thus the genome AA), the genus *Sitopsis* B (genome BB), the genus *Gigachilon* A and B (genome AABB), and the genus *Triticum* A, B, and D (genome AABBDD) (Löve 1982, 1984). The Triticeae are well suited for genome analysis because of their large chromosomes (Dewey 1982).

According to the genomic system of classification, species with different haplomes or combinations of haplomes are placed into different genera. The allotetraploid genus *Gigachilon* (genome AABB) is separated from the diploid *Crithodium* (AA) and *Sitopsis* (BB) *etc*. In this way Löve (1984) arrives at 37 genera within the Triticeae (Table 2). Other taxonomists accept a noticeably smaller number of genera. On the other hand, Stebbins (1956a) proposed to place all Triticeae into one single genus, due to their distinctive ability to hybridize – a viewpoint (Kellogg 1989) only recently described as theoretically justifiable but difficult to apply practically.

After writing his "Conspectus of Triticeae," Löve (1986), applying the genomic system of classification, established the new genera

| Genotype | | |
|---|---|---|
| 1 Crithodium (3) | AA,AAAA | |
| 2 Sitopsis (5) | BB | |
| 3 Gigachilon (4) | AABB,AAAABB | |
| 4 Orrhopyglum (1) | CC | |
| 5 Patropyrum (1) | DD | |
| 6 Triticum (1) | AABBDD | |
| 7 Comopyrum (1) | MM | |
| 8 Ambylopyrum (1) | ZZ | |
| 9 Chennapyrum (1) | LL | |
| 10 Kharapyrum (1) | UU | |
| 11 Aegilemma (2) | BBUU | |
| 12 Cylindropyrum (1) | CCDD | |
| 13 Aegilopodes (1) | CCUU | |
| 14 Gastropyrum (4) | DDMM,DDDDMM | |
| 15 Aegilonearum (1) | DDMMUU | |
| 16 Aegilops (6) | MMUU,MMMMUU | |
| 17 Secale (3) | RR | |
| 18 Dasypyrum (2) | VV,VVVV | |
| 19 Eremopyrum (9) | FF,FFFF | |
| 20 Henrardia (2) | OO | |
| 21 Pseudoroegneria (19) | SS | |
| 22 Elytrigia (8) | SSXXXX | |
| 23 Elymus (143) | SSHHYY | |
| 24 Pascopyrum (1) | SSHHJJNN | |
| 25 Psathyrostachys (9) | NN | |
| 26 Leymus (32) | JJNN,JJJJNN | |
| 27 Thinopyrum (17) | JJ-JJJJJJJ | |
| 28 Critesion 35) | HH-HHHHHH | |
| 29 Hordeum (1) | II | |
| 30 Hordelymus (1) | HHTT | |
| 31 Festucopsis (3) | GG | |
| 32 Australopyrum (1) | WW,WWWWW | |
| 33 Heteranthelium (1) | QQ | |
| 34 Crithopsis (1) | KK | |
| 35 Taeniatherum (2) | TT | |
| 36 Agropyron (3) | PP | |

Table 2. Overview of the genera of Triticeae according to the genomic system of classification after Löve (1984) and Dewey (1984), the numbers of the species of these genera (numbers in parenthesis; after Löve 1984), and the generic crossbreeding matrix. Matrix date from von Bothmer & Jacobsen (1989); von Bothmer et al. (1994); Dewey (1984); Frederiksen (1994); Frederikson & von Bothmer (1989, 1995); Jensen & Bickford (1992); Kihara (1937); Knobloch (1968); Lu & von Bothmer (1991); Lucas & Javier (1988); Morrison & Rajhatny (1959); Oehler (1935); Plourde et al. (1990); Sakamoto (1968, 1973); Schooler (1966); Sharma & Gill (1983); Wang (1986,1987); White (1940). Names in each case according to the nomenclature of Löve (1984). Capital letters: genome types. In accordance with Dewey (1984), in this table the genome types J and E are combined to J (Dewey marks them as J-E). In contrast to Löve (1984), Dewey does not justify a separation into two genome types. X and Y represent uncertain genome types (Dewey 1984); according to Löve (1984) Elytrigia possesses the genome types EEJJSS, EEJJJSS and EEEEJJJJSSSS, Elymus HHSS, HHHHSS, HHSSSS and HHHHSSSS. The division into the four subtribus Triticinae, Henrardiinae, Hordeinae and Agropyrinae after Löve (1984) is based on morphological and cytogenetic data.

*Trichopyrum* (including species formerly part of *Triticum* and *Elytrigia*) and *Psammopyrum* (including species formerly part of *Triticum* and *Elymus*). Other hitherto unknown genomes are the recently discovered genus *Kengyilia* (genome type PYS; Yen and Yang 1990; Baum et al. 1991a) and the genome type SYW (*Elymus scabrus*; Lu and von Bothmer 1993). Baum and Gupta (1990) propose to consider × *Triticosecale* as the new genus *Triticale*. Whereas not all species of Triticeae have been examined cytogenetically, it is to be expected that even more genera will be established. The following is based on the taxonomy of Löve (1984), as well as Dewey (1984) (Table 2).

Some authors reject the one-sided application of this criterion for classification. According to the opinion of von Bothmer and Jacobsen (1986), the crossbreeding results of representatives of the genera *Hordeum* and *Critesion* do not justify a separation into two genera. The separated species are not always morphologically distinguishable (Barkworth and Dewey 1985). Baum *et al.* (1987; comp. Baum *et al.* 1991b) believe that a system of classification should be based on as many characteristics as possible (comp. Seberg 1989). Basing it solely on cytogenetic results could be misleading because genome similarities would have to correspond to genetic, biochemical, and morphological similarities. Also, the absence of homologous pairing during meiosis could be due to the effect of individual genes and is, therefore, not necessarily an expression of missing homology. Furthermore, in some cases even non-homologous genomes were found to pair. Finally the genomic system of classification would be based on the assumption that the formation of species and genera follows different mechanisms: speciation would be based on variations of genomes without the genomic identity being lost; formation of genera would be based on the formation of "new" genomes.

According to Sakamato (1972) on the other hand, the hybridization of different genera is an important aid to recognize evolutionary parallelisms between non-related taxa. It is also remarkable that, in many cases, genera based on genomes also show the same mode of pollination (self-fertility, self-sterility) (Jensen *et al.* 1990).

In spite of these objections, the present work is based on the genomic system of classification because it seems most suitable to represent crossbreeding results, which after all are the primary focus of research on basic types. (A critical response to the arguments of Baum *et al.* [1987] against the genomic system of classification can be found in Jauhar and Crane [1989].) Because genera with different genome types can often be hybridized, one can assume that this is also true for the species within a genus (which must also possess the same genome types). This conclusion would not be possible if the classification were based on morphological criteria.

Indeed, even intergeneric crossbreeding is normally successful, as the following examples prove: von Bothmer and Jacobsen (1986) list 294 pairs of hybrid species within the genera *Hordeum* and *Critesion*. The annual Triticeae can normally be crossed artificially (Löve 1982). *Secale* species were just as successfully crossed with other *Secale* species as with species of the genus *Triticum* (*s.l.*; Kranz 1975). All species of the *Aegilops-Triticum-Secale*-group (*sensu lato* in each case) can be hybridized directly with each other (Oehler 1934). Successful crossbreedings occurred within *Eremopyrum* (Sakamoto 1973), *Taeniatherum* (Frederiksen and von Bothmer 1986), and *Elymus* (Lu and von Bothmer 1993). Until 1980, over 250 different hybrid combinations among *Agropyron, Hordeum,* and *Elymus* (Dewey 1980) were generated.

The most important differences between the genomic system and the classic taxonomy are: The diploid to hexaploid genus *Triticum* is divided in three genera (comp. Table 2): *Crithodium, Gigachilon,* and *Triticum.* The genus *Aegilops* is divided into a large number of mostly smaller genera: *Sitopsis, Orrhopygium, Patropyrum, Comopyrum, Amblyopyrum, Chennapyrum, Kiharapyrum, Aegilemma, Cylindropyrum, Aegilopodes, Gastropyrum, Aegilonearum,* and *Aegilops. Sitanion, Hystrix, Roegneria,* and *Asperella* are placed within *Elymus. Agropyron* is divided into several previously established genera (see above).

**6.2.2. Intergeneric Crossbreedings.** Table 2 gives an overview of the currently known generic hybrids, based on the genomic system of classification. The matrix contains 193 generic hybrid pairs of a total of 630 possible pairs (30.6%). Thirty-three of the 36 genera (after Löve's classification: 34 out of 37) are connected through hybridization.

This matrix might still be incomplete since it must be presumed that not all positive results are known to this author. Furthermore, it must be assumed that all possible combinations were not tested regarding their ability to hybridize. Hybridizations are normally attempted when connected to economic expectations (e.g., Plourde *et al.* 1990). Sakamoto (1973) believes it is possible to hybridize all genera of the Triticeae with one another. The genetic proximity of at least some different genome types is shown by the possibility of the mutual substitution of homologous chromosomes (Dvořák 1980).

Sharma and Gill (1983) consider Knobloch's (1968) summary, on which part of the crossbreeding matrix is based, to be "by no means complete and correct." These authors did not accept cases of the wheat-*Agropyron* and wheat-*Elymus* hybrids listed by Knobloch (1968) in which the seeds died off or cytogenetic data were lacking. A re-examination of the original literature quoted by Knobloch proved his statements to be correct as far as crosses within the tribe are concerned.

More detailed data regarding generic hybrids can be found in the sources listed in the legend for Table 2 and in Sando (1926), Katayama (1933), White (1940), Smith (1942), Myers (1959), Bowden (1967), and Franke *et al.* (1992).

Those genera that did not show any hybridizations with other genera are monotypic (*Crithopsis, Australopyrum*) or include only 3 species (*Festucopsis*). All subtribes (division according to Löve 1984) are interconnected.

Diploid genera with different haplomes were successfully crossed in 63 cases. Where systematic examinations are available, comparatively many hybrids are known. The genus *Crithodium* (genome AA) was

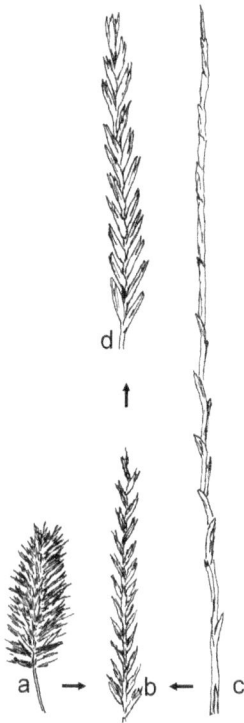

Figure 3. Spikes of *Eremopyrum orientale* (a), *Henrardia persica* (c), an F₁ hybrid from both (b), and an amphiploid (d); after Sakamoto 1972.

successfully hybridized with as many as 11 other haplome types (and thus with half of the possible hybrids), the genus *Secale* (genome RR) with 13.

A series of generic hybrids was created spontaneously, for example: *Agropyron* (PP) × *Elymus* (SSHHYY) (McIntyre 1988), *Elymus trachycaulis* (SSHH) × *Leymus mollis* (JJHH) (Löve 1982), *Leymus* (JJHH) × *Secale* (RR) (Heneen 1963), and *Critesion* (HH) × *Elytrigia* (SSXX) (McIntyre 1988). In North America, 18 naturally occurring generic hybrids are known within the perennial Triticeae (Barkworth and Dewey 1985).

Crossbreeding experiments between *Crithopsis* and other genera of the Triticeae have so far been unsuccessful (Sakamoto 1973). Whether intergeneric crossbreeding experiments were attempted with *Australopyrum* and *Festucopsis* is not known to this author.

On the basis of these results, one can conclude that *all* genera of the Triticeae can be connected through hybridization and, consequently, belong to the same basic type. Löve (1982, 1984) thinks that there is no doubt that all Triticeae genomes can be traced back to an original genome. Dvorák (1980) is of the same opinion.

**6.2.3. Observations on Hybrids.** Intergeneric and interspecific hybrids are usually morphological intermediates (e.g., Stebbins and Snyder 1956; Dewey 1980; von Bothmer and Jacobsen 1986; see Figure 3). In hybrids between species on different ploidal levels, normally the characteristics of the parent with the higher ploidal level prevails (Dewey 1975; Finch and Bennett 1980; von Bothmer and Jacobsen 1986). However, characteristics that none of the two parents possess can also be observed frequently (e.g., Stebbins and Snyder 1956; Dewey 1975; Figure 4).

Development of the hybrids may vary. In some cases it does not get beyond the embryonic stage so the seeds do not germinate. In other cases, one must help the young plants to survive the critical initial stage,

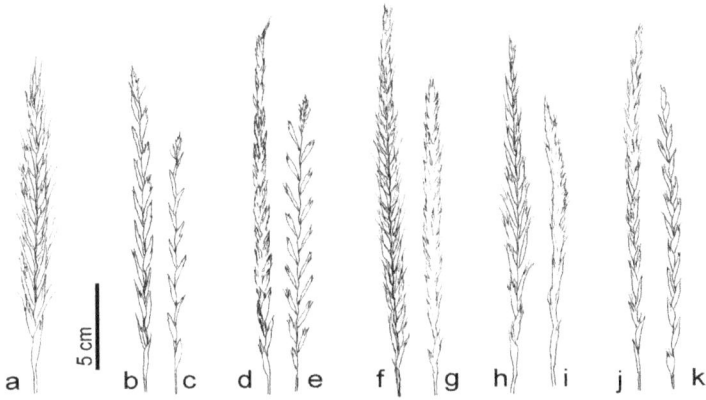

Figure 4. Spikes of *Agropyron drobovii* (a), *A. libanoticum* × *A. drobovii* (b), *A. libanoticum* (c), *A. dasystachium* × *A. drobovii* (d), *A. dasystachium* (e), *A. mutabile* × *A. drobovii* (f), *A. mutabile* (g), *A. ugamicum* × *A. drobovii* (h), *A. ugamicum* (i), *A. drobovii* × *A. leptourum* (j), and *A. leptourum* (k) (after Dewey 1980).

because the endosperm (second fertilization!) of the hybrids is often watery and incapable of sufficiently nourishing the young seedling. However, the hybrids are often vigorous, sometimes more so than the parental species (Stebbins and Snyder 1956; Dewey 1975, 1980; Finch and Bennett 1980).

Intergeneric hybrids are almost always sterile (Sharma and Gill 1983; Löve 1982). There are, however, some exceptions: some *Pseudoroegneria* (genome SS) × *Elymus* (SSHH) hybrids, *Pseudoroegneria* (SS) × *Elytrigia* (SSXX) hybrids, as well as *Elytrigia repens* (SSXX) × *Elymus trachycaulis* (SSHHYY), for example, are fertile (Dewey 1984). Interspecific hybrids are more frequent but are by no means always fertile.

The cause of the sterility is irregular meiosis. In generic hybrids, only autosyndetic chromosome pairing takes place, *i.e.*, only the homologous chromosomes of the parental species pair. (The genomic system of classification is based on such observations.)

The causes of the irregularities in meiosis, on the other hand, are only partially known. It is certain that chromosome rearrangements are among these causes. Whether other factors play a role is uncertain (Dvořák and McGuire 1981). For example, according to these authors, cytologically non-verifiable differences between the chromosomes of *Secale cereale* and *S. montanum* would have at least as big an effect on the meiotic pairing as would visible structural changes (comp. also Endo and Gill 1984). Also, introgressions from another genome and accumulations of large quantities of repetitive DNA prevent the pairing of homologues (Dvořák 1983).

## 6.3. The Delimitation of the Triticeae from Other Tribes of the Poaceae

### 6.3.1. Unsuccessful Intertribal Crossbreedings.
In his "Check List of Crosses in the Gramineae," Knobloch (1968) presents some intertribal hybrids. However, a re-examination of the original literature raised doubts in all cases. The reports either were mere suppositions (Nevski 1963) or lacked cytogenetic analysis. In the latter cases, because only seed set but no hybrid adult was achieved, self-pollination or parthenogenesis cannot be excluded (Hertzsch 1938; Cugnac 1949). This is also true for the hybrid between rice (*Oryza*) and wheat, which had rice-like anatomy, panicle-shaped blossoms, and composite starch grains, as reported by Wang and Tang (1982). Within the Triticeae, some cases of parthenogenetic development are known (Katayama 1933; Sakamoto 1968, 1974). In reference to some intertribal hybrids of Triticeae grasses, Terrell (1966) states that in hybrids "in which ovary stimulation or a few caryopses of some kind were produced there is no way of knowing whether these were actually successful cross-fertilizations instead of self-fertilization."

Hybrids between *Triticum aestivum* and *Avena sativa* (oats; tribe Aveneae; Kruse 1969), which were generated by exposing their heads to ultraviolet radiation, represent a special case of intertribal crossbreeding. The radiation served the purpose of inducing a breakdown of the sterility barriers. Eight hours of radiation were necessary to produce seeds capable of germination; four hours were not enough. The hybrids had the chromosome number 42 with eight hours of radiation and 41 with four hours. (The chromosome number of both parental species is also 42.) Notably, the hybrids showed a dominance of the maternal parent. During meiosis many multivalents were formed; however, disruptions of meiosis were less intense than one would have expected on the basis of the taxonomic distance of the species which were hybridized. Kruse concludes, on the basis of the dominance of the maternal parent in the hybrids and the behavior of the chromosomes during the meiosis, that the hybrid seeds develop through apomixis, which could be confirmed by a closer examination of embryogenesis. Proteins of both parental species were detected in the hybrid plants.

In view of the high radiation dose that was necessary to induce development, it must be questioned whether any real fertilization took place at all. High doses of radiation cause damage, so one cannot speak of normal parental species. It is possible that chromosome segments of the oat were installed into the wheat genome, which led to the formation of certain traits in the oats and caused a general dominance of the maternal parent. Due to these uncertainties, the results of Kruse are not interpreted as a "basic type connection" between *Triticum* and *Avena*.

Smith and Flavell (1974) later wrote that between oats, on the one hand, and wheat, rye and barley, on the other hand, no hybrids are known, which confirms the isolation of the tribes of the Aveneae from the other cereals.

Hybridization experiments between wheat and corn (Laurie and Bennett 1986, 1987, 1988; Laurie and Snape 1990) yielded conclusive results for the basic type biology. In all crossbreeding experiments, the corn genome was cast off completely during the first three cell division cycles, which led to the formation of haploid wheat plants open for further breeding experiments. The chromosome elimination takes place because corn chromosomes do not touch the spindle during mitosis. Thus, one of the conditions of the basic type definition is not met because this condition demands that embryogenesis proceeds under coordinated shaping of the paternal and maternal genetic material (comp. Scherer 1993, this volume). The basic type limit of the Triticeae is, thus, confirmed by these experiments. Endosperm (second fertilization) was either not formed at all or was manifested in a very abnormal way. In further crossbreeding experiments among distantly related taxa of the Poaceae, including genera of the Triticeae (Zenkteler and Nitzsche 1984), embryos developed only partially and perished in an early stage (6-10 days after the pollination). On that occasion no observations were made that would prove that the basic type criteria were met. This is in agreement with the closely examined hybrids between wheat and corn.

As a result, the enormous number of successful crossbreedings within the Triticeae outweighs the uncertain reports of intertribal hybridizations. Furthermore, without a knowledge of the hybrid plant it is impossible to ascertain whether a true hybrid exists. Therefore, extending the limit of the basic type beyond the Triticeae does not appear justified. This limit appears even more distinct as numerous hybrids are also known within the "neighboring tribe" of the Festuceae (e.g., Jenkin 1959; Barker and Stace 1986).

Since the preparation of the German edition of this paper, verified hybrids between the Triticeae and the small tribe Bromeae have been reported. These include *Agropyron* × *Bromus* hybrids (Gyulai *et al.* 1992), as well as the somatic hybrids *Bromus* × *Triticum* (Xiang *et al.* 1999). Therefore, the tribe Bromeae adjoins the Triticeae and should also be included in the basic type Triticeae + Bromeae.

**6.3.2. Other Criteria.** The question arises whether the basic type nature of the Triticeae can be confirmed on the basis of other characteristics beyond the criterion of hybridization. The answer is yes. MacFarlane and Watson (1982) list a number of anatomical and physiological characteristics of the genera of the subfamily Pooideae ( = Festucoideae) within the Poaceae. Because Triticeae are clearly different from other tribes in these characteristics, they state, "... in our numerical analyses they have behaved as a close-knit group, becoming subdivided... only at very low levels of the dendrogram" (p. 188). For example, the Triticeae

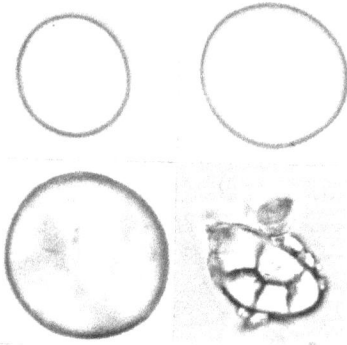

Figure 5. Simple starch grains of *Secale*, above left, *Hordeum*, above right, *Triticum*, below left, and composite starch grains of *Avena* (below right); the latter does not belong to the Triticeae. Enlargement: 1000x. Photo by B. van Cleve.

are characterized by simple starch grains (Figure 5).

Classical taxonomy of the grasses was based on diagnosis using simple morphological features, especially sturcture of the inflorescence (see above). This criterion, in the light of modern detailed analysis, is in accurate. Now the limits of the Triticeae are perceived using other morphological criteria that agree with cytogenetic traits. The genera *Brachypodium* (see Nevski 1963; Runemark and Heneen 1968) and *Lolium* (see Stebbins 1956a), for example, were formerly included in the Triticeae (as was *Nardus* originally by Linné). However, *Brachypodium* deviates from the chromosome number $x = 7$, which is strictly constant within the Triticeae (in this genus $x = 5$ and $x = 9$ also occur, see West et al. 1988). In *Lolium* the structure of the spike is "completely different... from the spike of typical Triticeae" (Stebbins 1956a). According to a study of Kellogg (1989), the genera *Brachypodium* and *Bromus* have to be separated from the Triticeae on the basis of morphological characteristics. Regarding the history of the delimitation of the Triticeae, see Holmberg (1926) and Barkworth (1992).

The quantities of repetitive DNA sequences were compared in the following genera: wheat (*Triticum*), rye (*Secale*), barley (*Hordeum*) and oats (*Avena*, the latter not belonging to the Triticeae). In these sequences, the differences between the genera increase, oat sand barley being more clearly different from each other than barley and rye, or rye and wheat (Flavell et al. 1977, 1980; Rimpau et al. 1978; comp. MacIntyre et al. 1988).

According to Bendich and McCarthy (1970), the DNA of oats seems to be less similar to the DNA of barley, rye, and wheat as the DNA of the latter three cereals compared the each other. The extent of DNA hybridization between oat DNA, on one hand, and wheat, rye, and barley DNA, on the other hand, is approximately equal in each case. Comparisons with DNA hybridizations among the three genera of Triticeae (wheat, rye and barley), on the other hand, reveal different affinities (Table 3; Rimpau et al. 1978). The comparisons show that one can distinguish seven groups of repetitive DNA sequences (Figure 6) in the four genera. Within the framework of the basic type concept, one can interpret these results in the following way: group VII is specific for the "oat" basic type, group II is specific for the "Triticeae" basic type, group I is common to both basic types, and the groups III through VI arose in the course of the diversification of the Triticeae. Each group can contain hundreds of non-related sequence families; each family consists of hundreds or thousands of identical or related sequences (Flavell et al. 1977).

| marked DNA | unmarked DNA | | | |
|---|---|---|---|---|
|  | wheat | rye | barley | oats |
| wheat | 74 | 58 | 32 | 22 |
| rye | 52 | 74 | 38 | 19 |
| barley | 42 | 44 | 71 | 20 |
| oats | 14 | 17 | 20 | 75 |

Table 3. Percent identity between the DNA of wheat, rye, barley and oats. The approximate equidistance of oats to the three genera of Triticeae is remarkable. (After Flavell et al. 1977)

Interestingly, group I, which according to the evolutionary perspective is the oldest group, shows no more divergence than the allegedly younger groups (Flavell et al. 1977). Within the basic type concept, this can be interpreted by assuming equal age and rapid diversification into the groups III through VI (rapid speciation).

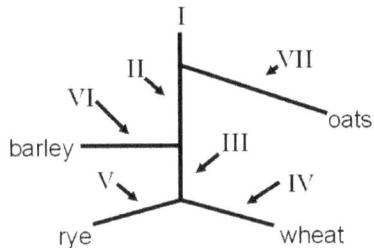

Figure 6. Sequence homologies of repetitive DNA-sequences. The distances represent the percentage of repetitive sequences, which at 60°C and 0.18 M Na+ are incapable of stable hybridization. The Roman numerals show different groups of repetitive sequences. (After Flavell et al. 1977)

Hamby and Zimmer (1988) compared the sequences of ribosomal RNA of different grasses. *Hordeum* and *Avena* (not a representative of the Triticeae, occasionally placed into a different supertribe), amazingly, possess greater similarities than *Hordeum* and *Triticum*. The sequence difference between the former is approximately 0.4%, between the latter is approximately 0.7%. All three genera belong to the subfamily Pooideae, which is regarded as a monophyletic group in this analysis. The authors believe that it is possible that additional data could change this viewpoint; the current database may be too small. In the framework of the basic type concept, one can assume that different basic types started their development with the same sequences, so that the results mentioned above are among the possible expectations.

## 6.4. Systematic Division of the Triticeae

The large number of synonyms mentioned above for virtually all species of Triticeae can, on the one hand, be an indication of the problems of establishing a system within the Triticeae, or on the other hand, they can be interpreted as an indication of the unity of this tribe (Runemark and Heneen 1968; Baum 1978b). Depending on the weight ascribed to criteria of hybridization, cytogenetics, morphology, anatomy, physiology, and biochemistry, one arrives at different divisions. Baum and Appels (1992) call the Triticeae a "very complex reticulate web," due to contradicting morphological and cytological results. West *et al.* (1988) call the systematic arrangement of the Triticeae "a phylogenetic puzzle;" "there are many aspects of their interrelationships and evolution that remain to be determined." Parallelisms (e.g., parallel reductions in the inflorescence), as well as introgressions and hybridizations, are thought to be among the causes of these difficulties (Runemark and Heneen 1968; Baum 1983; Kellogg 1989; Jauhar and Crane 1989). The result is a complicated network of relationships: "Evolution in the Triticeae is highly reticulate" (Jauhar and Crane 1989). According to the opinion of these authors, the use of cladistic algorithms is therefore inappropriate. Kellogg (1989) commented: "Widespread hybridization creates the problem of classification in the Triticeae; our classificatory system is hierarchical, but the history of the tribe is not."

**6.4.1. Numerical Analyses.** Baum (1978a, 1978b, 1983) did extensive numerical analyses of morphological and hybridization data for representative Triticeae species. He comes to the conclusion that there is a "clinal relationship" (gradual transitions) between the members of the Triticeae. According to him, it would be appropriate to assign certain species to several genera simultaneously. This would best correspond with biological reality. More recently, phenetic and cladistic analyses of the phylogeny within the Triticeae by means of DNA-DNA hybridization, DNA analysis of the *Nor* loci, sequencing of the 5S rDNA, and electrophoresis tests of isozymes were performed. In every analysis, approximately 15 species from all subtribes except the Henrardiinae (according to the division by Löve 1984) were investigated. McIntyre (1988) summarizes: "The various analyses do not give identical results in terms of the suggested relationship between the genomes." However, generally speaking, the genomic system of classification was confirmed by these results, although, at the same time noticeable contradictions were detected, which MacIntyre (1988) attributes to the complex evolutionary history of the Triticeae. For example, the so-called "350-family" DNA sequences, which are typical of the genus *Secale*, can also be found in *Argopyron*, which is taxonomically distant from *Secale*. Xin and Appels (1988) suspect, in this case, an independent parallel multiplication of this family of DNA sequences, but they also discuss the possibility of introgression into *Agropyron*. Such mechanisms make it difficult to reconstruct the phylogeny of the Triticeae (see below). A summary of several more recent papers on the systematics of the Triticeae can be found in West *et al.* (1988).

**6.4.2. Conclusions from Crossbreeding Results.** The genomic system preferred by Dewey and Löve does not always yield unambiguous results. There are, in part, more or less fluent transitions among the

| genome pair | number of species combinations | frequency of univalents |
|:-----------:|:------------------------------:|:-----------------------:|
| J, E | 1 | 2.89 |
| S, J | 1 | 4.34 |
| E,P | 1 | 6.42 |
| S,H | 1 | 7.12 |
| P,S | 3 | 7.80 |
| E,S | 2 | 8.14 |
| E,H | 1 | 10.07 |
| J,R | 1 | 11.05 |
| E,N | 2 | 11.22 |
| S,N | 1 | 12.04 |
| H,N | 1 | 12.18 |
| P,R | 1 | 12.64 |
| S,R | 2 | 12.97 |
| H,R | 6 | 13.10 |

Table 4. Chromosome combinations between different species of the Triticeae. (From Xin & Appels 1988, simplified)

haplome types. In Table 4 some results of the pairing of chromosomes during meiosis are summarized. They illustrate that the different genomes exhibit strong affinity to varying degrees. In some cases, it is apparently up to the individual author whether the haplomes of two species are assigned to two different haplome types (e.g., Zhang and Dvorák 1991). In many cases, the haplomes of different representatives of the same haplome type are subdivided further, for example, $S_1$, $S_2$ in *Pseudoroegneria*; $H_1$, $H_2$ in *Critesion* (Dewey 1984); $A_1$, $A_2$ in *Triticum*; and $R_1$, $R_2$, $R_3$ in *Secale* (Löve 1982). "It must be recognized, however, that each species has its unique version of a basic genome" (Dewey 1980).

Within the Triticeae, crossbreeding results are a suitable criterion for systematic purposes only under certain conditions. This can be seen in the fact that, in some cases, it is more difficult to hybridize certain species with other species of the same genus than with species from other genera (Sharma and Gill 1983; Morrison and Rajhathy 1959). Furthermore, chromosomal changes that prevent hybridization can occur rapidly even under breeding conditions (Snape et al. 1979; comp. also Katayama 1933). On the other hand, one cannot equate relative taxonomic isolation with a low potential for hybridization, as shown by the transgeneric crossbreeding results with the isolated genus *Erenecoyrum* (Sakamato 1968; comp. Katayama 1933).

Caution must be exercized when making conclusions from crossbreeding results also because the factors that prevent successful hybridization are not sufficiently known. Small causes can have large effects. It has long been known that chromosome 5 of the B genome prevents meiotic pairing in the hybrid genome (e.g., Sears 1976; Jampates and Dvorák 1986).

## 6.5. Speciation Mechanisms

**6.5.1. Variability.** The Triticeae show considerable variability. Spikelet arrangement varies from loose [spike interrupted, its internodes elongate] to dense [spike congested, its internodes barely developed]. The number of the spikelets per node varies between one and several, and the number of the florets per spikelet vary between one and three. Although most Triticeae are perennial, annual and biennual species also exist.

The degree of fertility includes the whole spectrum, ranging from complete self-fertility to complete self-sterility (Jensen et al. 1990). The amount of DNA varies between 4.2 and 9.5 pg (West et al. 1988), and within the genus *Secale* between 7.2 and 9.5 pg. An enormous range is seen in the quantity of repetitive DNA (McIntyre et al. 1988; see below) and in miscellaneous biochemical characteristics (for an overview see West et al. 1988).

It is assumed that all haplomes originated from the same set of chromosomes (Löve 1982; Dvorák 1980, 1983). What mechanisms generated this wide spectrum of variation and the numerous speciations?

**6.5.2. Chromosome Reorganizations.** In individual cases, this question often remains unsolved. "The molecular nature of mechanisms generating quantitative and qualitative change are not clearly understood" (West et al. 1988; this paper also contains a short overview of these mechanisms). The following processes are mentioned: translocations (exchange of parts of non-homologous chromosomes), deletions, inversions, introgressions (immigration of chromosomes or chromosome parts through interspecific hybridization and consecutive backcrosses), gene amplification that leads to repetitive sequences, increase of the heterochromatin, polyploidization, and cytologically non-verifiable smaller changes (e.g., Feldman 1966; Sears 1974; Rimpau et al. 1978; Flavell et al. 1979; Flavell 1982; Jones and Flavell 1982; Gustafson et al. 1983; Lukaszewsky and Gustafson 1983; Ehrendorfer 1984; Wang et al. 1985; Badaeva et al. 1986; Löve 1986; McIntyre 1988; Xin and Appels 1988).

Regarding the speed of such processes, Flavell (1982) thinks that "the evidence is now very substantial that gross changes, 'macromutations', accumulate very rapidly in the chromosome" (p. 318). A similar statement can be found in West et al. (1988). The contradiction described in the section "Systematic Division of the Triticeae" also allows the interpretation of a rapid radiation of the respective genera.

Kimber and Yen (1988) also ascribe the complexity of the taxonomy of the Triticeae to the frequency of so-called "pivotal-differential" evolution. This term refers to the evolution of different allotetraploids, each of which have one genome in common (the "pivotal genome") and differ in the other genomes (the "differential genomes"). Hybrids from such allotetraploids are often fertile, which is explained by the stabilizing effect of the common genome. This also facilitates the gene flow between distantly related species, thus veiling the phylogenetic path.

Introgressions can also cause inconsistencies between different dendrograms that were constructed by analyses of different parameters (comp. the section "Observations on Hybrids"). Also, the possibility of fluctuating evolutionary rates of different DNA or protein sequences has been discussed (McIntyre 1988). It must be assumed that morphological traits, isozymes, rDNA, cpDNA, *etc.* do not evolve "in concert."

**6.5.3. Polyploidizations.** Taxa with higher ploidy levels have presumably originated through auto- or (usually) allopolyploidization (Löve 1982). In some cases, the assumed path of formation could be either reproduced or at least made probable through experiments, for example, for *Pascopyrum* (genome SSHHJJNN; Dewey 1975, 1984) and for *Aropyron leptourum* (genome SSSSHH; Dewey 1972). Further data can be found in Sears (1948), Stebbins (1956b), Sakamoto (1967), Taylor and McCoy (1973), and Löve (1982). Undoubtedly the most famous example is the formation of common wheat (*Triticum aestivum*, genome AABBDD). In this case, a hybridization took place twice, according to Dvorák (1983; using classification in Table 2). The crossbreeding of *Crithodium monococcum* (AA) and a species of the genus *Sitopsis* (BB) first resulted in *Gragachilon polonicum* ssp. *turgidum* (AABB). The crossing of this with *Patropyrum* (DD) then resulted in common wheat. On the basis of biochemical and cytogenetic data, the donor of the BB genome is uncertain (Kimber and Athwal 1972; comp. Holm and Fröst 1979; Harborne *et al.* 1986; Kerby and Kuspira 1987; Talbert *et al.* 1991). Recently, *T. urartu* – formerly a single biospecies together with *T. monococcum* – was regarded as the A-donor (Dvorák *et al.* 1993).

The proof of a presumed path of polyploidization does not always succeed. Although it must be assumed that the genus *Elymus* originated from a combination of *Pseudoroegneria* (S) and *Critesion* (H), no natural hybrids between the proposed original genera are known, and only one hybridization between diploid species of these genera was successful (Dewey 1984). A greater potential of hybridization must be assumed in the past.

**6.5.4. Paths of Diversification.** On the basis of these results, it is no surprise that a phylogenetic tree for the Triticeae can hardly be constructed. Plus, the origin of the Triticeae is unknown (Sakamoto 1973). However, some suppositions regarding presumed paths of diversification of the Triticeae can be made.

Sakamoto (1973) regards the following traits as original: basic number of chromosomes $x = 7$, large chromosomes with mostly median centromeres; perennial growth-form; spike with three spikelets per node; and at least two florets per spikelet. Diploid forms exist in many genera. The fact that the perennial genera are older is confirmed by a series of results (West et al. 1988; comp. Löve 1982):

- The appearance of annual forms is regarded as advanced with respect to the perennial species;
- annual forms are more adaptable to changing environmental conditions;
- there are many self-fertilizing species among the annual forms, which is interpreted as an adaptation to changing environmental conditions;
- annual genera are relatively small, morphologically clearly distinguishable, and they populate only relatively small geographical areas;
- annual genera show a higher rate of polymorphic gene loci, as well as a larger percentage of heterozygotes than the perennial genera (McIntyre 1988); and
- compared with the perennial genera, the annual genera hardly form any intergeneric hybrids, which is attributed to swift differentiation and clear genetic isolation. Rapidly changing environmental conditions and resulting selection pressures in the region of the Near East are regarded as the causes for this observation.

Dvorák et al. (1984) view the hypothetical original type of the Triticeae as follows: "... was probably a cross-fertilizing perennial with long spikes that had several multifloreted spikelets per node" (p. 217). These are characteristics of *Psathyrostachys*, *Leymus* and some species of *Elymus*. If such forms are regarded as the starting points of the diversification of the Triticeae, then the annual species of *Triticum* (s.l.) must be the descendants of an evolutionary radiation that had to include a progressive loss of general qualities, as well as the formation of specialized characteristics. The species of the genus *Elytrigia* (*Thinopyrum sensu* Dewey), which differ from this original group in the loss of the multiple spikelets per node, would then represent the first step in this process (Dvorák et al. 1984).

Sakamoto (1973) distinguishes a younger Mediterranean group (including *Secale*, *Haynaldia* [= *Dasypyrum sensu* Löve], *Triticum s.l.*, *Aegilops s.l.*, *Eremopyrum*, *Henrardia*, *Heteranthelium*, *Crithopsis*, *Taeniatherum*, and *Psathyrostachys*) from an older arctic-temperate group (*Hordeum*, *Agropyron*, *Asperella* [which according to Löve belongs to *Elymus*], *Elymus* and *Sitanion* [which according to Löve also belongs to *Elymus*]). According to his statements, the differentiation of the perennial genera took place in the diploid state. The formation of the arctic-temperate group probably happened in the Upper Tertiary, and a rapid radiation of the Mediterranean group probably in the Quaternary. Finally, in each group the development of species took place through chromosomal differentiation, polyploidization in connection with natural hybridization of the diploid species, and adaptation to the respective local conditions. According to West et al. (1988), the fact that the distribution of many genera is limited to a single continent leads to the conclusion that the evolution of the genera took place mostly after continental drift.

Because taxa with an E haplome (*Thinopyrum* [comp. Table 2 where E and J are combined]) can be crossed easily with *Triticum* (ABD), Löve (1982) thinks that these taxa possibly represent a connection between the perennial and the annual taxa (comp. also Dewey 1984 and Dvorák 1980). Dvorák et al. (1984) regard the genus *Triticum* as a possible result of an adaptive radiation of a species complex with an E genome.

Due to the shape of the palea, as well as the heteromorphic 1- or seldom 2-floreted spikelets, which are located at the nodes in groups of 2 or 3, the subtribe Hordeinae can easily be distinguished from the other Triticeae. In the same way, the genera *Aegilops s.l.* (includes over 10 species according to Löve 1982), and *Triticum s.l.* are regarded as closely connected (Runemark and Heneen 1968; Sakamoto 1973; Flavell 1982).

The genus *Agropyron* (*sensu* Löve and Dewey) is relatively isolated, as only a few hybridizations were successful (Löve 1982; Dewey 1984). According to morphological and karyological criteria, the genus *Australopyrum* is a close relative of *Argopyron* (Hsiao et al. 1986).

Within the Triticeae, the genus *Henrardia* described by Hubbard (1946) shows the strongest degree of morphological reduction (Runemark and Heneen 1968). Almost all hybridizations with other Triticeae grasses failed (Sakamoto 1972). The genera *Heteranthelium* and *Crithopsis*, too, possess a very specific morphology (Sakamoto 1973, 1974).

Stebbins proposed degrees of karyotype symmetry as a criterion for the differentiation between younger and older genome types. Genera with asymmetrical karyotypes might be more advanced (Dvorák et al. 1984). This includes the following genera according to Löve's nomenclature: *Orrhopygium* (genome type CC), *Kiharapyrum* (UU), *Comopyrum* (MM) and *Chennapyrum* (LL). All those are without exception monotypic genera (see Table 2).

# References

Asay, K.H. 1992. Breeding potentials in perennial Triticeae grasses. *Hereditas* 116:167-173.

Badaeva, E.D., F.M. Shkutina, I.N. Bogdevich, and N.S. Badaev. 1986. Comparative study of *Triticum aestivum* and *T. timopheevi* genomes using C-banding technique. *Plant Systematics and Evolution* 154:183-194.

Barker, C.M. and C.A. Stace. 1986. Hybridization in the genera *Vulpia* and *Festuca* (Poaceae): Meiotic behaviour of artificial hybrids. *Nordic Journal of Botany* 6:1-10.

Barkworth, M.E. 1992. Taxonomy of the Triticeae: A historical perspective. *Hereditas* 116:1-14.

Barkworth, M.E and D.R. Dewey. 1985. Genomically based genera in the perennial Triticeae of North America: Identification and membership. *American Journal of Botany* 72:767-776.

Baum, B.R. 1978a. Taxonomy of the tribe Triticeae (Poaceae) using various numerical techniques, II: Classification. *Canadian Journal of Botany* 56:27-56.

Baum, B.R. 1978b. Taxonomy of the tribe Triticeae (Poaceae) using various numerical techniques, III: Synoptic key to genera and synopses. *Canadian Journal of Botany* 56:374-385.

Baum, B.R. 1983. A phylogenetic analysis of the tribe Triticeae (Poaceae) based on morphological characters of the genera. *Canadian Journal of Botany* 61:518-535.

Baum, B.R. and R. Appels. 1992. Evolutionary change at the 5S DNA loci of species in the Triticeae. *Plant Systematics and Evolution* 183: 195-208.

Baum, B.R, J.R. Estes, and P.K. Gupta. 1987. Assessment of the genomic system of classification in the Triticeae. *American Journal of Botany* 74:1388-1395.

Baum, B.R. and P.K. Gupta. 1990. Taxonomic examination of *Triticale* (× *Triticosecale*). *Canadian Journal of Botany* 68:1889-1893.

Baum, B.R., C. Yen, and J.L. Yang. 1991a. *Kengyilia habahenensis* (Poaceae: Triticeae)—a new species from the Altai mountains, China. *Plant Systematics and Evolution* 174:103-108.

Baum, B.R., C. Yen, and J.L. Yang. 1991b. *Roegneria*: its generic limits and justification for its recognition. *Canadian Journal of Botany* 69:282-294.

Bendich, A.J. and B.J. McCarthy. 1970. DNA comparisons among barley, oats, rye, and wheat. *Genetics* 65:545-565.

Bor, N.L. 1960. *The Grasses of Burma, Ceylon, India and Pakistan*. Pergamon Press, New York.

Bothmer, R. von and N. Jacobsen. 1986. Interspecific crosses in *Hordeum* (Poaceae). *Plant Systematics and Evolution* 153:49-64.

Bothmer, R. von and N. Jacobsen. 1989. Intergeneric hybridization between *Hordeum* and *Hordelymus* (Poaceae). *Nordic Journal of Botany* 9:113-117.

Bothmer R. von, B-R Lu, and I. Linde-Laursen. 1994. Intergeneric hybridization and c-banding patterns in Hordelymus (Triticeae, Poaceae). *Plant Systematics and Evolution* 189:259-266.

Bowden, W.M. 1959. The taxonomy and nomenclature of the wheats, barleys, and ryes and their wild relatives. *Canadian Journal of Botany* 37:657-684.

Bowden, W.M. 1967. Taxonomy of intergeneric hybrids of the tribe Triticeae from North America. *Canadian Journal of Botany* 45:711-724.

Cauderon, Y. 1979. Use of *Agropyron* species for wheat improvement. In: Zeven, A.C. and A.M.van Harten, eds. *Proceedings of a Conference on Broadening the Genetic Base of Crops*. Pudoc, Wageningen, Netherlands, pp. 175-786.

Conert, H.J. 1983. Unterfamilie Pooideae. In: *Hegi Illustrierte Flora von Mitteleuropa*. Volume 1, Part 3. Paul Parey, Berlin.

Crepet, W.L. and G.D. Feldman. 1991. The earliest remains of grasses in the fossil record. *American Journal of Botany* 78:1010-1014.

Cugnac, M.A. de. 1949. Sur la significance systématique de quelques expériences de croisement entre des Graminées appartenant à des genres éloignés dans la classification: *Elymus, Festuca, Lolium, Dactylis*. *Comptes Rendus des Séances de l'Academie des Sciences* 228:422-423.

Dewey, D.R. 1972. The origin of *Agropyron leptourum*. *American Journal of Botany* 59:836-842.

Dewey, D.R. 1975. The origin of *Agropyron smithii*. *American Journal of Botany* 62:524-530.

Dewey, D.R. 1980. Cytogenetics of *Agropyron drobovii* and five of its interspecific hybrids. *Botanical Gazette* 141:469-478.

Dewey, D.R. 1982. Genomic and phylogenetic relationships among North American perennial Triticeae. In: Estes, J.R., R.J. Tyrl and J.N. Brunken, eds. *Grasses and Grasslands*. University of Oklahoma Press, Norman, OK, pp. 51-81.

Dewey, D.R. 1984. The genomic system of classification as a guide to intergeneric hybridization with the perennial Triticeae. In: Gustafson, J.P., ed. *Gene Manipulation in Plant Improvement*. Plenum Press, New York, pp. 209-279.

Dvorák, J. 1980. Homoeology between *Agropyron elongatum* chromosomes and *Triticum aestivum* chromosomes. *Canadian Journal of Genetics and Cytology* 22:237-259.

Dvorák, J. 1983. The origin of wheat chromosomes 4A and 4B and their genome reallocation. *Canadian Journal of Genetics and Cytology* 25:210-214.

Dvorák, J, P. di Terlizzi, H.B. Zhang and P. Resta, 1993. The evolution of polyploid wheats: Identification of the A genome donor species. *Genome* 36:21-31.

Dvorák, J. and P.E. McGuire. 1981. Nonstructural chromosome differentiation among wheat cultivars, with special reference to differentiation of chromosomes in related species. *Genetics* 97:391-414.

Dvořák J, P.E. McGuire, and S. Mendlinger 1984. Inferred chromosome morphology of the ancestral genome of *Triticum*. *Plant Systematics and Evolution* 144:209-220.

Ehrendorfer, F. 1984. Experimentelle Grundlagen der Taxonomie: Neuere Methoden der Verwandtschaftsforschung bei höheren Pflanzen. *Verhandlungen der Deutschen Zoologischen Gesellschaft* 77:57-67.

Endo, T.R. and B.S. Gill. 1984. Somatic karyotype, heterochromatin distribution, and nature of chromosome differentiation in common wheat, *Triticum aestivum* L. em. Thell. *Chromosoma* 89:361-369.

Feldman, M. 1966. Identification of unpaired chromosomes in $F_1$ hybrids involving *Triticum aestivum* and *T. timopheevii*. *Canadian Journal of Genetics and Cytology* 8:144-151.

Finch, R.A. and M.D. Bennett. 1980. Mitotic and meiotic chromosome behaviour in new hybrids of *Hordeum* with *Triticum* and *Secale*. *Heredity* 44:201-209.

Flavell, R.B. 1982. Sequence amplification, deletion and rearrangement: Major sources of variation during species divergence. In: Dover, G.A. and R.B. Flavell, eds. *Genome Evolution*. Academic Press, London, pp. 301-323.

Flavell, R.B, M. O'Dell, and D. Smith. 1979. Repeated sequence and DNA comparisons between *Triticum* and *Aegilops* species. *Heredity* 42:309-322.

Flavell, R.B., J. Rimpau, and D.B. Smith. 1977. Repeated sequence DNA relationships in four cereal genomes. *Chromosoma* 63:295-222.

Flavell, R.B., J. Rimpau, D.B. Smith, M. O'Dell, and J.R. Bedbrook. 1980. The evolution of plant genome structure. In: Leaver C.J., ed. *Genome Organization and Expression in Plants*. Springer, Edinburgh, pp. 35-47.

Franke, R., R. Nestrowicz, A. Senula, and B. Staat. 1992. Intergeneric hybrids between *Triticum aestivum* L. and wild Triticeae. *Hereditas* 116:225-231.

Frederiksen, S. 1994. Hybridization between *Taeniatherum caput-medusae* and *Triticum aestivum* (Poaceae). *Nordic Journal of Botany* 14:3-6.

Frederiksen, S. and R. von Bothmer. 1986. Relationships in *Taeniatherum* (Poaceae). *Canadian Journal of Botany* 64:2343-2347.

Frederiksen, S. and R. von Bothmer. 1989. Intergeneric hybridization between *Taeniatherum* and different genera of Triticeae, Poaceae. *Nordic Journal of Botany* 9:229-240.

Frederiksen S. and R. von Bothmer. 1995. Intergeneric hybridizations with Eremopyrum (Poaceae). *Nordic Journal of Botany* 15:39-47.

Gill, B.S. and G. Kimber. 1974. Giemsa C-banding and the evolution of wheat. *Proceedings of the National Academy of Sciences* 71:4086-4090.

Gustafson, J.P., A.J. Lukaszewski, and M.D. Bennett. 1983. Somatic deletion and redistribution of telomeric heterochromatin in the genus *Secale* and in *Triticale*. *Chromosoma* 88:293-298.

Gyulai, G., J. Janovszky, E. Kiss, L. Lelik, A. Csillag, and L.E. Hezsky. 1992. Callus initiation and plant regeneration from inflorescence primordia of the intergeneric hybrid *Agropyron repens* (L.) Beauv. × *Bromus inermis* cv. *nanus* on a modified nutritive medium. *Plant Cell Reports* 11:266-269.

Hamby, R.K. and E.A. Zimmer. 1988. Ribosomal RNA sequences for inferring phylogeny within the grass family (Poaceae). *Plant Systematics and Evolution* 160:29-37.

Harborne, J.B., M. Boardley, S. Fröst, and G. Holm. 1986. The flavonoids in leaves of diploid *Triticum* species (Gramineae). *Plant Systematics and Evolution* 154:251-257.

Hegnauer, R. 1963. *Chemotaxonomie der Pflanzen, Volume 2: Monocotyledoneae.* Birkhäuser, Basel, Switzerland.

Heneen, W.K. 1963. Cytology of the intergeneric hybrid *Elymus arenarius* × *Secale cereale. Hereditas* 49:61-77.

Hertzsch, W. 1938. Art- und Gattungskreuzungen bei Gräsern. *Züchter* 10:261-263.

Holm, G. and S. Fröst. 1979. A preliminary contribution to the discussion of the origin of bread wheat. *Hereditas* 91:295-296.

Holmberg, O.R. 1926. Über die Begrenzung und Einteilung der Gramineen-Tribus Festuceae und Hordeae. *Botaniska Notiser* 79:69-80.

Hsiao, C., R.R.C. Wang, and D.R. Dewey. 1986. Karyotype analysis and genome relationships of 22 diploid species in the tribe Triticeae. *Canadian Journal of Genetics and Cytology* 28:109-120.

Hubbard, C.E. 1946. *Henrardia*, a new genus of the Gramineae. *Blumea* 3(Suppl.):10-21.

Hutchinson, J. 1973. *The Families of Flowering Plants.* Clarendon Press, Oxford, England.

Jampates, R. and J. Dvorák. 1986. Location of the *Ph1* locus in the metaphase chromosome map and the linkage map of the 5Bq arm of wheat. *Canadian Journal of Genetics and Cytology* 28:511-519.

Jauhar, P.P. and C.F. Crane. 1989. An evaluation of Baum *et al.*'s assessment of the genomic system of classification in the Triticeae. *American Journal of Botany* 76:571-576.

Jenkin, T.J. 1959. Fescue species (*Festuca* L.). In: Kappert, H. and W. Rudorf, eds. *Handbuch der Pflanzenzüchtung.* Volume IV. Parey, Berlin, pp. 418-434.

Jensen, K.B., Y.F. Zhang and D.R. Dewey. 1990. Mode of pollination of perennial species of the Triticeae in relation to genomically defined genera. *Canadian Journal of Plant Science* 70:215-225.

Jensen, K.B. and I.W. Bickford. 1992. Cytology of intergeneric hybrids between *Psathyrostachys* and *Elymus* with *Agropyron* Poaceae Triticeae. *Genome* 35(4):676-680.

Jones, J.D.G. and R.B. Flavell. 1982. The structure, amount and chromosomal localisation of defined repeated DNA sequences in species of the genus *Secale. Chromosoma* 86:613-641.

Katayama, Y. 1933. Crossing experiments in certain cereals with special reference to different compatibility between the reciprocal crosses. *Memoirs of the College of Agriculture, Kyoto Imperial University* 27:1-

75.
Kellogg, E.A. 1989. Comments on genomic genera in the Triticeae (Poaceae). *American Journal of Botany* 76:796-805.

Kerby, K. and J. Kuspira. 1987. The phylogeny of polyploid wheats *Triticum aestivum* (bread wheat) and *Triticum turgidum* (macaroni wheat). *Genome* 29:722-737.

Kihara, H. 1937. Genomanalyse bei *Triticum* und *Aegilops*, VII: Kurze Übersicht über die Ergebnisse der Jahre 1934-36. *Memoirs of the College of Agriculture, Kyoto Imperial University* 41:1-61.

Kimber, G. and R.S. Athwal. 1972. A reassessment of the course of evolution of wheat. *Proceedings of the National Academy of Sciences* 69:912-915.

Kimber, G. and Y. Yen. 1988. Analysis of pivotal-differential evolutionary patterns. *Proceedings of the National Academy of Sciences* 85:9106-9108.

Knobloch, I.W. 1968. *A Check List of Crosses in the Gramineae.* Self-published, East Lansing, MI.

Kranz, A.R. 1975. *Wildarten und Primitivformen des Roggens* (Secale L.) *Cytogenetik, Genökologie, Evolution und züchterische Bedeutung. "Fortschritte der Pflanzenzüchtung", Beiheft No. 3 der Zeitschrift fur Pflanzenzüchtung.* Paul Parey, Berlin.

Kruse, A. 1969. Intergeneric hybrids between *Triticum aestivum* L. (v. Koga II, 2n = 42) and *Avena sativa* L. (v. Stål, 2n = 42) with pseudogamous seed formation: Preliminary report. *Royal Veterinary and Agricultural College Yearbook* 1969:188-200.

Laurie, D.A. and M.D. Bennett. 1986. Wheat × maize hybridization. *Canadian Journal of Genetics and Cytology* 28:313-316.

Laurie, D.A. and M.D. Bennett. 1987. The effect of the crossability loci *Kr1* and *Kr2* on fertilization frequency in hexaploid wheat × maize crosses. *Theoretical and Applied Genetics* 73:401-409.

Laurie, D.A. and M.D. Bennett. 1988. The production of haploid wheat plants from wheat × maize crosses. *Theoretical and Applied Genetics* 76:393-397.

Laurie, D.A. and J.W. Snape. 1990. The agronomic performance of wheat doubled haploid lines derived from wheat × maize crosses. *Theoretical and Applied Genetics* 79:813-816.

Löve, A. 1982. Generic evolution of the wheatgrasses. *Biologisches Zentralblatt* 101:199-212.

Löve, A. 1984. Conspectus of Triticeae. *Feddes Repertorium* 95:425-521.

Löve, A. 1986. Some taxonomical adjustments in eurasiatic wheatgrasses. *Veröffentlichungen des Geobotanischen Institutes der ETH, Stiftung Rübel* 87:43-52.

Lu, B.R. and R. von Bothmer. 1991. Cytogenetic studies on the intergeneric hybrids between *Secale cereale* and *Elymus caninus*, *E. brevipes*, and *E. tsukushiensis*. *Theoretical and Applied Genetics* 81:524-532.

Lu, B.R. and R. von Bothmer. 1993. Meiotic analysis of *Elymus caucasicus*, *E. longearistatus*, and their interspecific hybrids with twenty-three

*Elymus* species (Triticeae, Poaceae). *Plant Systematics and Evolution* 185:35-53.

Lucas, H. and J. Javier. 1988. Phylogenetic relationships in some diploid species of Triticineae: Cytogenetic analysis of interspecific hybrids. *Theoretical and Applied Genetics* 75:498-502.

Lukaszewski, A.J. and J.P. Gustafson. 1983. Translocations and modifications of chromosomes in *Triticale* × wheat hybrids. *Theoretical and Applied Genetics* 64:239-248.

MacFarlane, T.D. and L. Watson. 1982. The classification of Poaceae subfamily Pooideae. *Taxon* 31:178-203.

McIntyre, C.L. 1988. Variation at isozyme loci in Triticeae. *Plant Systematics and Evolution* 160:123-142.

McIntyre, C.L., B.C. Clarke, and R. Appels. 1988. Amplification and dispersion of repeated DNA sequences in the Triticeae. *Plant Systematics and Evolution* 160:39-59.

Melderis, A., C.J. Humphries, T.G. Tutin, and S.A. Heathcote. 1980. Tribe Triticeae Dumort. In: Tutin, T.G., V.H. Heywood, N.A. Burges, D.M. Moore, D.H. Valentine, S.M. Walker, and D.A. Webb, eds. *Flora Europaea*. Volume 5. Cambridge University Press, Cambridge, England, pp. 190-206.

Merker, A. 1992. The Triticeae in cereal breeding. *Hereditas* 116:177-280.

Morrison, J.W. and T. Rajhathy. 1959. Cytogenetic studies in the genus *Hordeum*, III: Pairing in some interspecific and intergeneric hybrids. *Canadian Journal of Genetics and Cytology* 1:65-77.

Myers, W.M. 1959. The Wheatgrasses, *Agropyron spp.* In: Kappert, H. and W. Rudorf, eds. *Handbuch der Pflanzenzüchtung* Volume IV. Parey, Berlin, pp. 503-524.

Nevski, S.A. 1963. Tribe XIV: Hordeae Benth. In: Komarov, V.L., ed. *Flora of the U.S.S.R.* Volume II. Israel Program for Scientific Translations, Jerusalem. [English translation of Nevski, S.A. 1934. Tribe XIV: Hordeae Benth. In: Komarov, V.L., ed. *Flora of the U.S.S.R.* Volume II. Botanical Institute of the Academy of Sciences of the USSR, Leningrad], pp. 369-570.

Oehler, E. 1934. Untersuchungen über Ansatzverhältnisse, Morphologie and Fertilität bei *Aegilops-Secale*-Bastarden. *Zeitschrift für Inductive Abstammungs und Vererbungslehre* 67:317-341.

Oehler, E. 1935. Untersuchungen an *Aegilops-Haynaldia*- and *Triticum-Haynaldia*-Bastarden. *Zeitschrift für Inductive Abstammungs und Vererbungslehre* 68:187-208.

Plourde, A., G. Fedak, C.A. St-Pierre, and A. Comeau. 1990. A novel intergeneric hybrid in the Triticeae: *Triticum aestivum* × *Psathyrostachys juncea*. *Theoretical and Applied Genetics* 79:45-48.

Porter, C.L. 1967. *Taxonomy of Flowering Plants*. W.H. Freeman and Company, San Francisco.

Rimpau, J., D. Smith and R. Flavell. 1978. Sequence organization analysis of the wheat and rye genomes by interspecies DNA/DNA hybridization. *Journal of Molecular Biology* 123:327-359.

Runemark, H. and W.K. Heneen. 1968. *Elymus* and *Agropyron*, a problem of generic delimination. *Botaniska Notiser* 121:51-79.

Sakamoto, S. 1967. Cytogenetic studies in the tribe Triticeae,V: Intergeneric hybrids between the two *Eremopyrum* species and *Agropyron tsukushiense*. *Seiken Zihô* 19:19-27.

Sakamoto, S. 1968. Cytogenetic studies in the tribe Triticeae, VI: Intergeneric hybrid between *Eremopyrum orientale* and *Aegilops squarrosa*. *Japanese Journal of Genetics* 43:167-171.

Sakamoto, S. 1972. Intergeneric hybridization between *Eremopyrum orientale* and *Henrardia persica*, an example of polyploid species formation. *Heredity* 28:109-115.

Sakamoto, S. 1973. Patterns of phylogenetic differentiation in the tribe Triticeae. *Seiken Zihô* 24:11-31.

Sakamoto, S. 1974. Intergeneric hybridization among three species of *Heteranthelium*, *Eremopyrum* and *Hordeum*, and its significance for the genetic relationships within the tribe Triticeae. *New Phytologist* 73:341-350.

Sando, W.J. 1926. Hybrids of wheat, rye, *Aegilops* and *Haynaldia*. *Journal of Heredity* 26:229-232.

Scherer, S. 1993. Basic types of life. In: Scherer, S., ed. *Typen des Lebens*, Pascal, Berlin, pp. 11-30.

Schooler, A.B. 1966. *Elymus caput-medusae* L. crosses with *Aegilops cylindrica* Host. *Crop Science* 6:79-83.

Sears, E.R. 1948. The cytology and the genetics of the wheats and their relatives. *Advances in Genetics* 2:239-270.

Sears, E.R. 1974. The wheats and their relatives. In: King, R.C., ed. *Handbook of Genetics, Volume 2: Plants, Plant Viruses, and Protists*. Plenum Press, New York, pp. in 59-91.

Sears, E.R. 1976. Genetic control of chromosome pairing in wheat. *Annual Review of Genetics* 10:31-51.

Seberg, O. 1989. Genome analysis, phylogeny, and classification. *Plant Systematics and Evolution* 166:159-171.

Sharma, H.C. and B.S. Gill. 1983. Current status of wide hybridization in wheat. *Euphytica* 32:17-31.

Smith, D.B. and R.B. Flavell. 1974. The relatedness and evolution of repeated nucleotide sequences in the genomes of some Gramineae species. *Biochemical Genetics* 12:243-256.

Smith, D.C. 1942. Intergeneric hybridization of cereals and other grasses. *Journal of Agricultural Research* 64:33-47.

Snape, J.W., V. Chapman, J. Moss, C.E. Blanchard, and T.E. Miller. 1979. The crossabilities of wheat varieties with *Hordeum bulbosum*. *Heredity* 42:291-298.

Stebbins, G.L. 1956a. Taxonomy and the evolution of genera, with special reference to the family Gramineae. *Evolution* 10:235-245.

Stebbins, G.L. 1956b. Cytogenetics and evolution of the grass family. *American Journal of Botany* 43:890-905.

Stebbins, G.L. 1972. The evolution of the grass family. In: Youngner, V.B. and

C.M. McKell, eds. *The Biology and Utilization of Grasses*. Academic Press, New York, pp. 1-17.

Stebbins, G.L and L.A. Snyder. 1956. Artificial and natural hybrids in the Gramineae, tribe Hordeae, IX: Hybrids between western and eastern North American species. *American Journal of Botany* 43:305-312.

Talbert, L.E., G.M. Magyar, M. Lavin, T.K. Blake, and *S.I.* Moylan. 1991. Molecular evidence for the origin of the S-derived genomes of polyploid *Triticum* species. *American Journal of Botany* 78:340-349.

Taylor, R.J. and G.A. McCoy. 1973. Proposed origin of tetraploid species of crested wheatgrasses based on chromatographic and karyotypic analyses. *American Journal of Botany* 60:576-583.

Terrell, E.E. 1966. Taxonomic implications of genetics in ryegrasses (*Lolium*). *Botanical Review* 32:138-164.

Wang, C.P. and S.H. Tang. 1982. A new hybrid genus of Gramineae (Poaceae). *Acta Phytotaxonomica Sinica* 20:179-181.

Wang, R.R.C. 1986. Diploid perennial intergeneric hybrids in the tribe Triticeae, II: Hybrids of *Thinopyrum elongatum* with *Pseudoroegneria spicata* and *Critesion violaceum*. *Biologisches Zentralblatt* 105:361-368.

Wang, R.R.C. 1987. Diploid perennial intergeneric hybrids in the tribe Triticeae, III: Hybrids among *Secale montanum*, *Pseudoroegneria spicata*, and *Agropyron mongolicum*. *Genome* 29:80-84.

Wang, R.R.C, D.R. Dewey, and C. Hsiao. 1985. Intergeneric hybrids of *Agropyron* and *Pseudoroegneria*. *Botanical Gazette* 146:268-274.

West, J.G, C.L. McIntyre, and R. Appels. 1988. Evolution and systematic relationships in the Triticeae (Poaceae). *Plant Systematics and Evolution* 160:1-28.

White, W.J. 1940. Intergeneric crosses between *Triticum* and *Agropyron*. *Scientific Agriculture* 21:198-232.

Xiang, F.N., G.M. Xia, A.F. Zhou, H.M. Chen, Y. Huang, and X.L. Zhai. 1999. Asymmetric somatic hybridization between wheat (*Triticum aestivum*) and *Bromus inermis*. *Acta Botanica Sinica* 41(5):458.

Xin, Z.Y. and R. Appels. 1988. Occurrence of rye (*Secale cereale*) 350-family DNA sequences in *Agropyron* and other Triticeae. *Plant Systematics and Evolution* 160:65-76.

Yen, C. and J.L. Yang. 1990. *Kengyilia gobicola*, a new taxon from west China. *Canadian Journal of Botany* 68:1894-1897.

Zenkteler, M. and W. Nitzsche. 1984. Wide hybridization experiments in cereals. *Theoretical and Applied Genetics* 68:311-315.

Zhang, H.B. and J. Dvorák. 1991. The genome origin of tetraploid species of *Leymus* (Poaceae: Triticeae) inferred from variation in repeated nucleotide sequences. *American Journal of Botany* 78:871-884.

# 7. The Genera *Geum* (Avens), *Coluria*, and *Waldsteinia* (Rosaceae, Tribe Geeae)

REINHARD JUNKER

## Abstract

Currently known data from hybridization, morphology, anatomy, cytogenetics (basic number of chromosomes), and mycology (infestation by parasites) allow the preliminary conclusion that the tribe Geeae (Rosaceae), represented by the genera *Geum*, *Waldsteinia*, and *Coluria*, can be separated from the other tribes of the subfamily Rosoideae as a distinct basic type. Within the genus *Geum*, hybrids are easily obtained and frequently occur in nature. These hybrids are usually vigorous and—especially within the subgenus *Geum*—partially fertile. Most characteristics are intermediate on a polygenic basis; in some cases also, simple hereditary paths can be observed. Reasons for hybrid sterility are discussed. The present data are interpreted within the hypothesis of a basic type diversification from a genetically polyvalent ancestral form.

## 7.1. Introduction

In the year 1957, the Polish botanist W. Gajewski published an extensive monograph on the Rosaceae genus *Geum* L. (avens), along with the closely related genera *Coluria* and *Waldsteinia* (Gajewski 1957). He reports numerous crossbreeding experiments and gives an overview of morphological, cytological, genetic, geographical, and paleontological data for this genus. In the present contribution, the most important results of his work, as well as some newer investigations, are introduced and discussed within the framework of basic type biology (Junker 1989, this volume; Scherer 1993, this volume). Table 1 gives an overview of the taxonomy of the Geeae and those genera to which they show closest affinity. Gajewski (1957, 1959) places the genera *Geum*, *Waldsteinia*, and *Coluria* into a separate tribe and, together with the Dryadeae and the Cercocarpeae, into the subfamily Dryadoideae. Most taxonomists, however, do not give this group the status of a subfamily and place them—together with a number of other tribes—into the subfamily of the Rosoideae (El-Gazzar 1981; Challice 1974; and others).

Figure 1. The four avens native to Germany. Above left: *Geum urbanum* L. (wood avens), with enlarged blossom and fruit (fishhook type), is the most frequent indigenous species of aven and populates relatively dry locations near forest paths, walls, and in hedges. Height: approximately 30-50 cm and more. Above right: *G. rivale* L. (water avens, 25-40 cm high), with its fruit in the center of the drawing (fishhook type with hairs), loves wet meadows, ditches, and riversides of the lower and middle mountain areas. Below left: *G. montanum* L. (alpine avens, 10-15 cm high), with plumose fruit, grows on alpine meadows, primarily at 1500 to 2500 m altitude. Below right: *G. reptans* L. (creeping avens, 10-15 cm high), is usually found in areas over 2000 m altitude, on moraine hills and rocky slopes, especially in the silicate mountains. (Drawing by M. Häberle)

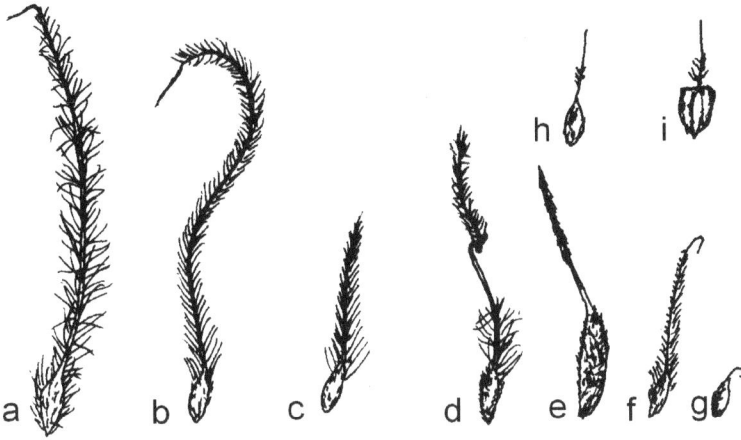

Figure 2. Style forms of the Geeae. a-h: subgenera of *Geum*: a: *Oreogeum*, b: *Erythrocoma*, c: *Acomastylis*, d: *Geum*, e: *Orthurus*, f, g: *Oncostylus*, h: *Woronowia*; i: *Waldsteinia*. (After Huber 1961 and Gajewski 1957)

## 7.2. Characteristics and Distribution of the Geeae

The Geeae are perennial herbs with a thick rhizome. They possess pinnate or pinnatified leaves (Figure 1). The genus *Waldsteinia* possesses "if you will, only the (rather large, palmate or palmately divided) terminal lobe" (Bolle 1933, p. 7; Figure 4). The plants have a basal leaf rosette and leafy, usually many-flowered stems. The petal and sepal whorls are usually pentameric, rarely hexa- to octomeric or tri- and tetrameric (Figure 1). At each calyx sinus, a bractlet occurs, in rare cases two (e.g., with *Geum reptans*). The flowers contain numerous (20 to over 100) stamens and separate carpels and are usually arranged in loose panicle-like cymes. The fruits are single-seeded nutlets.

The most variable characteristic is the style (Figures 1 and 2). It is more or less plumose and elongates after flowering. Upon reaching maturity, it remains connected to the simple fruit as a unit or is jointed with two parts. The upper part (stigmatic segment) detaches at a hook-shaped region at the apex of the rostrum (long lower part) before the fruit reaches maturity. In one case, the fruit is a plumed diaspore on a parachute, which is distributed by the wind in dry conditions (in wet conditions, distribution by animals through water adhesion is also possible); in the second case, a fish-hook or burr fruit is distributed by animals.

In addition, there are transitional forms (Bolle 1933, p. 12) and, in *Geum heterocarpum* (subgenus *Orthurus*), the so-called "harpoon" form with stiff, downward pointed hairs at the tip of the rostrum (Figure 2e). The species of the subgenus *Oncostylus* possess a style with a bent stigmatic segment, which does not detach from the rostrum (Figure 2f,g). In *Coluria* (Figure 3) and *Waldsteinia* (Figure 4), the style is constricted at

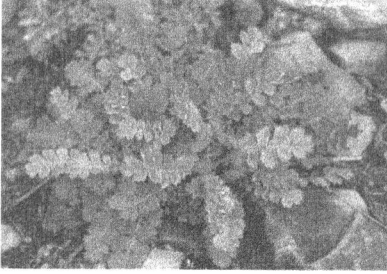

Figure 3. Basal leaf rosette of *Coluria geoides* (Pall) Leder. Native area: Altai Mountains.

Figure 4. *Waldsteinia fragarioides* (Michx.) Tratt. Native area: North America. (Photo by R. Sanders)

the basis and detaches completely (Figure 2i). On the basis of anatomical characteristics (number of stomates), it can be homologized with the stigmatic segment of the articulated style. Ants disperse the individual fruits of *Waldsteinia* (myrmecochory).

The different style forms and, hence, the associated dispersal mechanisms correspond to different habitats. The species with plumed fruits grow in open areas, as, for example, the indigenous alpine avens and the creeping avens (*Geum montanum* and *G. reptans*), which grow in higher, usually treeless areas, of the Alps. The genera with fish-hook/burr fruits live in open forests or along roadsides.

**7.2.1. Basic Chromosome Numbers**. The basic chromosome number of the genus *Geum* is $x = 7$, just as in *Coluria* and *Waldsteinia*. The genus *Dryas* (which due to the structure of its style is morphologically very closely related), on the other hand, possesses the chromosome number $x = 9$, just as all the other genera of the Dryadoideae (Table 1). Almost all cytologically examined species of *Geum* are polyploid (tetra- to dodecaploid); most are hexaploid.

Diploid representatives can be found only in *Geum* subgenus *Sieversia* and among the five *Coluria* and the four *Waldsteinia* species (Stebbins 1984). Within species, no ecotypes or races with different ploidy levels have been found thus far.

**7.2.2. Geographical Distribution.** The Geeae are widespread—commonly found between the arctic and tropic circles in the northern and southern hemispheres (Bolle 1933, p. 96). The 11 subgenera of *Geum* listed by Gajewski are in most cases geographically separated. For example, representatives of *Sieversia* (*s. str.*) grow only in northeast Asia and Japan; *Oreogeum* in the mountains of central Europe; *Andicola* in the Chilean Andes; *Oncostylus* in Australia, Tasmania, and South America; and *Erythrocoma* in North America. Only the species-rich subgenus, *Geum*, is widespread, populating large parts of the northern hemisphere and some parts of the southern hemisphere.

| | basic number of chromosomes | style form | uniformly pinnate basal leaves | oblong petals | nodding blossoms | carpels few | stolons | specially shaped leaves |
|---|---|---|---|---|---|---|---|---|
| Geeae | 7 | | | | | | | |
| Geum | 7 | | | | | | | |
| Sieversia (2) | 7 | plumose | | | | | | |
| Neosieversia (1) | 7 | plumose | x | | | | | |
| Oreogeum (3) | 7 | plumose | + | + | | | + | |
| Erythrocoma (4) | 7 | plumose | x | x | x | | | |
| Acomastylis (7) | 7 | plumose | | | | x | | |
| Andicola (1) | – | plumose | x | | | | | |
| Oncostylus (9) | – | fishhook | | | | | | |
| Geum (25) | 7 | fishhook | | | + | | | + |
| Stylipus (1) | 7 | fishhook | | | | | | + |
| Orthurus (2) | 7 | harpoon | | | | | | |
| Woronowia (1) | 7 | detaches | | | | | | |
| Waldsteinia (6) | 7 | detaches | | | | x | x | x |
| Coluria (5) | 7 | detaches | | | | | | |
| Dryadeae | 9 | plumose | | | | | | |
| Dryas (3) | 9 | plumose | | | | | | |
| Cowania (7) | 9 | plumose | | | | | | |
| Fallugia (1) | 9 | plumose | | | | | | |
| Purshia (2) | 9 | plumose | | | | | | |
| Cercocarpeae | 9 | plumose | | | | | | |
| Cercocarpus | 9 | plumose | | | | | | |

Table 1. The genera of the subfamily Dryadoideae, as well as the subgenera of *Geum* (after Gajewski 1957, 1959), their basic chromosome numbers and style form, as well as special characteristics; in parenthesis: number of the species of the *Geum* subgenera, as well as the genera *Waldsteinia* and *Coluria* and the genera of the *Dryadeae*. The subgenera *Sieversia* (s. str.), *Neosieversia*, *Oreogeum*, and *Erythrocoma* are frequently united as subgenus *Sieversia* (s. l.), which sometimes is separated from *Geum* as its own genus. + = in part, x = always

## 7.3. Hybridization Results

In spite of considerable morphological differences among the individual species, the genus *Geum* is characterized by a strong tendency towards interspecific hybridization. However, this tendency does not extend to transgeneric hybrids with *Coluria* and *Waldsteinia* as these have not yet been documented beyond doubt. Table 2 summarizes the results of the crossbreeding experiments of Gajewski (1957).

Besides these artificially-generated hybrids, natural hybrids are also known and occur frequently where the areas of distribution of the species overlap. Gajewski describes 23 natural hybrids, most of which were also generated artificially. Only 32 of the 56 species of *Geum* were included in the investigation of Gajewski.

Basic Types of Life

1 pentapetalum
2 selinifolium
3 glaciale
4 montanum
5 reptans
6 bulgaricum
7 peckii
8 radiatum
9 calthifolium
10 sikkimense
11 elatum
12 rossii
13 turbinatum
14 andicola
15 renifolium
16 uniflorum
17 divergens
18 parviflorum
19 albiflorum
20 leiospermum
21 pusilum
22 involucratum
23 lechlerianum
24 triflorum
25 canescens
26 ciliatum
27 campanulatum
28 rivale
29 capense
30 silvaticum
31 pyrenaicum
32 coccineum
33 quellyon
34 magellanicum
35 peruvianum
36 brevicarpellatum
37 riojense
38 laciniatum
39 latilobum
40 molle
41 canadense
42 aleppicum
43 virginianum
44 boliviense
45 macrophyllum
46 perincisum
47 oregonense
48 japonicum
49 fauriei
50 urbanum
51 roylei
52 hispidum
53 vernum
54 heterocarpum
55 kokanicum
56 speciosum

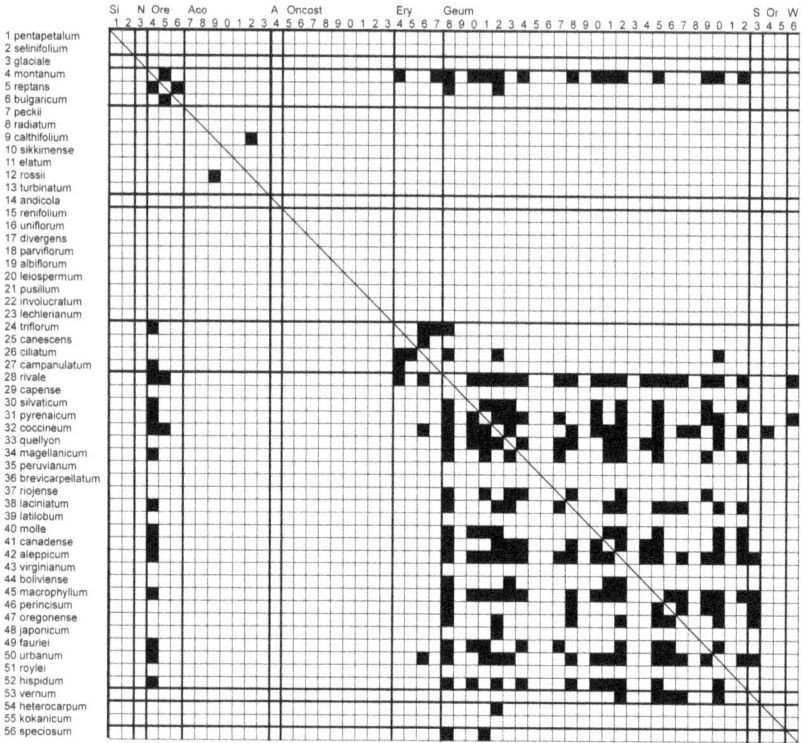

Table 2. Crossbreeding matrix of *Geum* L. After Gajewski (1957, 1959) and Bolle (1933). Abbreviations (subgenera after Gajewski 1957, 1959; comp. Table 1): Si = *Sieversia*, N = *Neosieversia*, Ore = *Oreogeum*, Acom = *Acomastylis*, A = *Andicola*, Ery = *Erythrocoma*, S = *Stylipus*, Or = *Orthurus*, W = *Woronowia*.

Hybridization attempts rarely failed. As far as examined, reciprocal crosses yield identical results (Gajewski 1957, p. 114). With one exception, all subgenera, which were included in the investigation, are interconnected through hybridization. The exception is the subgenus *Acomastylis*. Gajewski managed to conduct only a few crossbreeding experiments with one species from *Acomastylis*, *G. rossii*, and all were unsuccessful. However, Gajewski was convinced that other species of this subgenus can be crossed with representatives of other subgenera. The larger gaps in the crossbreeding matrix, thus, exist due to lack of experimental results. Natural hybrids between the subgenera are largely impossible because of geographical separation.

The species of the subgenera of *Geum* are particulary easy to crossbreed, even if different levels of ploidy exist (Table 3). In this case all hybridizations produced hybrids. Crosses (within the subgenus *Geum*) between dodecaploid species (n = 84) are sterile to almost fully fertile, between dodecaploid and decaploid (n = 70) almost sterile to partially fertile, between dodecaploid and hexaploid (n = 42) almost or

| No. | hybridized species from *Geum* | maximum conjugation | pollen fertility | seed fertility |
|-----|-------------------------------|---------------------|------------------|----------------|
| **in each case from different subgenera** | | | | |
| 6 | *rivale* (Ge, 42) × *triflorum* (Er, 42) | 14/14 | 0.3 | 0 |
| 7 | *coccineum* (Ge, 42) × *heterocarpum* (Ort, 42) | 7/2 | 11.2 | 0 |
| 10 | *urbanum* (Ge, 42) × *vernum* (St, 42) | 12/18 | 1.7 | 0.3 |
| 15 | *montanum* (Ore, 28) × *rivale* (Ge, 42) | 14/ 7 | 34.5 | 15.5 |
| 17 | *montanum* (Ore, 28) × *silvaticum* (Ge, 42) | 14/ 7 | 25.5 | 34.4 |
| 21 | *montanum* (Ore, 28) × *aleppicum* (Ge, 42) | 14/ 7 | 4.2 | 2.4 |
| 23 | *montanum* (Ore, 28) × *urbanum* (Ge, 42) | 14/ 7 | 0.45 | 0.6 |
| 24 | *montanum* (Ore, 28) × *macrophyllum* (Ge, 42) | 6/23 | 0.1 | 0.1 |
| 25 | *pyrenaicum* (Ge, 28) × *montanum* (Ore, 28) | 14/14 | 0.01 | 0.15 |
| **within subgenus *Oreogeum*** | | | | |
| 30 | *montanum* (28) × *reptans* (42) | 14/ 7 | 0.7 | 0 |
| **within subgenus *Geum*** | | | | |
| 31 | *magellanicum* (84) × *riojense* (84) | 42/ 0 | 61.9 | 27.3 |
| 34 | *fauriei* (84) × *riojense* (84) | 34/16 | 0.2 | 0 |
| 37 | *riojense* (84) × *quellyon* (70) | 35/ 7 | 36.4 | 24.4 |
| 55 | *fauriei* (84) × *rivale* (42) | 21/21 | 0.2 | 0 |
| 61 | *rivale* (42) × *quellyon* (70) | 21/14 | 0.1 | 0 |
| 67 | *rivale* (42) × *urbanum* (42) | 21/ 0 | 78.0 | 72.1 |
| 68 | *rivale* (42) × *coccineum* (42) | 21/ 0 | 66.6 | 94.5 |
| 75 | *rivale* (42) × *macrophyllum* (42) | 7/28 | 0.7 | 0.1 |
| 90 | *silvaticum* (42) × *aleppicum* (42) | 21/ 0 | 0 | 0 |
| 101 | *molle* (42) × *aleppicum* (42) | 21/ 0 | 0.2 | 0.01 |
| 103 | *molle* (42) × *macrophyllum* (42) | 6/30 | 0.2 | 0 |

Table 3. Maximum conjugation (number of bivalents / univalents), fertility of pollen and seeds in *Geum*-hybrids. The numbering corresponds to the numbers assigned to the hybrids generated by Gajewski (1957). Abbreviations, see Table 1; the numbers in parenthesis indicate the gametic chromosome number. (Data from Gajewski 1957)

fully sterile, between decaploid and hexaploid fully sterile, and between hexaploid species sterile to almost fully fertile. In the latter case, hybrids of geographically separated species are more often sterile than those between sympatric species.

*Geum montanum* (subgenus *Oreogeum*) hybridizes with species of subgenus *Geum* remarkably easily, with a degree of fertility of the seeds of up to 34% (Gajewski 1957, p. 155; see Table 3). On the other hand, hybrids between *G. montanum* and *G. reptans* (both belonging to *Oreogeum*) are almost completely sterile (they grow occasionally in contiguous habitats in the mountains).

Altogether, the conclusion of Hess et. al. (1970, p. 365) seems justified, that, most likely, all species of the genus *Geum* can hybridize with each other.

**7.3.1. Intergeneric Hybrids in the Geeae.** Gajewski (1957, p. 116) reports that, for example, in some hybrids between *Waldsteinia geoides* and *Coluria geoides,* as well as between *Geum rivale* and *Waldsteinia ternata,* the fruits began to develop but then degenerated. However, he thinks that viable seeds could be generated in extensive crossbreeding experiments using embryo cultures and a larger number of biotypes.

**7.3.2. Crossbreeding Experiments with Species of Other Genera of the Rosaceae.** Attempts to crossbreed species of *Geum* with species of *Potentilla* ("cinquefoil") failed completely (Gajewski 1957, p. 117). This author does not know of any other crossbreeding experiments.

### 7.4. Characteristics of the *Geum* Hybrids

**7.4.1. Degrees of Fertility and Vitality of the Hybrids.** Within the subgenus *Geum*, a majority of the hybrids is partially to almost fully fertile. Hybrids between the subgenera, on the other hand, are usually sterile or almost sterile, with the exception of a series of hybrids between *G. montanum* and species of the subgenus *Geum* (see above). In partially fertile hybrids, the degree of fertility of the pollen and seeds often fluctuates widely. The degree of fertility of the seeds of the hybrid between *G. montanum* and *G. rivale,* for example, lies between 3.5% and 32% (Gajewski 1957, 155).

Most hybrids are quite vigorous, often as much so as the parents, irrespective of the degree of fertility (Gajewski 1957, p. 334). In rare cases, however, puny or malformed hybrids occur, especially in the $F_2$. No connection to disruptions during meiosis could be detected (Gajewski 1957, p. 338). The variability of the vitality and fertility as well as morphological characteristics usually increases in the $F_2$ (see tables in Gajewski 1957, pp. 282-314). Gajewski says regarding the $F_2$ of *G. montanum* and *G. rivale*: "... it becomes easily understood that among the 98 $F_2$ plants no two were alike, and that the characters of both parental species appear in a variety of combinations" (1957, p. 269). He was able to show that some characteristics are inherited in blocks (dependent on each other [1957, p. 271]).

**7.4.2. Causes of Hybrid Sterility.** Concerning causes of the sterility of hybrids, Gajewski (1957, p. 331ff) names incompatibilities between pollen and stigmatic tissue, as well as disharmony among embryo, endosperm, and nucellar tissue. Different degrees of ploidy or non-homologies of homoploid species cause disruptions in meiosis and formation of univalents. However, there are a number of fertile hybrids between heteroploid species. On the other hand, many hybrids of homoploid species (as well as hybrids within the same subgenus) are sterile (see Gajewski 1957; Table pp. 341-343 and pp. 351-353). Table 3 contains an excerpt from the summary of data published by Gajewski and illustrates the following possible connections:

In many cases, the fertility of hybrids apparently depends on the extent of chromosome homology (comp. Table 6 below). However, there are notable exceptions (compare, for example, hybrid nos. 68 and 90, or nos. 17 and 23). Genic sterility could explain these results and the fact that, in numerous cases when three species are mutually crossed, varying proportions of fertility occur (Table 4; comp. Lönnig 1993, p. 170ff).

| | | |
|---|---|---|
| G. aleppicum | x G. coccineum | partially fertile |
| G. coccineum | x G. molle | fertile |
| G. molle | x G. aleppicum | sterile |
| G. aleppicum | x G. rivale | partially fertile |
| G. rivale | x G. urbanum | fertile |
| G. urbanum | x G. aleppicum | sterile |
| G. coccineum | x G. hispidum | fertile |
| G. hispidum | x G. canadense | partially fertile |
| G. canadense | x G. coccineum | sterile |

Table 4. Triple mutual crosses with varying degrees of fertility. (After Lönnig 1993 and Gajewski 1957)

"Sterility is sometimes caused by genes scattered among different species and does not have any phylogenetic meaning" (Gajewski 1959, p. 384). The fertility barriers are not correlated with morphological characteristics: "The conclusion which is to be drawn from these considerations is that incompatibility barriers often depend on factors which are not distributed within a genus in a manner strictly correlated with the degree of morphological differentiation between species" (Gajewski 1957, p. 334). Neither is there any strict correlation in the differences in degree of ploidy and fertility (compare, for example, hybrid nos. 17, 37 and 90).

Among the hexaploid species of subgenus *Geum*, four groups of species can be distinguished that crossbreed with almost full fertility but are more or less reproductively isolated from species from other groups. The differences in the fertility among these groups can be explained with chromosomal sterility only in the case of the North American group of *G. macrophyllum*, *G. oregonense*, and *G. perincisum*. In the other cases, genic sterility must be the cause because no considerable differences in the behavior of the chromosomes during meiosis can be detected. The three species *G. canadense*, *G. laciniatum*, and *G. boliviense* are fully fertile among each other, but are, in part, geographically separated (the first two are found in North America, the latter in South America). It follows that geographical separation, also, cannot be generally regarded as the cause of fertility barriers. "There is no doubt that the hexaploid *Eugeum*[1] species even from different continents still possess homologous loci in their chromosomes" (Gajewski 1959, p. 385; comp. also p. 387).

**7.4.3. Mode of Inheritance of Phenotypes in Hybrids.** Most characteristics exhibit intermediary inheritance, with continuous variation in the $F_2$, which indicates polygenic inheritance (Gajewski 1957, p. 263ff). Some qualities, like the color of the blossom (white, yellow, pink or red with different nuances), formation of anthocyan, emargination of the petals, formation of a petal claw [narrowed stalk-like base of the petal], formation of stolons, *etc.* are determined by one or a few loci, so that these characteristics—in accordance with Mendel's rules—divide in the $F_2$. In true cross-breeds, one therefore finds what

---

1 The subgenus *Geum* was formerly called *Eugeum*.

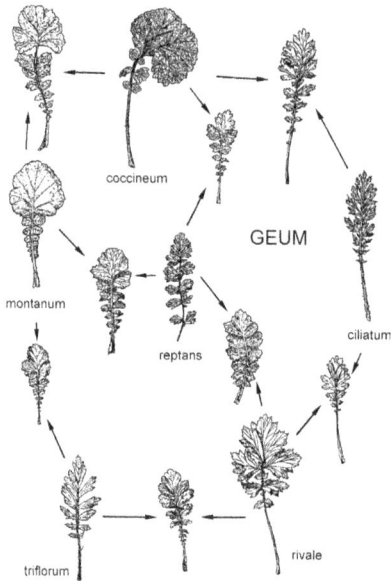

coccineum

GEUM

montanum

ciliatum

reptans

triflorum

rivale

Figure 5. Basal leaves of some species and hybrids of *Geum*. The arrows point from the parents to the hybrid.

is to be expected (vitality of the $F_1$, independence of vitality and fertility, high degree of variability of the $F_2$).

**7.4.4. Inheritance of Flower Colors.** As an example, the inheritance of flower colors and the style will be described.

Among the Geeae, there are species with red, pink, yellow, and white flowers (with color gradations). Based on the hybridizations between species with differently colored flowers, the following conclusions regarding the genetic basis can be made:

The flower color is determined by three loci (Gajewski 1957, p. 361). In absence of the allele C, the flowers are white. Allele C together with allele A is the basis for the pink floral color; allele B and allele C produce the red color; alleles A, B, and C produce yellow flowers (Table 5). Homo- and heterozygotes show different color nuances.

---

white: ccaabb, ccAabb, ccAAbb, ccaaBb, ccAaBb, ccAABb, ccaaBB, ccAaBB, ccAABB
pink: CCAAbb, CCAabb, CcAAbb, CcAabb
red: CCaaBB, CCaaBb, CcaaBB, CcaaBb
yellow: CCAABB, CCAABb, CCAaBB, CCAaBb, CcAABB, CcAABb, CcAaBB, CcAaBb

---

Table 5. Genotypes of the flower colors white, pink, red, and yellow. Explanations see text.

Thus, the hybrid of the red-blossomed *G. coccineum* (CCaaBB) and the pink-blossomed *G. rivale* (CCAAbb), for example, possesses yellow blossoms (CCAaBb). The hybrid of the white-blossomed *G. canadense* (ccAABB) and *G. coccineum* (CCaaBB) has orange-yellow blossoms (CcAaBB), *etc.* Gajewski (1957, 361) adds: "It undoubtedly is a substantial proof of the common phylogenetic origin of the whole group of species."

**7.4.5. Inheritance of the Style.** In reference to the inheritance of the style, hybrids between species with plumose styles and those with the fishhook form (Figure 1, 2, 6) are of particular interest. Between *G. montanum* and a series of species of the subgenus *Geum*, a number of fertile hybrids with this combination could be generated (see above). In the $F_1$ individuals, one can distinguish three different style forms with gradual transitions (Gajewski 1957, p. 147ff.; Gajewski 1964; comp. Figure 2 and 6): Form I: straight, more or less hairy, the stigmatic segment does not detach (*montanum*-like). Form II: partial

Figure 6. Fruit of the hybrid of *Geum montanum* and *G rivale*, showing style structures.

bend between rostrum and stigmatic segment; the stigmatic segment usually detaches. Form III: rostrum bent to a hook; the stigmatic segment detaches; the style is relatively short and has few hairs (*rivale*-like). The early development is generally *rivale*-like. The first flowers to develop possess rather *montanum*-like styles in their final shape (form I), the later flowers are *rivale*-like. This variability of the style could be observed in several populations of *G. montanum* hybrids. Some hybrids between *G. montanum* and various subgenera only develop styles of form I.

In different pairs of hybrids, different proportions of dominance exist. "These results indicate that the particular species of the subgenus *Eugeum* [*sic*] possess various sets of hereditary factors determining the expression of the achenes in these species. These sets when united with the set of genes from *G. montanum* may be in $F_1$ completely recessive, they may be partially dominant or even completely dominant as in the case of the dodecaploid species" (Gajewski 1957, p. 368). Gajewski then comes to the conclusion that the two homologous genomes, *i.e.*, the genome belonging to the species subgenus *Geum* and that of *G. montanum*, possess alleles that indicate the common origin of both subgenera.

In the $F_2$ of hybrids of subg. *Oreogeum* × subg. *Geum*, a high variability of style forms exists (Gajewski 1957, p. 268f); some plants simultaneously possess different style forms. Occasionally, the style quite strongly resembles that of one of the two original species. The fertility of the hybrids is independent of the style form or other morphological characteristics. "The segregation in $F_2$ and backcross families seem to indicate that... the structure of the styles probably depends on few genes with major effects and many modifying genes" (Gajewski 1959, p. 385).

Hybrids between *G. heterocarpum* (subgenus *Orthurus*), with a style of the "harpoon" form, and *G. coccineum* (subgenus *Geum*), with fishhook form, proved to be completely sterile. The fruits did not ripen completely so that a comparison with the styles of the parental species was possible only in a limited way. The downward pointed hairs were

| | Ge42 | Gm42 | Ge70 | Ge84 | Ort28 | St42 | Ore28 | Ore42 | Er42 |
|---|---|---|---|---|---|---|---|---|---|
| G. rivale (Ge42) | 21/0 | 7/28 | 21/14 | 21/21 | — | 14/14 | 14/7 | 14/14 | 14/14 |
| G. coccineum (Ge42) | 21/0 | 7/28 | 21/14 | 21/21 | 7/21 | 14/7 | 14/14 | 14/14 | |
| G. montanum (Ore28) | — | — | — | — | — | — | — | — | 7/21 |
| | Ge42 | Gm42 | Ge70 | Ge84 | Ort28 | St42 | Ore28 | Ore42 | Er42 |
| G. rivale (Ge42) | 3 | 1 | 3 | 3 | — | 2 | 2 | 2 | 2 |
| G. coccineum (Ge42) | 3 | 1 | 3 | 3 | 1 | — | 2 | 2 | 2 |
| G. montanum (Ore28) | — | — | — | — | — | — | — | — | 1 |

Table 6. Maximum conjugation (number of bivalents/univalents) in hybrids between species of different subgenera of *Geum* or within the subgenus *Geum* (Ge). In the lower part, the numbers of homologous sets of chromosomes are listed. Other abbreviations: Er = *Erythrocoma*, Gm = *G. macrophyllum* of the subgenus *Geum*, Ore = *Oreogeum*, Ort = *Orthurus*, St = *Stylipus*; the numbers following the abbreviations give the gametic number of chromosomes. (Data from Gajewski 1957)

missing in young styles. The top of the rostrum was only slightly bent (Gajewski 1957, 132f.).

**7.4.6. Chromosome Homologies.** Chromosome homologies (Table 6) were determined on the basis of the maximum chromosome conjugations during meiosis in hybrids of *G. rivale*, as well as *G. coccineum* (of the subgenus *Geum*), with different species of the subgenera *Geum*, *Oreogeum*, *Erythrocoma*, *Orthurus*, and *Stylipus*. In all hybrids considerable differences between maximum and minimal conjugation, and thus considerable variability, were observed. The above-mentioned chromosome homologies are therefore not complete. In Table 7, some extreme examples are given (a complete list of all 121 successfully crossed pairs can be found in Gajewski 1957, p. 351ff).

From Table 6 and 7, the following conclusions can be drawn: The three sets of chromosomes of the hexaploid species of the subgenus *Geum* are homologous. The low degree of pairing between *G. macrophyllum* and *G. aleppicum*, on the one hand, and other hexaploid species of the subgenus *Geum*, on the other hand, (with a maximum of 7 bivalents) is attributed by Gajewski (1959, p. 384) to premature desynapsis. *Geum montanum* and species of the subgenus *Geum* usually have two homologous sets of chromosomes, which is also true for the pair *G. montanum / G. reptans* (both in subgenus *Oreogeum*); the third set of chromosomes of *G. reptans* is not homologous to the third set of chromosomes of the subgenus *Geum*. The subgenera *Stylipus* and *Geum* have at least two, only partially, homologous sets of chromosomes. Subgenus *Orthurus* (*G. heterocarpum*) probably shares only one set of chromosomes with subgenus *Geum*. *Geum montanum* shares at most one set of chromosomes with subgenus *Erythrocoma*. Thus, the subgenus *Geum* probably shares at least one set of chromosomes with all subgenera examined thus far (in some cases with only partial homology).

| No. | species crosses in the genus *Geum* | conjugation | |
|---|---|---|---|
| | | maximum | minimum |
| 3 | *urbanum* (Ge, 42) × *ciliatum* (Er, 42) | 14/14 | 7/28 |
| 7 | *coccineum* (Ge, 42) × *heterocarpum* (Ort, 28) | 7/21 | 2/31 |
| 11 | *macrophyllum* (Ge, 42) × *vernum* (St, 42) | 8/26 | 0/42 |
| 44 | *magellanicum* (Ge, 84) × *hispidum* (Ge, 42) | 21/42 | 14/35 |
| 57 | *fauriei* (Ge, 84) × *macrophyllum* (Ge, 42) | 21/21 | 13/37 |
| 77 | *rivale* (Ge, 42) × *oregonense* (Ge, 42) | 9/24 | 0/42 |
| 112 | *aleppicum* (Ge, 42) × *boliviense* (Ge, 42) | 21/ 0 | 13/16 |

Table 7. Maximum and minimal extent of chromosome conjugation during the meiosis (number of bivalents/univalents); abbreviations see table 6. (Data from Gajewski 1957)

## 7.5. Are the Geeae a Basic Type?

**7.5.1. Hybrids.** Based on current knowledge this question cannot be answered with certainty. Crossbreeding results of some subgenera are still missing (no experiments were made). The results of crossbreeding *Coluria, Waldsteinia,* and *Geum* suggest an affiliation with the same basic type (see above). Gajewski (1957, p. 330) unites all species of *Geum* in a so-called "comparium." This comparium includes coenospecies, which can produce sterile interspecific hybrids. Between two comparia, no hybrids can be generated. (Coenospecies are characterized by partial fertility of the hybrids).

This definition is remniscent of the definition of the basic type, but is less distinct. Gajewski thinks that *Coluria* and *Waldsteinia* are to be excluded from the comparium of *Geum,* although he considers it possible that hybrids could be generated (see above).

Intertribal hybrids are not known and were probably hardly ever tried (exception: crossbreeding experiments with *Potentilla,* see above). On the basis of the crossbreeding data, there is, therefore, no reason to extend the boundaries of the basic type beyond the tribe Geeae.

Table 1 lists genera that are taxonomically close to the Geeae, according to the classification of Gajewski (1957, 1959). However, not all taxonomists follow the uniting of *Geum, Coluria* and *Waldsteinia* into the tribe Geeae, as proposed by Gajewski (El-Gazzar 1981; Challice 1974; Hutchinson 1973). In the next section, therefore, certain data will be discussed that suggest a delimitation of the Geeae on the basis of other criteria, as well as other possible connections.

**7.5.2. Other Criteria.** Besides the crossbreeding results, the chromosome numbers (only Geeae with n = 7 within the Dryadoideae, Table 1), as well as the susceptibility to rust fungi, suggest a delimitation of the Geeae from the remaining Dryadoideae (El-Gazzar 1981; Table

| (sub)genus of host | genus of rust fungus | | |
|---|---|---|---|
| | a | b | c |
| Waldsteinia (4) | — | + | — |
| Geum (9) | + | + | — |
| Sieversia (2) | — | + | — |
| Orthurus (1) | + | — | — |
| Acomastylis (1) | — | + | — |
| Dryas (1) | — | — | + |

Table 8. The hosts of some rust fungi, and (in parenthesis) the number of species within those host genera that were examined. a: Phragmidium, b: Puccinia, c: Kuehneola (from El-Gazzar 1981).

8). "*Dryas* shares no parasites with any other rosaceous genus to my knowledge... These fungi reflect the isolated position of *Dryas*. The genus bears no rust that might indicate its affinity" (Savile 1979, p. 417). Still, the same author states: "The genera and subgenera of Geeae are linked by a seemingly natural group of *Puccinia* species."

He also mentions a connection between *Waldsteinia fragarioides* and *Geum triflorum*—both are parasitized by the monotypic smut fungus *Ustacystis* (p. 419)— and concludes, "It seems well, from mycological and other evidence, to regard Geeae as a reasonably typical tribe of Rosoideae, and Dryadeae as an atypical one with pronounced climatic adaptations" (p. 420f; the author uses the same taxonomy as Gajewski). Furthermore, only the Geeae have pinnate basal leaves, *Waldsteinia* representing a borderline case (only the end leaflet is developed; Figure 4). Plus, the leaves of *Dryas* are extremely xeromorphic (Huber 1961), and *Dryas* does not possess any calyx bractlets, this also being the case in only one species of *Geum*—*G. vernum*, which actually is distant from *Dryas*. El-Gazzar (1981) mentions another group of anatomical characteristics that separate the Geeae from *Dryas*. All these characteristics – with the exception of the basic number of chromosomes – admittedly do not strongly support a delimitation of the Geeae.

Thus, Bolle (1933, p. 21; however with a comparatively narrower empirical basis) considers a delimitation of *Dryas* and *Sieversia* (*s. str.*), on one hand, and the other subgenera of *Geum*, on the other hand, as possible. However, Gajewski (1957, p. 8) thinks that *Sieversia*—with reference to the number of petals, the form of the calyx, and the shape of the leaves – so clearly differs from *Dryas* and so very much resembles *Geum* that its placement into *Geum* is more convincing. This opinion is confirmed by more recent findings. However, the subgenus *Sieversia* is similar to *Dryas* in its espalier-like shrubby growth and the lack of known infections by rust fungi — in contrast with the other subgenera of *Geum* (Savile 1979, p. 419). Crossbreeding experiments are needed to resolve this conflict.

Regarding the structure of the style, *Dryas* and some subgenera of *Geum* closely resemble each other (Figure 7). The plumose style can, however, also be found in completely different plant families— for

Figure 7. *Dryas octopetala* (white mountain-avens) in fruit (left), compared with the plumose fruit of *Geum montanum*. The genus *Dryas* is morphologically closest to the Geeae. It grows in higher region of the Limestone Alps.

example, with *Pulsatilla* among the Ranunculaceae. There it must be interpreted as an analogy, which is also possible in the case of *Dryas / Geum*.

Paleontological data are almost completely absent.

**7.5.3. Taxonomy Within the Geeae.** The two genera *Coluria* and *Waldsteinia* can easily be separated morphologically from *Geum* on the basis of the nature of the style. Korotaeva (1983), on the basis of anatomical structures of the fruits, detected a sharp boundary between *Coluria* and *Waldsteinia*, on one hand, and *Geum*, on the other. Within *Geum*, Gajewski distinguishes the above-mentioned 11 subgenera by morphological and geographical criteria. Bolle (1933) awards the status of genera to most of these subgenera. Due to easy hybridization, however, Gajewski does not consider this as justifiable. The relationships within *Geum* are unclear. Depending on which traits the classification is based on, different divisions and different affinities emerge among the subgenera (Gajewski 1957, p. 5; Bolle 1933, p. 101ff; Král 1966; Table 1).

### 7.6. Mechanisms of Basic Type Diversification

The difficulty of reconstructing a phylogenetic tree can be compared to the uncertainties of establishing a taxonomic system within the Geeae tribe (comp. Gajewski 1959, p. 385ff; Bolle 1933, p. 105). The subgenus *Geum* could have originated through amphiploidy between a tetraploid species of *Oreogeum* and an unknown diploid species (perhaps closely related to *Waldsteinia*). The dodecaploid species could have originated in a similar way. Amphidiploidization also succeeded twice in crossbreeding experiments (Gajewski 1959, p. 387).

**7.6.1. Evolutionary Lines or a Flexible Ancestral Form?** According to the basic type concept, ancestral forms with a high potential of variation stand at the beginning of evolution. The following questions result from this postulate:

1. **Can the present data be interpreted in the sense of descent from a genetically polyvalent, *i.e. a* variable and flexible ancestral population?** First, the high degree of hybridization can certainly be interpreted in this sense. The manner of heredity of the flower color (see above), as well as a number of other characteristics, suggests a close genetic relationship and can also easily be interpreted in this sense. In addition, the fact that, based on different characteristics, varying affinities exist between the individual subgenera also fits this picture. Here are some examples:

- The subgenus *Neosieversia* with the single species *G. glaciale* shows different relationships to other species or subgenera, depending on which characteristic is in view: "... equally sized leaflets of the basal leaves is reminiscent of *Erythrocoma, Geum reptans,* and *Acomastylis-Euacomastylis*; the large quantity of the carpels with their plumose structure points to *Geum reptans*; the hairy filaments and base of the stamen, to *Erythrocoma* and *Acomastylis*" (Bolle 1933, p. 101). Due to other characteristics, however, *Neosieversia* has to be separated from these subgenera.

- An oblong form of the petals is found in *G. reptans* (subgenus *Oreogeum*), *G. andicola* (subgenus *Andicola*), and the species of *Erythrocoma*.

- The flowers in *Erythrocoma* and *G. rivale* are pendulous.

- Similar calyx formation occurs in *Waldsteinia* and the species of *Acomastylis* "which often also contains only a few carpels in one flower, as is also the case in *Waldsteinia*" (Bolle 1933, p. 101).

- *Coluria* has "closest connections" with *Acomastylis*, "because the only essential characteristic, that which separates *Coluria elegans* from *Acomastylis elata*, particularly its var. *shensina*, is the articulation of the style" (Bolle 1933, p. 102). Stolons are found in the species of *Waldsteinia* and in *G. reptans* (Bolle 1933, 15).

- The style of *G. speciosum* (subgenus *Woronowia*) closely resembles the style of *Waldsteinia. Geum speciosum* includes characteristics of *Geum,* as well as of *Waldsteinia* and *Coluria* (Gajewski 1957, pp. 83, 377, 392).

- *Waldsteinia*-like leaves can also be found in species of *Geum*—for example, sometimes in *G. vernum*, *G. canadense* and others (Bolle 1933, p. 7). The cauline leaves of *G. urbanum* have only "terminal leaflets", just like *Waldsteinia*.

Other examples can be found in Bolle (1933, p. 102f) and Gajewski (1957, 86ff).

2. **What did such an ancestral form look like?** Its appearance must remain an issue of speculation. A variable diploid ancestral form seems probable. The different subgenera of *Geum* would then have originated through multiple polyploidization. Through reorganization of the chromosomes or due to polyploidization of various ancestral forms, differences could have developed in the individual subgenera and could have stabilized due to separation processes (comp. Junker 1993, this volume).

**7.6.2. The Style as an Example.** Among the 65 species of the Geeae, the large palette of style structures (Figure 1, 2, 6, 7) makes possible the diverse dispersal mechanisms of the fruits (fishhooks of various construction, plumose styles, harpoon-shaped structures). According to the basic type concept, an ancestral population would have to "store" all these possibilities, both in the genome and the gene pool. In the course of the diversification of basic types, these different possibilities would then have to have been "distributed" to different biospecies while, at the same time, the genetic potential was decreased or certain genes were "switched off", so to speak. This would be a way to interpret convergences in the style (comp. Gajewski 1957, p. 384).

Gajewski (1957, 367), on the other hand, postulates a gradual transformation of the style during the evolutionary history of the genus (comp. Gajewski 1964) and, with Bolle (1933), pleads for an evolutionary path from the plumose style to the fishhook (Gajewski 1957, p. 387). However, the only argument they use is similarity (similarities between the plumose style of *Geum* and other species of the Dryadoideae). According to Bolle, the strange hook-shaped style is not as drastic as one might think at first glance because he regards it only as an advanced result of a torsion, which can also be found in subgenus *Oreogeum*. An intermediate form is found in *G. capense*.

Within the individual subgenera, Bolle distinguishes original and more advanced forms. The above-mentioned relationships (Table 1) exist between the original forms. From these, the following three evolutionary lines supposedly originated (Bolle 1933, p. 105) from an unjointed style: in one line, the end of the style was bent and rolled together (*Oncostylus*-line); in another line, the style was bent doubly in the middle and was cut at the point of inflection (subgenus *Geum* and others); finally, in the

*Waldsteinia*-line the style was constricted at the base and then severed. Regarding other characteristics, these three lines of evolution show fewer differences. Furthermore, all three lines have many evolutionary trends in common—for example, the abundant foliage of the stem and plentiful branching. We are therefore confronted with two different interpretations of the current distribution of style forms. According to the evolutionary interpretation, new qualities must have partially convergent origins, as, for example, the independent formation of fishhook styles in the subgenera of *Geum* and *Oncostylus*. According to an interpretation within the framework of the basic type concept (genetically polyvalent ancestral forms), a convergent *reduction* of an originally extensive gene pool took place (comp. also Gajewski 1957, p. 384).

In what respect can either of these contrary viewpoints be favored on the basis of the known results? The basic type model offers a satisfying explanation for the many, often fertile, hybrids between species with different style forms with highly variable $F_1$ and $F_2$. Maybe the variable $F_1$ of subg. *Geum* × subg. *Oreogeum* (see above) can be considered a model for a complex ancestral form. Still, the hook-shaped curvature of the style occurs both in unjointed and jointed styles, as does the plumose structure and the shortening of the style (Bolle 1933, p. 12; comp. Table 1). These characteristic combinations can be assessed as "remains" of a genetically polyvalent ancestral form.

Both interpretations lead to different questions:

a. The evolutionary framework raises the question of the mechanisms of the formation of new structural elements. Gajewski (1957, 1959) and Bolle (1933) do not address this question.

b. Within the framework of the basic type concept, the question must be asked whether a variable ancestral population with a rich gene pool makes sense. After all, the present-day species of the Geeae have adapted to different means of dispersal (see above). Actually *Geum rivale*, with its hairy hook-shaped style, presents a clear model, at least for the idea that both epizoochory as well as anemochory can be realized simultaneously, although, not in a unilaterally specialized form. (In addition, it is remarkable that *Geum rivale*—in contrast to other *Geum* species—is not at all specialized, as far as the altitudinal and geographic range of its distribution is concerned. Its distribution is widespread.) One can further assume that the second of the fishhook forms (in subg. *Oncostylus*, comp. Figure 2), as well as the prerequisites for a dispersal by ants, existed latently in an ancestral population. A loss of the torsion and the joint could secondarily have led to the plumose structure. Similar alterations could have resulted in the harpoon-form style and the style forms of the subgenus *Oncostylus*.

A second speculative interpretation is based on the postulate of "programmed variability:" different realizations in the regulatory realm are accessible through mutations. For this concept, there are models in other plants: the "water crowfoot" (*Ranunculus aquatilis*) possesses three different types of leaves (submerged leaves, floating leaves, aerial leaves), which consequently require three genetic programs, which can be called upon depending on environmental stimulation. Similarly, one could imagine a selection of programs in a genetically polyvalent ancestral form that in present-day species are either partially missing or "masked" through mutations. The observation that plants in altered surroundings suddenly (not only after an evolution phase!) display surprising abilities seems to support this idea.

Only detailed genetic examinations can give answers to the questions raised by both models. The analyses of Gajewski proved polygenic heredity for the style structure (Gajewski 1964), resulting in a highly variable $F_1$ (see above). In either case, the concept of flexible ancestral forms, therefore, does not appear unrealistic.

The three species of *Geum, G. rivale, G. coccineum,* and *G. silvaticum*, are closely related to *G. montanum*, given the fertility of their hybrids, as well as morphological traits (Gajewski 1957). They also show an especially large number of hybrids with other species of other subgenera and, therefore, could be regarded as an ancestral group of a larger part of the genus *Geum*, not only the subgenus *Geum*, as Gajewski assumes.

Do phyto-geographical data help? Apart from *Geum*, all subgenera are geographically isolated (see above). Forerunners of the present-day species of *Geum* could have attained their current disjunct distribution only through land bridges. Gajewski (1957, p. 385) considers other explanations are untenable. With respect to the above-mentioned question, these data do not help significantly. Within the interpretative framework of the basic type concept, one can assume that through an early processes of separation different "genetic raw material" for speciation (genetic drift) became available in the subsequently disjunct areas of distribution.

Similar questions emerge regarding the descent of other structures. Thus, different flower structures require different floral visitors. Insects that prefer flat extended blossoms, as in *G. urbanum*, cannot pollinate a nodding blossom with cup-like whorl of petals and sepals, as in *G. rivale*.

**7.6.3. The Example of the Adaptations of *Geum reptans*.** The creeping avens (*G. reptans*) grows in the Alps on moraines and rocky slopes. It has a powerful main root and stolons with which it can cross large stones (Gajewski 1957, p. 170) – a special adaptation to its habitat.

While blossoming, the flower stalk measures hardly more than 10 cm but grows to a length of 25 cm when the fruit reaches maturity exposing its plumose structure to the wind. According to the basic type concept, these qualities must have existed in the postulated ancestral form. The genetic potential of an organism will show when certain conditions provoke the realization of already existing programs.

## 7.7. Conclusions

The present data allow the conclusion that the three genera *Geum*, *Waldsteinia*, and *Coluria* belong to the same basic type. The observable variability of their traits can be explained by micro-evolutionary processes and does not require the evolution of completely new systems. The hypothetical ancestral form had to be relatively unspecialized and genetically polyvalent.

# References

Bolle, F. 1933. Eine Übersicht über die Gattung *Geum* L. und die ihr nahestehenden Gattungen. *Feddes Repertorium Beiheft* 72:1-119.

Challice, J.S. 1974. Rosaceae chemotaxonomy and the origins of the Pomoideae. *Botanical Journal of the Linnean Society* 69:239-259.

El-Gazzar, A. 1981. Chromosome numbers and rust susceptibility as taxonomic criteria in Rosaceae. *Plant Systematics and Evolution* 137:23-38.

Gajewski, W. 1957. A cytogenetic study in the genus *Geum* L. *Monographiae Botanicae* 4:1-416.

Gajewski, W. 1959. Evolution in the genus *Geum*. *Evolution* 13:378-388.

Gajewski, W. 1964. The heredity of seed-dispersing mechanisms in *Geum*. *Proceedings XIth International Congress of Genetics* 1964:423-430.

Hess, H.E., E. Landolt and R. Hirzel. 1970. *Flora der Schweiz*. Volume 2. Birkhäuser, Basel, Switzerland.

Huber, H. 1961. Rosaceae. In: *Gustav Hegi Illustrierte Flora von Mitteleuropa*. Volume 4, Part 2. Paul Parey, Berlin.

Hutchinson, J. 1973. *The Families of Flowering Plants*. Clarendon Press, Oxford, England.

Junker, R. 1989. Grundtypkonzept und Mikroevolution in der Schöpfungsforschung. *Praxis der Naturwissenschaften Biologie* 38(8):23-27.

Junker, R. 1993. Prozesse der Artbildung. In: Scherer, S., ed. *Typen des Lebens*, Pascal, Berlin, pp. 31-45.

Korotaeva, E.I. 1983. Fruit structure in some representatives of the subtribe Geinae (Rosaceae). *Botanicheskij Zhurnal* 68:1367-1373.

Král, M. 1966. Die Begrenzung der Gattung *Parageum* Nakai et Hara. *Preslia* 38:151-153.

Lönnig, W.E. 1993. *Artbegriff, Evolution und Schöpfung*. Naturwissenschaftlicher, Cologne, Germany.

Savile, D.B.O. 1979. Fungi as aids in higher plant classification. *Botanical Review* 45:378-503.

Scherer, S. 1993. Basic types of life. In: Scherer, S., ed. *Typen des Lebens*, Pascal, Berlin, pp. 11-30.

Stebbins, G.L. 1984. Polyploidy and the distribution of the arctic-alpine flora: New evidence and new approach. *Botanica Helvetica* 94:1-13.

# 8. Relationships among the Genera and Species of the Pome-Fruited Plants (Rosaceae, Subfamily Maloideae)

HERFRIED KUTZELNIGG

## Abstract

The Maloideae are well separated from other subfamilies of the Rosaceae by the pome fruit, the basic chromosome number $x = 17$ and the exclusive presence of trees and shrubs. They are comprised of about 15-30 genera including some 950 species. Their main distribution is in temperate zones of the northern hemisphere. Recently it has been suggested to consider this group of plants as subtribe Pyrinae, which along with three other genera comprises tribe Pyreae of subfamily Spiraeoideae.

Some genera are still undergoing speciation. Heterobathmy—the co-existence of primitive and advanced characteristics—is observed frequently. The variability of morphological and chemical characters within the subfamily is reviewed, with special emphasis on taxonomic significance.

The combination of characters within the genera does not allow for a taxonomic division of the Maloideae. On the contrary, the various genera are closely related to each other and show a "networking of characters." In addition, intergeneric sexual hybrids provide evidence for relationships between genera that do not seem to be closely related according to classical taxonomic approaches. Intergeneric grafting affinities show similar results.

It is suggested that all genera of the Maloideae have arisen from a genetically polyvalent basic population, without processes of macroevolution. Accordingly, the Maloideae are represented by a phylogenetic shrub, rather than by a phylogenetic tree.

Recent morphological and molecular data place four genera, formerly members of Spiraeoideae, next to the Maloideae, i.e., the woody plants *Kageneckia* ($x = 17$), *Lindleya* ($x = 17$), and *Vauquelinia* ($x = 15$) and even the herbaceous genus *Gillenia* ($x = 9$) which is thought to be an ancestor of the group.

The data known allow for the attribution of the taxonomic rank of a basic type to the Maloideae. It is discussed whether the genera *Kageneckia*, *Lindleya*, and *Vauqueliana*, as well as *Gillenia* could be members of the same basic type.

## 8.1. Introduction

When I was confronted with the basic type concept for the first time in about 1980, I was working on a revision of the pome-fruited or pomaceous plants (Maloideae) for *Hegi's Illustrated Flora of Central Europe*. In this way, I had the great opportunity to simultaneously look at this group of genera from the viewpoint of the basic type biology. It soon became obvious that the Maloideae are a good example, not the least because of their numerous hybrids.

In this work, the more important results and connections will be represented in a way that can be understood even by those readers who are not familiar with the group of plants. Whoever wants to become more familiar with this matter may refer to Schneider (1906), Rehder (1940), Petrides (1973), Krüssmann (1976-1978), Little (1980), Dirr (1998), Elias (2000), Hanelt (2001), Kutzelnigg (1994), and Kalkman (2004). A modern taxonomic revision of the Maloideae has been started by several North American teams (Robertson *et al.* 1991; Phipps *et al.* 1991; Morgan *et al.* 1994; Campbell *et al.* 1995; Evans and Campbell 2002; Campbell *et al.* 2007; Potter *et al.* 2007).

## 8.2. Systematic Position of the Maloideae Within the Family of the Rosaceae

The pome-bearing plants belong to the diversified rose family (Rosaceae), which besides the roses also includes strawberries, raspberries, spiraeas, cherries and almonds, to name a few examples. Altogether there are approximately 100 genera. They have much in common, such as the presence of a hypanthium, a double perianth, usually more than one free carpel, and alternate leaves usually bearing stipules. They are so well separated from other plant families that it is hardly possible to name a family closely related to the Rosaceae (other than the two smaller families Neuradaceae and Chrysobalanaceae).

Within the Rosaceae, there are notable differences, particularly regarding the fruits (Figure 1). For example, spiraea in fruit possesses separated follicles, cinquefoil and avens have numerous free nutlets, and cherry has a single stone fruit. In other cases, such as strawberries and apples, an increased growth of sterile floral tissue leads to the formation of aggregate fruits despite the fact that the carpels themselves are free (at least to a high degree). Accordingly, the systematic arrangement within the Rosaceae is also based primarily on the structure of the fruits. The Rosaceae are usually divided into the 4 subfamilies Rosoideae, Spiraeoideae, Prunoideae (= Amygdaloideae), and Maloideae. While the Rosoideae and Spiraeoideae include both herbs and woody plants, the Prunoideae (stone-fruited plants) and the Maloideae include only woody plants. As far as some vegetative characteristics are concerned (for

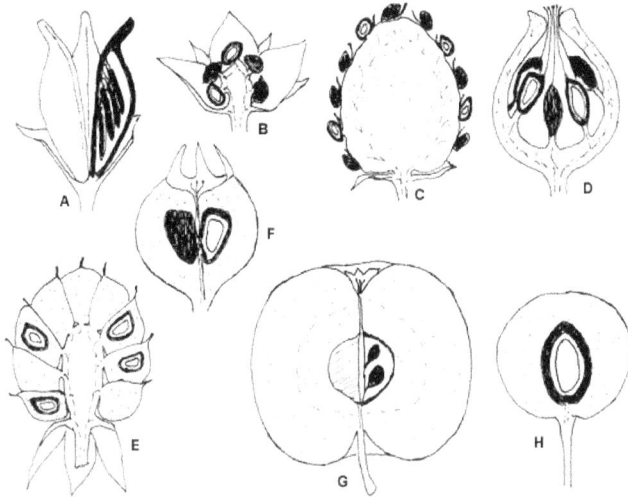

Figure 1. Longitudinal sections of the fruits of some representatives of the Rosaceae (rose family). A Subfamily Spiraeoideae: A *Spiraea*; B-E subfamily Rosoideae: B *Potentilla* (cinquefoil), C *Fragaria* (strawberry), D *Rosa* (rose), E *Rubus* (blackberry); F-G subfamily Maloideae: F *Mespilus* (medlar), G *Malus* (apple); H subfamily Prunoideae: H *Prunus* (cherry). Pulp stippled, hard layers of fruit wall blackened, papery layer in G lightly hatched, seeds unshaded oval shapes except in A and G showing whole seeds, blackened. (modified from Strasburger 1983).

example, the anatomy of the wood and the morphology of the thorns), the two latter subfamilies basically agree; however, notable differences are found, for example, in their flowers and fruits. In the Prunoideae, a stone fruit originates from a superior ovary, which is formed from only one single carpel (Figure 1H). In the Maloideae, the ovary is inferior. The typically five free carpels are indirectly connected through tissue proliferations of the floral cup as the fruit matures, so that an accessory-aggregate fruit, called a pome, is formed (Figure 1 F-G).

In contrast to the heterogenous Rosoideae and Spiraeoideae, the Prunoideae (=Amygdaloideae) and Maloideae in each case are homogeneous and clearly separated from all other genera of the Rosaceae. The Maloideae can be characterized as follows: The ovary is inferior. The fruit is a pome. The basic chromosome number is uniformly 17. They are exclusively woody plants.

The pomaceous plants are usually given the taxonomic rank of a subfamily (Maloideae Weber 1964, *J. Arnold Arbor.* 45:164). Some authors assign to them the rank of family (Malaceae), others only to tribe.

For some time it was unclear whether the genus *Dichotomanthes* (from Southern China) belongs to the Maloideae or – as Gladkova (1969) suggested – should be placed into the independent subfamily Dichotomanthoideae. Considerations that reflect the special position of the genus include the single carpel connected to the floral cup only in

| | Number of Species | Carpels becoming stony | Number of ovules | Ovules with only 1 integument | Number of carpels | Carpels with false partition | Carpels free above | Pinnate leaves | Leaves evergreen | Only shrubs (no trees) | Thorns | Tetraploid groups exist | Tendency of apomixis | Dihydrochalcones occur | Flavone glycosides occur | Prunasin (cyanogen) occurs |
|---|---|---|---|---|---|---|---|---|---|---|---|---|---|---|---|---|
| Cydonia (quince) | 1 | – | 20-24 | – | 5 | – | – | – | – | – | – | – | – | – | – | + |
| Docynia | 2-6 | – | 3-10 | – | 5 | – | – | – | (+) | – | – | | – | + | – | |
| Chaenomeles (flowering quince) | 3 | – | 20-24 | – | 5 | – | – | – | – | + | + | ● | | – | – | + |
| Pseudocydonia | 1 | – | 20-24 | – | 5 | – | – | – | (+) | – | – | – | – | – | – | |
| Pyrus (pear) | 20-30 | ● | 2 | – | 2-5 | – | – | – | (+) | – | (+) | ● | (+) | – | – | – |
| Malus (apple) | 50 | – | 1-6 | – | 2-5 | (+) | – | – | (+) | – | (+) | + | (+) | + | – | – |
| Sorbus s.l. (mountain ash) | 200 | ● | (1)-2 | – | 2-5 | – | (+) | (+) | – | – | – | + | + | – | (+) | + |
| Micromeles (dwarf apple) | 25 | – | 2 | – | 2-5 | – | – | – | – | – | – | | – | + | | |
| Aronia (chokeberry) | 3 | – | 2 | – | 5 | – | – | – | – | + | – | | – | + | | |
| Stranvaesia (laurel medlar) | 5-14 | – | 2 | – | 5 | – | (+) | – | (+) | – | – | | – | – | | |
| Photinia (shiny medlar) | 40-60 | ● | 2 | – | 2-5 | – | (+) | – | (+) | – | (+) | (+) | | – | – | + |
| Rhaphiolepis (grape apple) | 15 | – | 2 | (+) | 2 | – | – | – | + | + | – | | – | – | – | |
| Eriobotrya (loquat) | 10-15 | – | 2 | – | 2-5 | – | (+) | – | + | + | – | | – | – | – | |
| Amelanchier (serviceberry) | 6-10 | – | 2 | – | 5 | + | ● | – | – | – | – | + | (+) | | – | + |
| Peraphyllum (sqaw apple) | 1 | – | 2 | – | 2-4 | + | – | – | – | + | – | | – | – | | |
| Malacomeles (false serviceberry) | 2 | – | 3 | – | 3-5 | + | – | – | + | + | – | | – | + | | |
| Cotoneaster (cotoneaster) | 100-180 | + | 2 | – | 2-5 | – | (+) | – | (+) | – | – | + | + | – | – | + |
| Chamaemeles | 1 | + | 2 | (+) | 1-2 | – | – | – | + | – | – | | – | + | – | |
| Pyracantha (firethorn) | 6-10 | + | 2 | – | 5 | – | ● | – | + | + | + | – | | – | + | – |
| Hesperomeles (Audean apple) | 15 | + | 1 | – | 5 | – | – | – | + | – | (+) | | – | + | | |
| Mespilus (medlar) | 1-2 | + | 2 | – | 4-5 | – | – | – | – | (+) | – | | – | – | – | |
| Crataegus (hawthorn) | 200-240 | + | 2 | – | 1-5 | – | – | – | – | (+) | + | (+) | – | | + | (+) |
| Osteomeles (stone apple) | 4-5 | + | 1 | – | 5 | – | – | + | – | – | – | | – | + | | |
| Dichotomanthes | 1 | + | 2 | – | 1 | – | + | – | – | – | – | – | – | – | – | + |

Table 1. Distribution of selected characteristics in the genera of the Maloideae. + = applies, (+) = applies in part, ● = rudimentary, – = does not apply, empty field = result still open. The delimitation line marks the border of the classic division into the tribus Maleae (*Cydonia* to *Malacomeles*) and Crataegeae (*Cotoneaster* to *Dichotomanthes*).

the lower part and the style is attached laterally. In accordance with Robertson *et al.* (1991) Zhang and Baas (1992), and Campbell *et al.* (2007), *Dichotomanthes* is submerged in the Maloideae.

After the publication of the German edition of this paper the internal taxonomy of Rosaceae has been revised (Potter *et al.* 2007). The authors recognize three subfamilies. All genera previously assigned to Maloideae and Prunoideae are included in Spiraeoideae, the former as subtribe Pyrinae of tribe Pyreae, the latter as tribe Amygdaleae. Subfamily Rosoideae is divided into Rosoideae *s.str.* and a small new subfamily Dryaoideae.

Figure 2. above left: domestic apple (*Malus domestica*) blooming branch; above right: snowy mespilus (*Amelanchier ovalis*) blooming branch; below left: oneseed hawthorn (*Crataegus monogyna*), fruit-bearing branch; below right: domestic pear (*Pyrus communis*), appearance of a blooming tree. Photos: R. Wiskin.

## 8.3. Number of Genera and Species, and their Distribution

The Maloideae include about 950 species, the numbers ranging between 200 and 2000 species, depending on the particular author. They are divided into some 15 to 30 genera. The different figures reflect the different opinions about the delimitation of the species and genera. Currently, apart from smaller deviations, the Maloideae essentially include the 24 genera listed in Table 1. (In the modern treatment of Potter et al. (2007) there are only few changes. The authors split *Eriolobus* from *Malus*; *Heteromeles* from *Photinia*; *Aria*, *Chamaemespilus*, *Cormus* and *Torminalis* from *Sorbus*, but include *Micromeles* in *Aria*). The number of species and the common names (if available) can be found in this table. Figure 2 shows some typical representatives.

By far the most Maloideae exist in the northern hemisphere, mainly in the temperate zone and higher latitudes, with a definite center of distribution in Asia, followed by Europe and North America. Only two genera reach the southern hemisphere (*Photinia* in Java, *Hesperomeles* in South America). Almost all genera, often with multiple species,

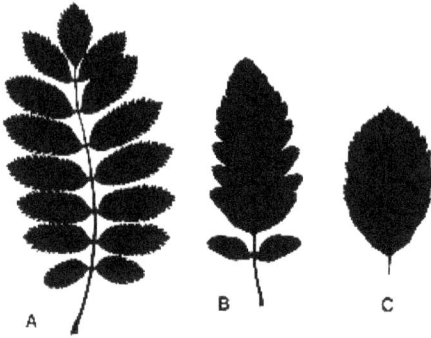

Figure 3. Photocopies of the leaves of the mountain ash (*Sorbus aucuparia* L.) (A), whitebeam (*Sorbus aria* (L.) Crantz) (C) and their hybrid, the Thuringian hybrid mountain ash (*Sorbus* × *pinnatifida* (Smith) Düll) (B).

are grown as decorative plants. Examples are the cotoneasters, ornamental quinces (*Chaenomeles*), firethorn (*Pyracantha*), ornamental and crab apples (*Malus*), and serviceberries (*Amelanchier*) (see Rehder 1940; Petrides 1973; Krüssmann 1976-1978; Dirr 1998; Elias 2000). Several species have considerable economic importance as fruit trees, especially the commercial apple (*Malus domestica*), pear (*Pyrus communis*), and quince (*Cydonia oblonga*), but also foreign species (see Hanelt 2001), as for example the loquat (*Eriobotrya japonica* (Thunb.) Lindl.).

## 8.4. Specification Processes in the Maloideae

In some Maloideae, the process of speciation seems to have already been completed. This is true for genera like *Chamaemeles, Cydonia,* or *Mespilus,* as well as *Sorbus* subgenus *Cormus* and subgenus *Chamaemespilus*. In these cases one speaks of phylogenetically old species.

On the other hand, there are genera like *Malus, Pyrus, Cotoneaster,* and *Crataegus,* as well as *Sorbus* subgenus *Aria* and subgenus *Aucuparia,* where intense speciation processes are apparently still taking place. They are regarded as phylogenetically young. The species are often highly variable and are usually connected with related species through hybridization. Since the hybrids are normally fully fertile, all possible intermediate forms and an intensive exchange of genetic material occur in the overlapping areas. For taxonomists this has always caused considerable problems (and disagreements) as far as the delimitation of the individual species is concerned because often whole species groups are connected in this way. It is remarkable that certain authors, after they had proposed or acknowledged a certain number of species, expressed the opinion that all the species of a particular genus could just as well be regarded as one single species (*e.g.,* Klotz 1982 for *Cotoneaster*).

With *Crataegus,* the special situation in North America resulted in the description of over 1000 species, most of which have since been dismissed. Today, only approximately 200 species are recognized.

In the above-mentioned genera, a common process may be differentiation through selection and isolation, combined with partial genetic exchange through hybridization. With *Crataegus*, this is the main mechanism of microevolution.

Additionally in *Cotoneaster*, apomixis (asexual reproduction through seeds) is common, producing genetically fixed populations. In this case, there is no genetic recombination (due to the absence of sexuality) so the descendants are rather homogeneous and therefore taxonomically well demarcated. This phenomenon has become widely known through the genus *Rubus* (blackberries) of the subfamily Rosoideae. Since sexual reproduction is usually not completely repressed, there are still possibilities of adaptation within certain limits. Incidentally, apomixis is usually combined with polyploidy, in most cases, probably allopolyploidy.

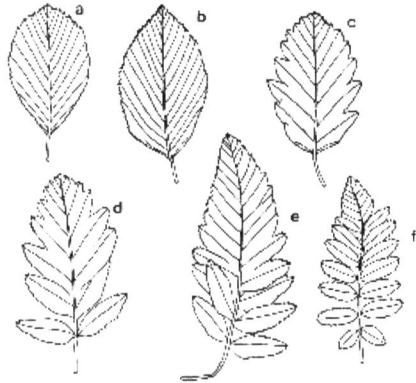

Figure 4. Leaves of the whitebeam (*Sorbus aria* (L.) Crantz) (a, b) and of its hybrids with the mountain ash (*Sorbus aucuparia* L.): c = *Sorbus pseudothuringiaca* Düll, constant hybrid apomict; d, e, f = variants (nothomorphs) of the primary sexual hybrid *Sorbus* × *pinnatifida* (Smith) Düll. (After Düll 1961)

Such processes are also partially detected in *Sorbus*. In this case, however, an additional process can also take place (see Düll 1961). Species, which developed in such different ways that they were placed into different subgenera under the current system, have in many cases apparently met again and produced hybrids. Figure 3 shows an example of this. The phenomen of wide hybridization is called reticulate evolution and has turned out to be a rather common event in microevolution.

Since the parents are not pure breeds, heterozygosis can generate different phenotypes (see Figure 4 d, e, f). Being sterile, the primary hybrids rarely persist in nature. Crucial for the case concerned, apomixis has allowed some of the well-adapted phenotypes to reproduce. In *Sorbus* a whole assortment of such apomictic forms behave as independent species with their own (though often small) areas of distribution. This shows that genetic exchange between species, already developed in different directions, in certain situations offers selective advantages over the parental species.

## 8.5. Variability and Distribution of Selected Characteristics

In view of a group with so many species like the pome-bearing plants the question arises as to what extent a taxonomic system is even possible. That is, it has to be determined whether there are any characteristics that can be regarded as systematic criteria. The systematic position in question here is between the level of subfamily and genus, *i.e.*, tribe. In order to identify taxonomically suitable criteria, it is necessary to analyze the extent of variation and the distribution of these characteristics in the various genera. This will be done by means of selected examples. Table 1 contains an overview.

**Growth form** can be shrub-shaped or tree-shaped. Some genera grow only as shrubs. For taxonomic purposes, this criterion then is only of subordinate importance since, on one hand, the growth within the same species, depending on the location, can be either shrub- or tree-shaped and, on the other hand, the majority of genera include both trees and shrubs. As extreme adaptations, one can even find prostrate dwarf shrubs and espalier-like shrubs. Exotic ornamental species often grow differently in their transplanted area than in their home country, and different strains can differ from each other in their growth.

Shoots developed as **thorns** are characteristic of several genera, while others lack them completely or partially. The taxonomic usefulness of this trait is limited because thorny and thornless plants may occur within the same species (e.g., the medlar, *Mespilus germanica*).

The **leaf blade** is usually undivided (with a straight or lobed edge); there are, however, two genera with pinnate leaves. Despite the visibility and simplicity of this trait, it is taxonomically far less significant than one might assume. The genus *Sorbus* for example includes representatives with undivided as well as pinnate leaves (see Figure 3).

As far as the **lifespan of the leaves** is concerned, most species are deciduous, rarely also persistent or evergreen (e.g., the firethorn [*Pyracantha*]), but again there are transitional forms, influences of local conditions, as well as coexisting variants within the same species or genus (e.g., cotoneaster).

Among the reproductive characteristics, most authors until about 1980 ascribe high taxonomic value to the nature of the **carpel wall** of the individual fruits at the time of maturity: It is either parchment-like (e.g., the apple; Figure 1 G) or gristly (pear) or stony (e.g., the medlar; Figure 1 F). The fruits of the latter type are called "stone apples" [translation of German *Steinäpfel*], the fruits of the former "core apples" [translation of German *Kernäpfel*]. There are intermediate forms in *Photinia* as well as partial transitions in certain species of the genera *Pyrus* (pear) and *Sorbus*, in which the stone cells of the pulp accumulate in a bony endocarp-like fashion in proximity of the carpels.

The **number of the carpels** is usually five. However, lower numbers also occur, often several different numbers in the same genus, (e.g., *Crataegus* [hawthorns]). Only rarely do the carpels possess false septa, namely in *Amelanchier* (serviceberries), *Peraphyllum* (squaw apples), and *Malacomeles* (false serviceberries). As an exception, they can also be found in the genus *Malus* (apple and crab-apples). The pulp usually surrounds the carpels to their tops; however, there are also species in which the carpel tops are exposed.

There are usually two **ovules** in the carpels (Figure 1 G). Among the species with core apples, *Cydonia* (quinces) and related genera possess numerous (20-24) ovules, and *Docynia* possesses ten. *Malus* usually has two per carpel but can occasionally possess up to six. The number of ovular integuments, which become the seed coat, is usually two, while *Rhaphiolepis* and *Chamameles* possess only one integument.

The **pollen grains** are all very similar and are hard to distinguish from those of other Rosaceae. They are normally tricolporate, possessing three longitudinal furrows with one germination pore each. A certain variability can be observed in the exine (structure of the surface of the pollen), but the intraspecific variability can be greater (e.g., among different cultivars) than the intergeneric variability. A smooth pollen surface can be found in genera as different as *Cydonia, Chaenomeles,* and *Cotoneaster.* In *Sorbus,* some species have a smooth pollen surface, others a sculptured exine.

The **chromosome complement** is usually diploid. In several genera, there are, beside diploid, also tetraploid (more rarely triploid or pentaploid) species and even individuals (cytotypes). An especially high percentage of polyploid species can be found in the genera *Cotoneaster, Crataegus,* and *Sorbus.*

The tendency towards **apomixis**—*i.e.,* asexual (parthenogenetic) reproduction without meiosis—varies. In *Sorbus* and *Cotoneaster,* it is rather high (see Jankun and Kovanda 1987). In *Crataegus* on the other hand—at least in the Central European species—it is rare and could be observed only in *Pyrus* after artificial induction. In *Malus* there are species with and without apomictic tendencies.

**Cyanogenic substances**—which emit hydrocyanic acid when the plant is injured—were detected in the seeds of almost all genera (exceptions: *Mespilus, Pyrus,* and *Rhaphiolepis*). Their presence in the remaining plant parts is variable. Their quantity can fluctuate strongly, depending on environmental conditions, and can occasionally be missing completely.

Among **phenolic substances**, two groups are of special taxonomic interest (see Challice 1981): 1. Dihydrochalcone (phloridzine) can only be found in *Malus* and *Docynia.* 2. Flavon-C-glycosides—*i.e.,* substances with a C-C-linkage instead of an oxygen-linkage were found in several

| | Cydonia | Docynia | Chaenomeles | Pseudocydonia | Pyrus | Malus | Sorbus | Micromeles | Aronia | Stranvaesia | Rhaphiolepis | Photinia | Eriobotrya | Amelanchier | Peraphyllum | Malacomeles | Cotoneaster | Chamaemeles | Pyracantha | Hesperomeles | Mespilus | Crataegus | Osteomeles | Dichotomanthes |
|---|---|---|---|---|---|---|---|---|---|---|---|---|---|---|---|---|---|---|---|---|---|---|---|---|
| Cydonia | | | P | | X | S | P | S | | | | | | P | | | | | | | | P | | |
| Docynia | | | | | | P | | | | | | | | | | | | | | | | | | |
| Chaenomeles | P | | | | P | P | | | | | | | | | | | | | | | | P | | |
| Pseudocydonia | | | | | | | | | | | | | | | | | | | | | | | | |
| Pyrus | X | | P | | | | X | X | S | | | | | P | | | | | | | | P | X | |
| Malus | S | P | P | | X | | | X | S | P | | P | | | | | P | | | | | P | P | |
| Sorbus | P | | | | X | X | | S | S | | | | | | | X | S | | | S | | S | X | |
| Micromeles | S | | | | S | S | S | | | | | | | | | | | | | | | | | |
| Aronia | | | | | | P | S | | | | | | | | | | | | | | | P | | |
| Stranvaesia | | | | | | | | | | | | | | | | | S | | | | | | | |
| Rhaphiolepis | | | | | | | | | | | | S | | | | | | | | | | | | |
| Photinia | | | | | | P | | | | | | | | | | | | | | | | | | |
| Eriobotrya | P | | | | | | | | | | S | | | | | | | | | | | | | |
| Amelanchier | | | | | | P | X | | | | | | | | | | | | | | | P | | |
| Peraphyllum | | | | | | | | | | | | | | | | | | | | | | | | |
| Malacomeles | | | | | | | | | | | | | | | | | | | | | | | | |
| Cotoneaster | | | | | | P | S | | | | | | | | | | | S | | | | P | | |
| Chamaemeles | | | | | | | | | | | | | | | | | | | | | | | | |
| Pyracantha | | | | | | | S | | | S | | | | | | | | | | | | | S | |
| Hesperomeles | | | | | | | | | | | | | | | | | | | | | | | | |
| Mespilus | | | | | | P | P | S | | | | | | P | | | | | | | | | | |
| Crataegus | P | | P | | | X | P | X | P | | | | | | | | P | | X | | | | | |
| Osteomeles | | | | | | | | | | | | | | | | | | | | S | | | | |
| Dichotomanthes | | | | | | | | | | | | | | | | | | | | | | | | |

Table 2. Sexual generic hybrids (S) and intergeneric grafting affinities (P) within the Maloideae, as far as currently known. X = both cases documented. (According to Weber 1964; Browicz 1970; Vitkovskiy and Korovina 1983 (for *Micromeles*); Robertson *et al.* 1991; Hanelt 2001; Coombes *et al.* 2007 (for *Eriobotrya*); and others). The delimitation line marks the boundary of the classic division into the tribus Maleae (*Cydonia* to *Malacomeles*) and Crataegeae (*Cotoneaster* to *Dichotomanthes*).

genera. In *Sorbus* there are groups of species with and without these substances. For the *Sorbus aria*-group, species include both individuals possessing flavon-C-glycosides and those lacking it, depending on their geographical origin.

In summary, it can be said that, based on the distribution of characters, there is no reason to divide the Maloideae into two or more tribes. Many characteristics show a high degree of variability within species and genera. Plus, different combinations of two or more characteristics often lead to contradictory arrangements. These observations indicate multiple connections between the different genera. One could describe this situation with the term "network of characters".

### 8.6. Conspicuous Character Combinations in Different Species and Genera of the Maloideae

Some characteristics of the above-mentioned network of genera are quite impressive, when one looks at their distribution across the subfamily. They often appear in diverse species from seemingly dissimilar genera. Some examples will illustrate this:

Various North American species of hawthorn are strikingly similar to certain central European mountain ashes, not only in their leaves, but

Chamaemeles

Cotoneaster

Crataegus

Dichotomanthes

Hesperomeles

Mespilus

Osteomeles

Pyracantha

Stranvaesia

Sorbus

Rhaphiolepis

Pyrus

Pseudocydonia

Photinia

Peraphyllum

Micromeles

Malus

Malacomeles

Eriobotrya

Docynia

Cydonia

Chaenomeles

Aronia

Amelanchier

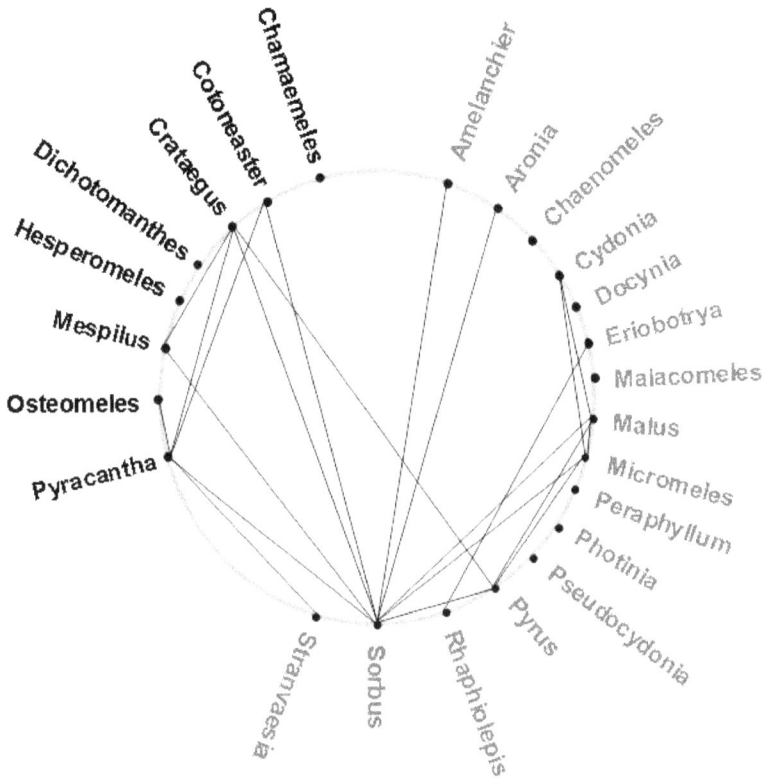

Figure 5. Intergeneric hybrids in Rosaceae, subfamily Maloideae. Left side: tribe Crataegeae, right side: tribe Maleae. For details see table 2.

also in the appearance of the inflorescence. A certain species of *Aronia* has leaves like the domestic pear. The hybrid × *Malosorbus florentina* (Zucc.) Browicz (*Malus sylvestris* (European crab apple) × *Sorbus torminalis* (wild service tree)) from the Mediterranean has leaves like the native European oneseed hawthorn (*Crataegus monogyna*). *Malus trilobata* (Poir.) Schneider has ones like *S. torminalis*. The fruits of *Sorbus domestica* L. (true service tree) from certain locations at first glance look like small domestic apples, and the bark of older *Sorbus domestica* looks like that of the pear tree.

Burgerstein determined as early as 1895 and 1896 that *Crataegus* and *Malus* closely resemble each other in their wood anatomy though divergent otherwise.

One can say about the monotypic genus *Chamaemeles* that the growth habit is like the flowering quince, the leaves remind one of firethorn, and the fruits of hawthorn. Otherwise the plant is close to *Cotoneaster*.

The monotypic genus *Pseudocydonia* from China possesses strong similarities with both *Cydonia* and *Chaenomeles*, as well as with *Pyrus*. However, it cannot be assigned to any of these genera (Weber 1964). *Pseudocydonia* therefore presents a taxonomic problem and is listed by different authors under different genera.

**8.6.1. Heterobathmy.** In phylogenetic research, the term heterobathmy refers to the simultaneous occurrence of original and advanced characteristics. It is assumed that some characteristics remained unchanged during the evolution, while others differentiated strongly.

The Rosaceae as a whole have long been regarded as an example of heterobathmy. The free carpels, for example, are regarded as primitive, whereas the above-mentioned feature of ongoing speciation is considered to be an advanced trait.

Within the Maloideae, heterobathmy is widespread. However, evaluation of heterobathmy depends on whether traits are considered to be primitive or advanced. Noteworthy opinions are divergent. Kalkman (1988), for example, regards more than two ovules per carpel in the Rosaceae as primitive and two or one as advanced. At the same time, he admits that others are of the opposite opinion. The same is true for leaf pinnation: according to Kalkman, it is ancestral; according to Robertson *et al.* (1991), it is an advanced condition. However one decides, numerous cases of heterobathmy always result. Thus, the genera of the Maloideae with numerous ovules would be primitive in the sense of Kalkman's definition, but regarding the undivided leaves they would be advanced (see Table 1). If, on the other hand, one considers the pinnate leaves to be an advanced trait, it is an astonishing fact that the genus *Osteomeles* (with pinnate leaves) possesses five carpels, which is universally regarded as an ancestral trait. Many other examples could be cited at this point.

Gabrieljan (1978), after intensive studies on the taxonomy of *Sorbus,* states that heterobathmy in this genus is frequent, and she writes that this complicates the systematic and phylogenetic studies. This is certainly true for other genera as well.

**8.7. Sexual Generic Hybrids**

An important indication of close relationship is the occurrence of hybrids. The genetic make-up of both parents must harmonize in the hybrids, so that the complex events of the individual development and the formation of characteristic traits can take place (Scherer 1993, this volume). In the pome-fruited plants, numerous hybrids which cross the generic boundaries are known. The results are summarized in Table 2 and Figure 5.

Among the 24 genera, 15 are connected through hybridization according to current knowledge, corresponding to 21 intergeneric hybrid combinations. Vitkovskiy and Korovina (1983) list some more combinations, but they are excluded from Table 2 and Figure 5 because they are not fully documented. The fact is that no extensive crossbreeding experiments have been performed up to now, which means that the observed hybrids are largely the products of coincidence. The high number of hybridizations is already an amazing result. The pome-bearers are quoted again and again as one of the classic examples of hybridization between genera— not the least because of their popularity.

A notable fact about these hybridizations is that genera that, according to some criteria, would be regarded as distant were crossbred many times. The gaps in the crossbreeding matrix are not very serious because many cases involve genera that are combined anyway by some authors (e.g., *Docynia* and *Pseudocydonia* into *Cydonia*).

Figure 6. The sexual intergeneric hybrid × *Sorbocotoneaster pozdnjakovii* Pojark., which originated from the crossbreeding of *Cotoneaster melanocarpus* Lodd. and *Sorbus aucuparia* L. subsp. *sibirica* (Hedl.) Krylov. Branch of a *Sorbus*-like plant. (Drawing by R. Sanders).

The most popular and more important genera are all directly or indirectly connected through hybridization (single exception: *Chaenomeles*, Japanese flowering quince).

There are, for example, common descendants of the whitebeam (*Sorbus aria*) and pear (*Pyrus communis*) or from a cotoneaster and the Sibirian mountain ash. The latter case (Figure 6)—as also the above-mentioned subgeneric hybrids of *Sorbus*—is an impressive example of what an intermediate between small, round, smooth-edged leaves, and large, pinnate leaves would look like.

The observed generic hybrids are usually viable, and often they are even able to reproduce by means of their own seeds—a situation one would not necessarily expect in generic hybrids.

As far as the fertility of the hybrids is concerned, there are apparently three possibilities:

1. Meiosis (division of the chromosomes) is undisturbed, and the seed set is good. Until now, this situation has been described expressly only in the hybrids between *Sorbus* and *Aronia*.

2. The hybrids are largely sterile, due to difficulty of chromosome pairing during meiosis. The fruits usually still develop, but only a few or no seeds are produced capable of

germinating. One example of this situation has been known for approximately 400 years: × *Sorbopyrus auricularis* (Knoop) Schneider (*Pyrus communis* (L.) Crantz × *Sorbus aria* (L.) Crantz), which since then has been propogated vegetatively. Accordingly, the fruits are intermediate between those of pear and whitebeam. The plant is known under the name "Bollwyller Pear" and has long been used as a fruit tree.

3. The problems of meiosis are avoided by apomixis. The seed set is good. In these cases, the genetic make-up of the hybrid can be transmitted for many generations.

With respect to the third possibility, three spontaneously formed combinations should be mentioned, whose descendants became naturalized, dispersed, and integrated into the spontaneous vegetation, thus, behaving as independent species of hybrid origin. The one that has been known for the longest time is × *Malosorbus florentina* (Zucc.) Browicz (*Malus sylvestris* Mill. × *Sorbus torminalis* (L.) Crantz), from the Mediterranean (Browicz 1970). Although its hybrid nature was recognized rather early, the plant even today is often listed as a species of the genus *Malus* (under the name *Malus florentina* (Zucc.) Schneider). Further to be mentioned are: × *Amelasorbus jackii* Rehder (*Amelanchier florida* Lindl. × *Sorbus scopulina* Greene) in North America, and × *Sorbocotoneaster pozdnjakovii* Pojarkova (*Cotoneaster melanocarpus* Lodd. × *Sorbus aucuparia* L. subsp. *sibirica* (Hedl.) Krylov) in Siberia (Figure 6).

Successful crosses with representatives of other subfamilies of the Rosaceae have never been observed.

The subgenera of *Sorbus* should again be mentioned. They are more different from each other than is the case with other subgenera of the Maloideae. Their difference is reflected in the fact that such different names as "mountain ash" or "rowan" (*Sorbus*) (in the narrow sense), "whitebeam," (*Aria*), "service tree" (*Cormus*) and "false medlar" (*Chamaemespilus*) are being used for them. In the course of time, the subgenera of *Sorbus* have been assigned repeatedly to the level of genera. At the moment it looks like this concept will become generally accepted in the future (Robertson *et al.* 1991; Phipps *et al.* 1991, Campbell *et al.* 2007, Potter *et al.* 2007). One of the consequences of such a change would be an elevated number of known intergeneric hybrids (*Aria* × *Aronia*, *Aria* × *Chamaemespilus*, *Aria* × *Pyrus*, *Aria* × *Sorbus*, *Aria* × *Torminalis*).

## 8.8. Intergeneric Grafting Affinities

In the context of systematic relationships, the phenomenon of grafting is of particular interest. Genetically different tissues are hereby united in one and the same individual, so that, for example, the "stock" (root + base of shoot) may come from the hawthorn, whereas the upper part of the plant ("scion") may come, for example, from the medlar. Even if there are groups of plants with a high tolerance, as far as the stock is concerned (e.g., beans and sunflower; see Kollmann and Glockmann 1985), in

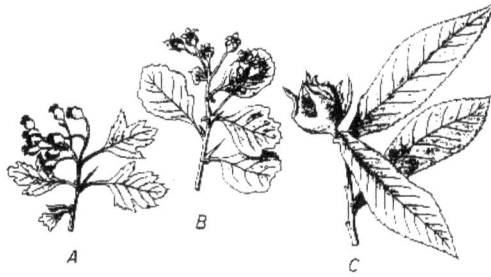

Figure 7. The grafting hybrid +*Crataegomespilus dardari* Bellair `Jules d' Asnière´ (B) between the parents: oneseed hawthorn (*Crataegus monogyna* Jacq.) (A) and medlar (*Mespilus germanica* L.) (C). (After Baur 1910).

most cases grafting succeeds only to a very limited extent. This is true also for the Maloideae. Apparently, a certain agreement between the regulative systems of the two partners must exist. How intense such interrelations can be is seen by the fact that the grafting stock often influences the characteristics of the graft. It has been known for a long time that pears are especially aromatic if they come from plants that were grafted on a stock of quinces.

Since grafting plays an important role in breeding and since the pomaceous plants include many commercial strains, numerous grafting combinations have been attempted (see Weber 1964; Hanelt 2001).

The corresponding data are listed in Table 2 as a supplement to the sexual generic hybrids. It should also be noted that successful grafting in both directions is an exception (until now, only the connection of the hawthorn and pear has succeeded, whereas an apple cultivar, for example, becomes united with the grafting partner only if another apple is used as a stock).

Some grafting combinations form especially close relationships. At the graft union the outer layers of the apical meristem of one partner grow over the inner layers of the other partner. This is called a periclinal chimera. As the bud grows, it produces the actual grafting hybrid. A good example is the grafting hybrid pictured in Figure 7 between its parents, which originated from a union of hawthorn (*Crataegus monogyna* Jacq.) and medlar (*Mespilus germanica* L.). The outermost layer of the apical meristem (and with it the epidermis) is medlar tissue; the remaining layers come from the hawthorn. The hybrid carries the name +*Crataegomespilus dardari* Bellair `Jules d'Asnières.' This plant has been known since 1898 and has been reproduced over and over through cuttings and can still be admired in some arboretums even today.

Beyond the boundaries of the pomaceous plants—e.g., with cherries or plums as representatives of the stone-fruited plants—grafting has never succeeded.

## 8.9. The Maloideae—A Systematically Uniform Group

Taxonomic results and hybridization compatibility are strong arguments to regard the Maloideae (despite their diversity) as a uniform systematic group with closely connected species and genera.

For a long time the division into the two tribes Maleae and Crataegeae proposed by Koehne (1890) was followed (e.g., it was used in Engler's syllabus of plant families [Schulze-Menz 1964] and by Kalkman 2004). The taxonomic criterion is the nature of the carpel walls at the time of fruit maturity. In the Maleae, it is parchment-like, in the Crataegeae stony. However, both the above-mentioned transitions as well as the incompatibility in the remaining characteristics speak against this seemingly clear division.

It is interesting that although Robertson and co-workers at first retained this tribal division, they later abandoned it (Robertson et al. 1991: "At the beginning of our research we saw little reason to question the recognition of the two tribes... However, more recent analysis indicates that the use of the single character of core texture is artificial, and tribes are not distinguished in this paper.") As we saw, other taxonomic criteria cannot be used for a meaningful system of the Maloideae. If one would use combinations of characters (as is often successfully done in systematics) one would arrive at insurmountable contradictions within the Maloideae (see Table 1).

Among others, it was Challice (1981) who found himself confronted with the incompatibility of different taxonomic criteria after he and his co-workers had spent years performing chemotaxonomic experiments in the Maloideae. He aptly says: "Apparently the chemotaxonomical data are inconsistent with classical methods and any compromise appears impossible." One could apply this statement to other taxonomic data as well. Kovanda (1965), for example, writes in a similar sense: "... the system of genera based on morphological, anatomical and distributional characters is contradictory to the results of the cytological and genetic approach."

The close relationship of the genera of the Maloideae is also reflected in statements of authors who investigated various aspects of the subfamily. For example, "... it is merely a question of discretion, whether one assigns the totality of all pomaceous plants to several genera, or to only one genus" (see Burgerstein 1896; Sax 1931; Weber 1964). Similarly, one could easily regard all species of *Cotoneaster* as a single species (Klotz 1982).

Last, but not least, the numerous intergeneric hybrids also indicate the reticulate relationships among genera. Although one cannot conclude from the occurrence of a common hybrid that the respective genera are more closely related with each other than other genera (Robertson et al. 1991), it is certainly justifiable to consider the occurrence of many hybrids as an indication of a very close relationship among all the Maloideae. This is further reinforced by the fact that there are no hybrids beyond the boundary of the subfamily.

## 8.10. The Evolution of the Maloideae

An attempt will be made to draw a hypothetical picture of the evolution of the Maloideae, based on the different aspects of their taxonomy.

It looks like all present-day forms of this group can be traced back to a common ancestral population. This population was genetically polyvalent, that is, it essentially contained the genetic amalgam of all the characters observed today, including their various alleles. The amalgam, which, for example, resulted in the development of thorns and the synthesis of the cyanogenous prunasin, apparently already existed. The characters did not necessarily have to be expressed, since they might have been covered by recessivity, or certain conditions of regulatory genes might have prevented their realization.

According to this idea, the different genera (perhaps also subgenera or sections) would have derived directly from the ancestral population via different combinations of genes and genetic complexes of the same gene pool. Indeed, all observations confirm that, from the time of hypothetical ancestral population or ancestral form until the formation of present-day groups, no macroevolutionary processes took place but that the variation within the original characteristics was sufficient.

One could also say it like this: The phylogenetic relationships of the Maloideae do not look like a tree but like a shrub. The differentiation of these relationships can be explained exclusively by microevolutionary processes.

One of the advantages of this hypothesis is that it can easily explain some phenomena of the taxonomy of the Maloideae that are difficulties for the evolutionary tree model.

The disjunct occurrence of rare characteristics (see Table 1), for example, would require a high frequency of independent and, therefore, convergent origins of similar structures if one follows the evolutionary tree model. It is hard to conceive of mechanisms capable of doing this.

The occurrence and existence of generic hybrids and the cooperation among the totality of genes suggested by such hybrids gives little room for independent development of genes or genetic complexes.

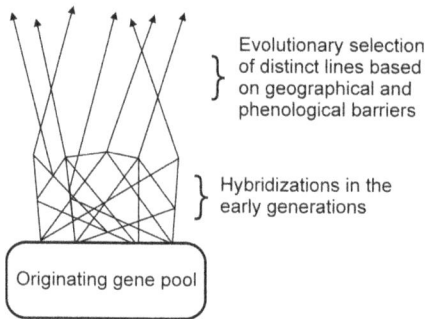

Evolutionary selection of distinct lines based on geographical and phenological barriers

Hybridizations in the early generations

Originating gene pool

Figure 8. Model of the evolution of the Maloideae (after Phipps *et al.* 1991).

Taxonomically difficult cases like the genus *Pseudocydonia*, which combines the qualities of several genera, fit poorly into the branches of an evolutionary tree but can easily be explained through an immediate descent from a polyvalent ancestral population.

The evaluation of qualities as "primitive" or "advanced" – as required by the evolutionary tree model – leads, as we have seen, to considerable problems. Depending on the particular author, the assigned polarities can be directly opposite. Plus, it is difficult to find an explanation why so many Maloideae simultaneously possess "primitive" and "advanced" traits (heterobathmy). If one assumes that pairs of traits like "leaves non-pinnate vs. pinnate" existed already in the ancestral population of the Maloideae side by side (polarity-neutral), the above-mentioned problems are solved.

After the completion of the first draft of the 1993 German editon of this paper, Phipps *et al.* (1991) published a graphic display of the evolution of the Maloideae (Figure 8) that essentially confirms the concept presented here. According to the opinion of these authors, the original gene pool can be traced back to alloploidy, a hypothesis which will be discussed later. The close relationships among the Maloideae and their polytomous (shrub-like) evolution was meanwhile also pointed out by Zhang (1992) on the basis of anatomical studies of the wood. In addition, molecular approaches of recent years (Morgan *et al.* 1994; Campbell *et al.* 1995; Evans and Campbell 2002; Campbell *et al.* 2007; Potter *et al.* 2007) confirm the uniformity of the group. The internal branches of the phylogenetic trees are very short indicating near connections of the genera.

## 8.11. The Pomaceous Plants as a Basic Type

In the introduction it was said that the Maloideae present a good example of a basic type. Throughout this paper, this has been demonstrated again and again with different lines of evidence, without expressly referring to it. Therefore, the results will once again be presented under the aspect of a basic type concept:

The subfamily of the pome-bearing plants can be understood as a basic type in the sense of Marsh (1976) since it complies with all necessary conditions: 1) It is clearly separated from other taxa, in this case from the other subfamilies (or tribes) of the rose family. 2) It represents a systematically uniform group despite its diversity. 3) There is a high degree of hybridization within the Maloideae, both between species and between genera. 4) Hybrids with representatives of other subfamilies (or tribes) are unknown. 5) The hybrids, including the intergeneric hybrids, always show qualities of both parents. They usually make it beyond the embryonic stage and in many cases even grow into vigorous plants.

## 8.12. The Taxonomic Position of *Kageneckia, Vauquelinia, Lindleya* and *Gillenia,* Formerly Placed in Spiraeoideae: Are They Members of the Basic Type?

Beginning with cytotaxonomic studies of Goldblatt (1976) it became more and more evident by morphological and molecular studies that the woody American genera *Kageneckia* (x = 17), *Vauquelinia* (x = 15) and *Lindleya* (x = 17), formerly placed in subfamily Spiraeoideae (bridlewreath subfamily), show close affinities to Maloideae (Morgan *et al.* 1994; Campbell *et al.* 1995; Evans and Campbell 2002; Campbell *et al.* 2007; Potter *et al.* 2007). They differ from Maloideae by dry, dehiscent fruits (follicles) instead of pome fruits, but the anatomy of the ovules and the early development of the carpels is pomoid. Evans and Campbell (2002) studied the DNA sequence of the gene for the granule-bound starch synthase (GBSSI). This gene exists in four copies in Maloideae and the three genera, and in two copies in other Rosaceae. The sequences of the gene copies are very similar within Maloideae, closely followed by *Kageneckia, Vauquelinia,* and *Lindleya*. Additional support for the close relationship comes from the fact that in both groups (as well as in *Gillenia,* see below) two copies of GBSSI have a long first intron and missing sixth intron, and that the introns can easily be aligned; introns are noncoding parts of a gene and generally much more variable than coding parts. As a consequence of all these data Evans and Campbell (2002) include the three genera in Maloideae. The new tribe Pyreae (= Maloideae *s.l.*) of his subfamily Spiraeoideae *s.l.* in the classification of Potter *et al.* (2007) is composed by *Kageneckia, Vauquelinia,* and *Lindleya* plus subtribe Pyrinae = Maloideae *s.str.*

These data make it necessary to ask if the three new genera should be included in the basic type discussed here. From the new insights into morphological and molecular data this seems probable. In addition it is no problem to bridge the obvious difference between dry follicles and typical pome fruits by imagining that the groups of follicles were surrounded by floral tissue to build a pomaceous fruit. But it must be recognized that differences exist and, most importantly, that crossings within the genera have not been tested to date.

## 8.13. *Gillenia* and the Early History of Maloideae

For a long time it has been suggested that Maloideae with their basic chromosome number $x = 17$ could be of allopolyploid origin, possibly by fusion of a member of Spiraeoideae ($x = 9$) and of a member of Prunoideae ($x = 8$) (wide-hybridization hypothesis) or by fusion of two members of Spraeoideae followed by the loss of one chromosome. Indeed, the allopolyploid nature was confirmed by isoenzyme studies. Evans and Campbell (2002) falsified the wide-hybridization hypothesis with evidence that the Maloideae as well as *Kageneckia*, *Vauquelinia* and *Lindleya* evolved from within a lineage that contained the ancestors of *Gillenia* ($x = 9$), a small herbaceous genus of eastern North America. *Gillenia* (= *Porteranthus*, Bowman's root or indian-physic) has only two copies of the GBSSI gene, but the structure and the position of the introns is very similar to those of the Maloideae *s.l.* Potter *et al.* (2007) place *Gillenia* as a single separate genus outside their tribe Pyreae but within the same supertribe Pyrodae. Nonmolecular support for a *Gillenia*-origin hypothesis includes association with the fungus *Phragmidium*, floral morphology, and fossil flowers (genus *Paleorosa*).

The hypothesis that the greatest part of Maloideae arose from early hybridization (and polyploidization) is congruent with our section "The evolution of the Maloideae." Therefore the question is if *Gillenia*, though herbaceous and without pome fruits, belongs to the basic type Maloideae *s.l.*, too. For a decision of this question, it is necessary to cross *Gillenia* with genera of Maloideae on the one side and with *Kageneckia*, *Lindleya* and/or *Vauqueliana* on the other side, an attractive task for further research that will likely prove difficult to accomplish.

# References

Baur, E. 1910. Pfropfbastarde. *Biologisches Zentralblatt* 30:497-514.

Browicz, K. 1970. "*Malus florentina*": Its history, systematic position and geographical distribution. *Fragmenta Floristica et Geobotanica* 16:61-83.

Burgerstein, A. 1895. Vergleichend-histologische Untersuchungen des Holzes der Pomaceen. *Sitzungsberichte Akademie der Wissenschaften Wien, Mathematisch-Naturwissenschaftl Klasse* (Abteilung 1) 104:723-772.

Burgerstein, A. 1896. Weitere Untersuchungen über den histologischen Bau des Holzes der Pomaceen, nebst Bemerkungen über das Holz der Amygdaleen. *Sitzungsberichte Akademie der Wissenschaften Wien, Mathematisch-Naturwissenschaftl Klasse* (Abteilung 1) 105:552-582.

Campbell, C.S., M.J. Donoghue, B.G. Baldwin,and M.F. Wojciechowski. 1995. Phylogenetic relationships in Maloideae (Rosaceae): evidence from sequences of the internal transcribed spacers of nuclear ribosomal DNA and its congruence with morphology. *American Journal of Botany* 27:903-918.

Campbell, C.S, R.C Evans, D.R. Morgan, T.A. Dickinson, and M.P. Arsenault. 2007. Phylogeny of subtribe Pyrinae (formerly the Maloideae, Rosaceae): Limited resolution of a complex evolutionary history. *Plant Systematics and Evolution* 266:119-145.

Challice, J.S. 1981. Chemotaxonomic studies in the family Rosaceae and the evolutionary origins of the Maloideae. *Preslia* 53:289-304.

Coombes, A., K. Robertson, D. Potter, and S. Still. 2007. × *Rhaphiobotrya*, A New Intergeneric Hybrid in Rosaceae and the Identity of Coppertone Loquat. Illinois Natural History Survey Research Projects, In the Context of the Survey's Strategic Plan, Fiscal Year 2008. http://www.inhs.uiuc.edu/resources/annualreports/06_07/Projs07.pdf

Dirr, M.A. 1998. *Manual of Woody Landscapes Plants: Their Identification, Ornamental Characteristics, Culture, Propogation and Uses*, 5th ed. Stipes Publishers, Champaign, IL.

Düll, R. 1961. Die Sorbusarten und ihre Bastarde in Bayern und Thüringen. *Berichte der Bayerischen Botanischen Gesellschaft* 34:11-65.

Elias, T.S. 2000. *The Complete Trees of North America: Field Guide and Natural History*, Chapman & Hall, Norwell, Massachusetts.

Evans, R.C. and C.S. Campbell. 2002. The origin of the apple subfamily (Maloideae; Rosaceae) is clarified by DNA sequence data from duplicated GBSSI genes. *American Journal of Botany* 89:1478-1484.

Gabrieljan, E. 1978. *The genus* Sorbus *in western Asia and the Himalayas*. Akademija Nauk Armjanskoj S.S.R., Erevan, Armenia.

Gladkova, V.N. 1969. On the systematic position of the genus *Dichotomanthes* Kurz. *Botanical Zhurnal* 54:431-436.

Goldblatt, P. 1976. Cytotaxonomic studies in the tribe Quillajeae (Rosaceae). *Annals Missouri Botanical Garden* 6:200-206.

Hanelt, P., ed. 2001. *Mansfeld's Encyclopedia of Agricultural and*

*Horticultural Crops (Except Ornamentals)*. 1. Engl. ed. Springer, Berlin.

Jankun, A. and M. Kovanda. 1987. Apomixis and origin of *Sorbus bohemica*. *Preslia* 59:7-116.

Kalkman, C. 1988. The phylogeny of the Rosaceae. *Botanical Journal of the Linnean Society* 98:37-59.

Kalkman, C. 2004. Rosaceae. In: Kubitzki, K. ed. *The Families and Genera of Vascular Plants, vol. 6, Flowering plants - Dicotyledons: Celastrales, Oxalidales, Rosales, Cornales, Ericales*. Springer, Berlin, pp. 43-86.

Klotz, G. 1982. Synopsis der Gattung *Cotoneaster* Medikus (Part 1). *Wissenschaftliche Beiträge der Friedrich-Schiller-Universität Jena* 10:7-82.

Koehne, E. 1890. *Die Gattungen der Pomaceen. Wissenschaftliche Beilage 95 zum Programm des Falk-Realgymnasium*, Berlin.

Kollmann, R. and C. Glockmann. 1985. Studies on graft unions, I: Plasmodesmata between cells of plants belonging to different unrelated taxa. *Protoplasma* 124:224-235.

Kovanda, M. 1965. On the generic concepts in the Maloideae. *Preslia* 37:27-34.

Krüssmann, G. 1976-1978. *Handbuch der Laubgehölze*. Second edition, Volume 3. Parey, Berlin.

Kutzelnigg, H. 1994. Unterfamilie Maloideae: Allgemeiner Teil sowie Gattungen *Cydonia, Pyrus, Malus, Sorbus, Mespilus* und *Cotoneaster*. In: Scholz, H. ed. *Gustav Hegi, Illustrierte Flora von Mitteleuropa* Second edition, Volume 4, Part 2B. Blackwell, Berlin.

Little, E.L. 1980. *National Audubon Society Field Guide to North American Trees: Eastern Region*. Chanticleer Press, New York.

Marsh, F.L. 1976. *Variation and Fixity in Nature*. Pacific Press, Mountain View, CA.

Morgan, D.R., D.E. Soltis, and K.R. Robertson. 1994. Systematic and evolutionary implications of *rbcL* sequence variation in Rosaceae. *American Journal of Botany* 81:890-903.

Petrides, G.A. 1973. A Field Guide to Trees and Shrubs: Northeastern and North-Central United States and Southeastern and South-Central Canada. 2nd ed. Houghton Mifflin Co., Boston.

Phipps, J.B., K.R. Robertson, J.R. Rohrer, and P.G. Smith. 1991. Origins and evolution of subfam. Maloideae (Rosaceae). *Systematic Botany* 16:303-332.

Potter, D., and 10 others. 2007. Phylogeny and classification of Rosaceae. *Plant Systematics and Evolution* 266:119-145.

Rehder, A. 1940. *Manual of Cultivated Trees and Shrubs Hardy in North America*. Second edition. The Macmillan Company, New-York.

Robertson, K.R., J.B. Phipps, J.R. Rohrer, and P.G. Smith. 1991. A synopsis of genera in the Maloideae (Rosaceae). *Systematic Botany* 16:376-394.

Sax, K. 1931. The origin and relationships of the Pomoideae. *Journal of the Arnold Arboretum* 12:3-22.

Scherer, S. 1993. Basic types of life. In: Scherer, S. ed. *Typen des Lebens*, Pascal, Berlin, pp.11-30.

Schneider, C.K. 1906. *Illustriertes Handbuch der Laubholzkunde.* Volume 1. Fischer, Jena.

Schulze-Menz, G.K. (1964) Rosaceae. In: Melchior, H., ed. A. *Engler's Syllabus der Pflanzenfamilien. Volume 2: Angiospermen.* Twelfth edition. Gebrüder Borntraeger, Berlin, pp. 11-30.

Strasburger, E., founder. 1983. *Lehrbuch der Botanik für Hochschulen.* 32nd edition. Gustav Fischer, Stuttgart.

Vitkovskiy, V.L. and O.N. Korovina, eds. 1983. Pome Fruits (apple, pear, quince). Vol. 14. In: Likhonos F.D., A.S. Tuz, and A.J. Lobachev, eds. *Flora of Cultivated Plants.* Moscow, Russia.

Weber, C. 1964. The Genus *Chaenomeles* (Rosaceae). *Journal of the Arnold Arboretum* 45:161-206, 302-345.

Zhang, S.Y. 1992. Systematic wood anatomy of the Rosaceae. *Blumea* 37:81-158.

Zhang, S.Y. and P. Baas. 1992. Wood anatomy of trees and shrubs from China, III: Rosaceae. *IAWA Bulletin New Series* 13:21-91.

# 9. "Living Stones" And Their Kin (Aizoaceae: Ruschieae) — A New Basic Type And Further Example Of Exceptionally Rapid Evolution

HERFRIED KUTZELNIGG

## Abstract

The "living stones" of the ice-plant family, together with their relatives, form a group of plants that is adapted to extreme aridity, native primarily to southern Africa. With over 1500 species, the plants on the one hand are very diverse and on the other are clearly united by their morphological and molecular similarities. On the assumption that the associated habitat originated only about 5 million years ago by conventional chronology, this group of plants is known in the literature as an example of a radiation that is unsurpassed both in extent and speed. Among the genera there are numerous hybridization connections, so that altogether the group can be understood as a basic type.

## 9.1. Introduction

Among the most impressive natural phenomena belong the plants that have become well known under the name "living stones" (as well as "stone plants" and "pebble plants"). They are water-storing (i.e., succulent) species, which in extreme cases barely protrude from the soil and are so camouflaged that, at first glance, they appear to be pebbles (Figure 1). They consist mainly of a single pair of leaves and are adapted to severe aridity and high solar radiation. This is because photosynthesis takes place in the protection of the soil, and the necessary sunlight arrives through a window (Figure 2). When the "stones" bloom (Figure 3), they bear flowers characteristic of the ice-plant family (Aizoaceae), which can sometimes be found

Figure 1: Typical representatives of the "living stones" (genus *Lithops*). Only the tips of the leaf pairs jut out from the rocky soil. Photo: A. Kutzelnigg.

in gardens throughout the world. The decorative flowers open most widely in the sun. With their very numerous, narrow petals and usually numerous, crowded stamens, they are somewhat reminiscent of the compound flower heads of the Compositae (= Asteraceae), but in contrast with the latter they are actually single flowers.

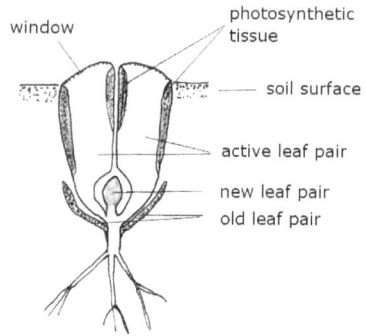

Figure 2: Longitudinal section through Lithops as a distinctive representative of the living stones. The plant consists mainly of two water-storing leaves that are almost completely hidden in the soil and where photosynthesis takes place. Light enters on the upper side of the leaves, which acts like a window, and passes through the glassy transparent leaf interior to the laterally located chlorophyll-containing tissue.

## 9.2. The Diversity Of The "Living Stones" And Their Kin (Ruschieae)

With about 101 genera and 1563 species, the group closely related to the "living stones" is extremely varied, so that the forms described above represent only an extreme example. The multitude of known forms of this plant group is related to the group's habitat, which readily accommodates different ecological niches due to its soil structure that varies over small areas. This environment is known as the Succulent Karoo, an arid zone in the area of western South Africa / Namibia with moderate temperatures and rainfall restricted to wintertime. Some species also grow in the neighboring areas, but beyond that there are few occurrences. The vegetation of the Succulent Karoo is

Figure 3: A relative of the living stones (genus Glottiphyllum) in flowering condition. Photo: A. Kutzelnigg.

particularly diverse with approximately 5000 species. Leaf succulents predominate, with 1750 species provided by the Aizoaceae alone, which comprise about 90% of the biomass of this region.

The "living stones" and their relatives are perennial and succulent. Indeed most are leaf succulents with triangular or rounded leaf cross-sections, but stem succulents with or without periodic leaf fall also occur. At their base, the paired leaves are often almost completely grown together. The dominant life form is the low subshrub, i.e., perennial plants in which the lower part becomes somewhat woody. However, fully woody, low or even larger shrubs are known, as well as lianas and tree-like forms. Not infrequently, depending on the season, different leaf

| Family | Subfamilies | Tribes |
|---|---|---|
| | | **Ruschieae** 101: 1563 |
| | Ruschioideae 110: 1585 | Apatesieae 7: 11 |
| Aizoaceae | Mesembryanthemoideae 9: 95 | Dorotheantheae 2: 11 |
| 130: 1845 | Aizooideae 6: 100 | |
| | Sesuvioideae 5: 37 | |

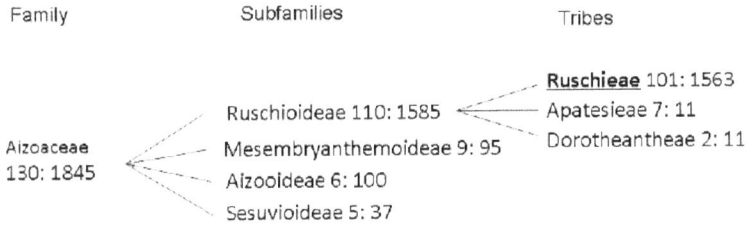

Figure 4: Systematic position of the "living stones" and their relatives (tribe Ruschieae) within the ice-plant family (Aizoaceae). The first number is that of the included genera, the second is of the included species.

forms develop (heterophylly). The flowers are white or yellow to red. The corresponding pigments—universal in the ice-plant family—are not the anthocyanins and flavonoids typical elsewhere in the plant kingdom, but the red betalains and yellow betaxanthins, respectively, well known in the red beet. The nectaries are crests of nectar-secreting tissue. The petals are transformed stamens. Like many other succulent plants, the species follow the CAM pathway of photosynthesis. This is an adaptation to the aridity of the habitat whereby $CO_2$ is absorbed at night then supplied to photosynthesis during the day. In this way, the stomata are closed by day so that water loss of the plant is greatly reduced. The fruits are usually many-seeded capsules that usually open with moisture and close during drought. This adaptation allows the germination of seeds during the short rainy periods. The chromosome number of $x = 9$ occurs uniformly.

Biosystematically the plants constitute the tribe Ruschieae, a characteristic part of subfamily Ruschioideae (Figure 4), which has recently been classified in the broadly delimited family of fig-marigolds or ice-plants (Aizoaceae) and earlier as part of the Mesembryanthemaceae ( "mesembs" used as a short vernacular). The name ice-plant refers to a typical family characteristic, namely, vesicular cells in the epidermis, which look like ice crystals in many cases. As shown by the number of species in Figure 4, the Ruschieae constitute the largest part of the family. The two remaining tribes of the subfamily (Apatesieae and Dorotheantheae) differ from the Ruschieae by the fact that, among other things, their leaves are flat and not strongly succulent, the species are annuals (to herbaceous perennials), and fruit splits into mericarps (nutlets) when ripe.

## 9.3. The Uniformity Of The Ruschieae

Despite the described diversity, the Ruschieae (in the broad sense of, for example, Schwantes 1971) comprise a uniform and outwardly well-defined group because of numerous similarities. Historically interesting is that, until the beginning of the 20th century, its representatives (as

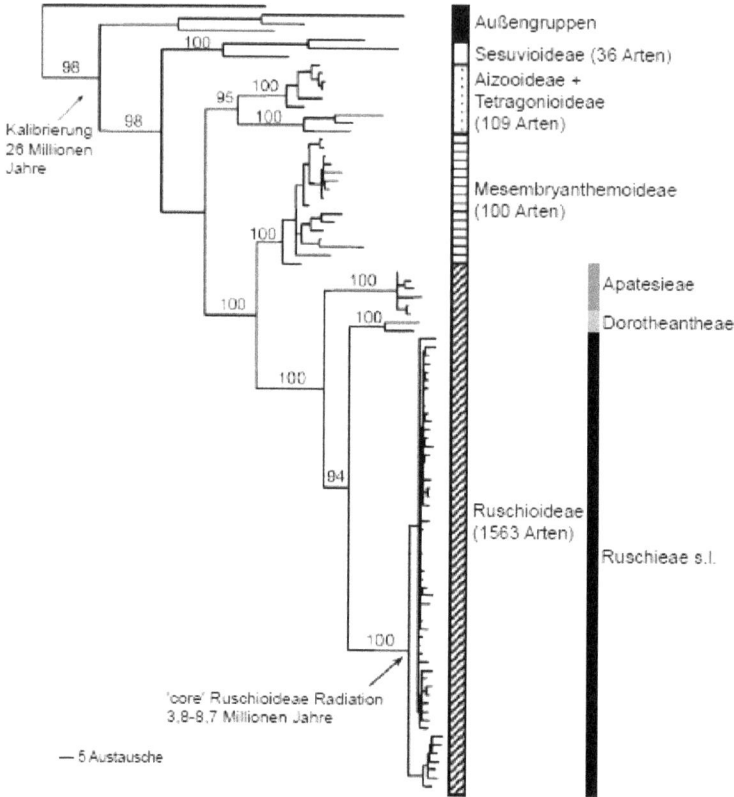

Figure 5: Molecular tree diagram (phylogram) of the ice-plant family (Aizoaceae) based on comparisons of DNA sequences from two regions of chloroplast DNA. The horizontal length of the branches corresponds to the number of changes in the DNA. One can see that the representatives of Ruschieae s.l., despite their high number of species, show almost no differences in DNA sequences. After Klak et al. 2003, 2004.

well as some further species of the family) were placed in the genus *Mesembryanthemum*, which was excessively large but has since been split apart. The uniformity of the plant group is demonstrated by various facts:

1. In terms of morphological criteria, the genera are so closely linked that it is not yet possible to establish a generally accepted subdivision of Ruschieae. Attempts to classify focused on characteristics of fruit structure and life forms. Schwantes (1971) distinguished 22 subtribes. Later Hartmann (1998, 2001) arranged the genera into 10 groups that were only partially congruent with the division of Schwantes and were not formally described as subtribes because of the tentativeness of the data. On the basis of the fine-structure of the nectaries, Chesselet et al. (2002) recently proposed to split the group into two tribes: the Ruschieae *s.str.* with nectaries

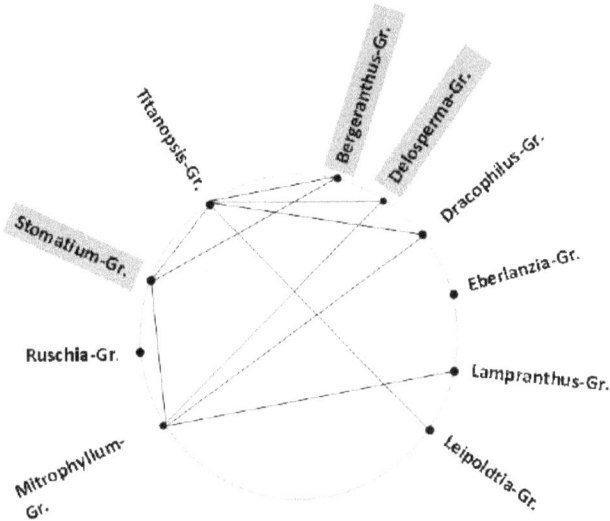

Figure 6: Hybridization connections among the generic groups of the Ruschieae. The three groups (*Bergeranthus, Delosperma* and *Stomatium* Groups) belonging to the Delospermeae, segregated by Chesselet et al. (2002), are highlighted in gray. The genus *Drosanthemum* (Tribe Drosanthemeae), recently segregated from the remainder, is included as part of the *Delosperma* Group in this diagram.

united into a ring and a newly formed tribe Delospermeae with unfused nectaries, comprising 3 of the 10 groups of Hartmann (*Bergeranthus, Delosperma* and *Stomatium* Groups) (see Figure 6). Generally, this classification has not been followed; especially, it is not reflected in the molecular data (see below). The separation of genera from each other is very difficult. This is a phenomenon often encountered in basic types, as well as very uniform taxa, in addition to the described characteristic reticulations.

2. The uniformity of Ruschieae was most impressively demonstrated by molecular comparisons of two plastid DNA sequences from representatives of the family Aizoaceae (Klak et al. 2003, 2004). The genetic variation among the representatives of Ruschieae (referred to as the core 'Ruschioideae' by the authors) is extremely limited (Figure 5), whereas the other two tribes of Ruschioideae (and the other subfamilies) can be clearly distinguished from the Ruschieae. Particularly noteworthy is that the three groups only recently separated as Tribe Delospermeae on morphological grounds are not supported by molecular data, so Chesselet et. al. (2004) have retracted the segregation of Delospermeae.

However, a few species form their own group within the Ruschieae (Figure 5, at the very bottom). This involves all six species of the genus *Drosanthemum* included in the investigation and a species that was recently transferred to the genus *Delosperma* and is now reincorporated into *Drosanthemum*. The genus *Drosanthemum* includes about 120 species and is morphologically very heterogeneous. It has been seen as part of Delosperma Group (which in the tree belongs to the large neighboring branch) and is morphologically difficult, in part, to separate from other genera of this group. It is remarkable that, among other things, species of the genus have leaves that are not fused at the base, which does not occur in the remaining Ruschieae, and the leaves usually are covered with large, water-filled vesicular cells. Parts of the genus inhabit somewhat different regions from the other genera. Thus the *Drosanthemum*-branch could actually form a separate evolutionary line, as suggested by the tree in Figure 5, but the genetic differences from the rest of the Ruschieae are very small. However, Chesselet et al. (2003) take the findings as an opportunity to separate at the rank of tribe this branch from the rest of the Ruschieae *s.str.* as the Drosanthemeae.

3. Landrum (2001) discovered a new type of xylem cells, known as wide-band tracheids, in some families of Caryophyllales, including the Aizoaceae. This type of tracheid is characterized by thick secondary walls and probably serves to prevent collapse of the cells under conditions of water stress. As far as concerns the 89 genera of Aizoaceae that have been examined, such tracheids are limited to leaves of members of the Ruschieae.

4. In a representative sample of 69 species of Aizoaceae Thiede et al. (2007) determined that all representatives of the Ruschieae (including *Drosanthemum*) differ from the remaining subgroups of the family by the absence of introns in plastid gene *rpo*C1.

Thus altogether, there are numerous observations that confirm on the one hand the internal uniformity of the other Ruschieae and on the other its clear demarcation from outside groups.

## 9.4. Rapid Evolution In The Ruschieae

Because of their variety, the living stones and their kin exhibit an impressive example of adaptive radiation, i.e., a microevolutionary splaying out with remarkable adaptations in response to very different living conditions. Given the high species number in the family with about 2,000 species in the Succulent Karoo environmental "meganiche," Ihlenfeldt (1994) conjectured that their evolution took place very quickly, indeed at a speed without parallel among the angiosperms. He justified this by saying that the associated diverse habitat arose as an ecological meganiche only about 5 million radiometric years ago. Klak et al. (2004) took up these ideas again in a separate publication in connection with the above sequence analysis but, instead, confined the radiation to the Ruschieae ('core' Ruschioideae) and tried to make it concrete on the basis of the molecular clock concept. Accordingly, the radiation should have taken place only during a relatively short time (essentially 3.8-8.7 million years ago by conventional dating methods) with speciation rates that until now were unknown from the animal and plant kingdoms.

## 9.5. The Ruschieae—A Basic Type

Like many other succulent plant groups, the Ruschioideae also enjoy great popularity with plant enthusiasts, as well as professional botanists. This situation has resulted from the past achievement of numerous artificial hybrids, which go beyond the otherwise usual chance products, as well as the intentional search for spontaneous hybrids in their natural habitats. Hammer and Liede (1990) provided a comprehensive report on the situation. But since then, numerous other hybrids have been added (see International Plant Names Index: www.ipni.org), so that now the relatively high number of 30 intergeneric hybrids is known. Of these, at least four were also observed in their natural habitat. Figure 6 shows the known crossing connections based on the groups following Hartmann (1998). It is clear that almost all groups are interconnected. Exceptions are the *Ruschia* and *Eberlanzia* Groups, of which many representatives are not maintained in culture. Nevertheless, one can assume that the related species, in principle, would be crossable and, therefore, belong to the same basic type because they do not fall outside its boundary by other criteria, especially comparisons of DNA sequences. Particularly interesting, of course, is the question of whether the Delospermeae (*Delosperma*, *Stomatium*, and *Bergeranthus* Groups), segregated from the Ruschieae by Chesselet et al. (2002) on the basis of unfused nectar glands, are also separated by crossability restrictions. Figure 6 clearly shows this is not the case. The question remains of how the genus *Drosanthemum* behaves. Since its species are rarely found in culture, there are unfortunately no corresponding hybrids, so that at present it must remain open whether this genus with its special position belongs to the same basic type. However, the genus *Delosperma* which cannot be well separated from *Drosanthemum* taxonomically is, as well as *Gibbaeum* of the *Delosperma* Group, connected by hybridization with the Ruschieae *s.str.*

So in summary, it can be established here from the hybridization connections, as well as from the morphological and biomolecular data, that the relationship of the "living stones" represents a basic type, possibly excluding of the genus *Drosanthemum*.

# References

Chesselet P, Smith GF & Van Wyk AE (2002) A new tribal classification of Mesembryanthemaceae: evidence from floral nectaries. Taxon *51*, 295-308.

Chesselet P, Van Wyk AE & Smith GF (2004) Notes on African plants. Mesembryanthemaceae. A new tribe and adjustments to infrafamilial classification. Bothalia *34*, 47–51.

Hammer SA & Liede S (1990) Natural and artificial hybrids in the Mesembryanthemaceae. South African J. Bot. *56*, 356-362.

Hartmann HEK (1998) Groupings in Ruschioideae (Aizoaceae). Mesemb Study Group Bull. *13*, 35-36.

Hartmann HEK (Ed.) (2001) Illustrated Handbook of Succulent Plants. Aizoaceae A-E; Aizoaceae F-Z. Berlin etc.: Springer.

Ihlenfeldt H-D (1994) Diversification in an Arid World: The Mesembryanthemaceae. Annu. Rev. Ecol. Syst. *25*, 521-546.

International Plant Names Index (IPNI): www.ipni.org (last accessed 06.2009)

Klak C, Khunou A, Reeves G & Hedderson T (2003) A phylogenetic hypothesis for the Aizoaceae (Caryophyllales) based on four plastid DNA regions. Amer. J. Bot. *90*, 1433-1445.

Klak C, Reeves G & Hedderson T (2004) Unmatched tempo of evolution in Southern African semi-desert ice plants. Nature *427*, 63-65.

Landrum JV (2001) Wide-band tracheids in leaves of genera in Aizoaceae: the systematic occurrence of a novel cell type and its implications for the monophyly of the subfamily Ruschioideae. Plant Syst. Evol. *227*, 49-61.

Schwantes G (1971) The classification of the Mesembryanthemaceae. In Herre H (Ed.) The genera of the Mesembryanthemaceae. Cape Town, South Africa: Tafelberg.

Thiede J, Schmidt SA & Rudolph B (2007) Phylogenetic implication of the chloroplast rpoC1 intron loss in the Aizoaceae (Caryophyllales). Biochem. Syst. Ecol. *35*, 372-380.

# 10. The Basic Type of Ducks, Swans, and Geese (Anseriformes, Anatidae): Biological and Paleontological Aspects[1]

SIEGFRIED SCHERER

In their species richness, the ducks, swans, and geese are so variagated and multiform that it is as if, by some exuberant creative whim, a whole cornucopia of fantasy has been poured out.

Klös and Klös (1975, p. 243)

**Abstract**

Among the Anseriformes, more than 400 anatid hybrids are known involving 85% of the anatid species. More than half of the hybrids are intergeneric, 31% of the hybrids are intertribal. Thus far, no hybrids between the Anatidae and other avian families are known. A number of hybrids possess morphological and ethological characteristics, that cannot be observed in either parent but do occur in other species of this family. In several cases, hybrids between two different species show similarities to a third species.

Several ethological and morphological characteristics occur in completely different representatives of this group, whereas it is difficult to place aberrant members into a specific tribe because they possess characteristics from various tribes. Whereas the attempt to use molecular taxonomy does not yield satisfying results for a systematic arrangement within the Anatidae, molecular data apparently do provide sufficient cause to separate this family from other avian taxa. The known data indicate clear differences between the magpie goose (*Anseranas semipalmata*) and the other Anatidae. Moreover, since no hybrids between Anseranas and other representatives of the Anseriformes are known, it is obvious that this species should be separated from the Anatidae.

---

[1] This paper was written in 1992 and was the attempt of a hobby ornithologist to interpret a synopsis of various data within the framework of a model of anatid origins from genetically polyvalent ancestral forms. The author now works in another field and did not have the possibility to update this paper. The body of data concerning the hybrids is still up to date (a few hybrids could be added), and the conclusion that hybridization allows for a very clear distinction of the Anatidae from all other bird families is as valid as it was in 1992. The rest of the paper reports data and reflects insights more than 17 years old and should be considered an early attempt to describe the basic type of the Anatidae . It may well be that current data would require divergent interpretations in many respects.

Thus far, no fossil connections between the Anatidae and other bird groups could be established; the ducks appear in the fossil record without transition. Difficulties to interpret the Eocene bird *Presbyornis* as a "missing link" in the evolution of the anatids are discussed. Since specialized forms occur in early strata, the evolution of the Anatidae from unknown ancestors is unclear.

Based on these results, the ducks, swans, and geese (Anatidae, excluding the Anhimidae and Anseranatidae) are for now considered to belong to the same basic type. The difficulties of an intra-anatid classification (which are common to all taxonomic systems of this family), the characteristics observed in hybrids, and the fossil record are interpreted within the framework of a potentially rapid radiation of this basic type during the Oligocene. The ancestral form already may have been genetically polyvalent with a high potential of variability.

## 10.1. Introduction

Ducks, swans, and geese belong to one of the best-known families of birds, the Anatidae, which include terrestrial as well as aquatic birds. Their nostrils connect with each other, they have two pairs of muscles located between the sternum and the trachea, they have ten to eleven primaries, their fifth secondary is absent, and the lower mandible has a long angular process. All species, with the exception of the Hawaiian goose (*Branta sandvicensis*) and the magpie goose (*Anseranas semipalmata*), possess fully developed webs. The diagnostic character of the Anseriformes is the skin-covered bill with a tip reinforced by a "nail" and lateral horn lamellae, which, depending on the feeding habits, can be coarse or fine. The basic systematic work was done by Delacour and Mayr (1945). The taxonomy suggested by them has been accepted without major modifications by more recent authors (e.g., Johnsgard 1978; Table 1).

Within the Anatidae a large number of interspecific hybrids are known. Therefore this family is exceptionally well suited to test the basic type concept suggested by Marsh (1976). Part of the material presented here has been published elsewhere (Scherer and Hilsberg 1982; Scherer and Sontag 1986) and is presented in abbreviated and updated form. The large amount of literature concerning the Anatidae and the wide range of these publications allow for only a small selection to be cited.

## 10.2. Data...[2]
### 10.2.1. Hybridization within the Anatidae.
#### 10.2.1.1. Formation of Hybrids. It has been known for over 100 years that within the Anatidae numerous hybrids have been produced (for an extensive synopsis see Gray 1958, as well as Johnsgard 1960). Without claiming completeness and without considering subspecies, Gray lists 361 hybrids, to which Scherer and Hilsberg (1982) added another 57.

The species within each tribe differ considerably when it comes to the forming of hybrids. Even closely related forms which produce fertile hybrids show differences in hybrid frequency (e.g., greylag goose (*Anser anser*) and Ross's goose (*A. rossi*); Canada goose (*Branta canadensis*) and red-breasted goose (*B. ruficollis*); common shelduck (*Tadorna tadorna*) and paradise shelduck (*T. variegata*); mallard (*Anas platyrhynchos*) and Meller's duck (*A. melleri*)). The probable reason for this is unequal opportunities to cross-breed. The smallest tribes, which mostly consist of only one species, do not show any hybrids.

Figure 1 presents, for the sake of a better visual summary, a doubled (diagonally symmetrical) cross-breeding matrix. In this matrix, patterns

O  Anseriformes
 F  Anhimidae (screamers 2/3)
 F  Anatidae
   S  Anseranatinae
     T  Anseranatini (magpie goose 1/1)
   S  Anserinae (40/148)
     T  Dendrocygnini (whistling ducks 2/9)
     T  Anserini (geese, swans 3/22)
     T  Cereopsini (Cape Barren goose 1/1)
     T  Stictonettini (freckled duck 1/1)
     T  Tadornini (shelducks, sheldgeese 5/15)
     T  Tachyerini (steamer ducks 1/3)
     T  Cairinini (perching ducks 9/13)
     T  Merganettini (torrent duck 1/1)
     T  Anatini (dabbling ducks 4/39)
     T  Aythyini (pochards 3/16)
     T  Mergini (sea ducks 8/20)
     T  Oxyurini (stiff-tail ducks 3/8)
--------------------------------
O  Anhimiformes
 F  Anhimidae (screamers 2/3)
O  Anatiformes
 F  Anseranatidae
   S  Anseranatinae (magpie goose 1/1)
 F  Anatidae
   S  Dendrocygninae (whistling ducks 2/9)
   S  Stictonettinae (freckled duck 1/1)
   S  Anserinae (geese, swans 11/23)
   S  Cereopsinae (Cape Barren goose 1/1)
   S  Tadorninae (shelducks, sheldgeese 9/19)
   S  Plectropterinae (spur-winged goose 1/1)
   S  Anatinae (dabbling ducks 48/101)

Table 1. Two different taxonomies of Anatidae and Anhimidae according to Johnsgard (1978), upper part, and Wolters (1975-1982), lower part. O = Order, F = family, S = Subfamily, T = Tribe. In parentheses the number of genera (left of slash) and species (right of slash) within the group is indicated.

---

2    Origins research as an historic science cannot claim objectivity; results depend to a high degree on a chosen metaphysic frame of interpretation (Steinbrunner 1991). Although even the classical natural sciences cannot be objective in the strictest sense, the results of empirical science will be considered to be "objective" data in this paper. Such data (in the first part of the paper) will be distinguished from origins interpretations (in the second part).

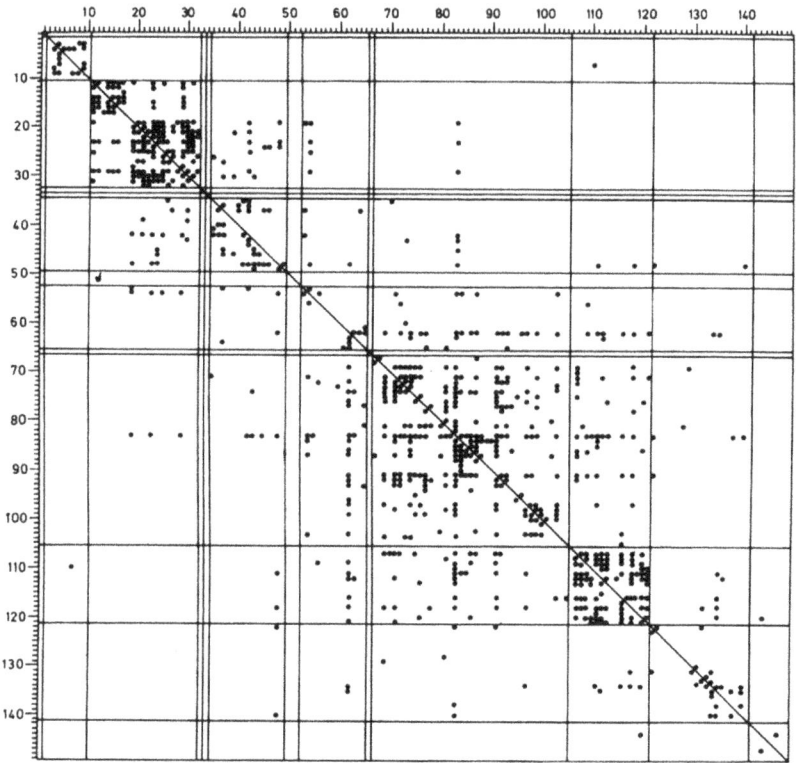

Figure 1. Cross-breeding matrix of the Anatidae. Numbers indicate 149 species according to Johnsgard (1978). Horizontal and vertical lines indicate the following tribes: 1: Anseranatini (magpie goose); 2-10: Dendrocygnini (whistling ducks); 11-32: Anserini (swans and geese); 33: Ceropsisni (Cape Barren goose); 34: Stictonettini (freckled duck); 35-49: Tadornini (sheldgeese and shelducks); 50-52: Tachyerini (steamer ducks); 53-65: Cairinini (perching ducks); 66: Merganettini (torrent duck); 67-105: Anatini (dabbling ducks); 106-121: Aythyini (pochards); 122-141: Mergini (sea ducks); 142-149: Oxyurini (stiff-tailed ducks). Reprinted from Scherer & Hilsberg (1982). Refer to this paper for species names.

surface immediately. As is to be expected, hybrids occur primarily along the diagonal line between closely related species. However, there are notable exceptions, especially within the Cairinini and Mergini.

Except in the Dendrocygnini and Anatini, the percentage of intergeneric hybrids is relatively high. A remarkable 52% of all known hybrids are intergeneric; 31% are intertribal (documentation in Scherer and Hilsberg 1982).

**10.2.1.2. Fertility of the F$_1$ Generation.** In many cases it cannot be decided for sure whether a hybrid is fertile or sterile. Intrageneric hybrids are generally considered to be fertile; although, this has not been tested in every case. Fertile intergeneric and, more rarely, fertile intertribal hybrids do occur. In some cases it is possible to breed hybrids

| Parents | Resulting similarity | Reference |
|---|---|---|
| Aythy fuligula × A. ferina | Aythya affinis | Perrins 1961 |
| A. fuligula × A. nyroca | A. baeri | Gillham et al. 1966 |
| A. fuligula × A. marila | A. affinis | Voous1955 |
| A. ferina × A. nyroca | A. nyroca | Gillham et al. 1966 |
| A. valisneria × A. americana | A. ferina | Weller1957 |
| Anser anser × Branta canadensis | Anser caerulescens | Harrison & Harrison 1966 |
| Anas clypeata × A. crecca | Anas formosa | Harrison 1953 |
| A. clypeata × A. penelope | A. formosa | Harrison 1959 |
| A. clypeata × A. cyanoptera | A. rhynchotis | Harrison & Harrison 1965 |
| A. acuta × A. crecca | A.*formosa | Harrison & Harrison 1971 |
| A. penelope × A. sibilatrix | A. americana | Harrison & Harrison 1968 |
| Mergus merganser × Tadorna tadorna | A. platyrhynchos | Lind & Paulson 1963 |

Table 2. It is often found that hybridization of two species produces offspring that are similar to another species.

over several generations (Lorenz 1958; von de Wall 1963; Sharpe and Johnsgard 1966). Kolbe (1984) mentions several three-species hybrids in captivity. The hybrid from two species was crossed with a third species. Whether hybrid populations in their natural environment break down after the $F_2$ generation (as is the case in other groups of animals; comp. Mayr 1967, Dillon 1978), has not been investigated systematically within the Anatidae. Fertile individuals were observed in 20% of the hybrids listed in Figure 1 (Scherer and Hilsberg 1982). In reality the percentage of fertile hybrids is certainly much higher because in many cases fertility tests are simply lacking. Of the 90 fertile hybrids, 59 were intrageneric, 18 intergeneric, and a remarkable 13 intertribal.

**10.2.1.3. Morphology and Behavior of Hybrids.** As is to be expected, hybrids generally exhibit intermediate states of parental characters, or the parental characters are exhibited unchanged. However, a number of hybrids do not resemble their parents, but rather, other species (Table 2; e.g., Harrison and Harrison 1963; Sage 1966). This similarity can go so far that hybrids in the open are actually mistaken for a third species, especially within the Aythyini (Perrins 1961; figures in Scherer and Hilsberg 1982).

In some cases, this similarity to a third species is only partial. An example is the cross between the cinnamon teal (Anas cyanoptera) and the northern shoveler (A. clypeata), investigated extensively by Harrison and Harrison (1965). The hybrids hardly show any sexual dimorphism and resemble mostly the Australian shoveler (A. rhynchotis). On the other hand, the reddish-brown dotted rear portion of their body and the absence of white feathers is reminiscent of the Cape shoveler (A. smithii) and the red shoveler (A. platalea). Plus, a white ring around the neck can show up, which is a character of the mallard (A. platyrhynchos). Hybrids of two species, thus, show characters of four other species, which in this case are fairly closely related to their parents.

Figure 2. Hybrid between *Anas acuta* (pintail, left) and *Anas crecca* (common teal, right). The hybrid exhibits a "bimaculated face pattern" which is not present in the parents. Artwork by Thomas Hilsberg.

This phenomenon, however, also occurs in hybrids of parents from different tribes. An example of this is the hybrid between the common merganser (*Mergus merganser*) and the common shelduck (*Tadorna tadorna*). Besides parental characters which can be exhibited fully or intermediately, the "single secondaries are coloured partly as in Goosander partly as in Shelduck but considered as a whole colour pattern is quite different from the two parent species and similar to the speculum of the Mallard (*A. platyrhynchus*)" (Lind and Poulson 1963).

It is particularly interesting that the "bimaculated face pattern" (see the section "Morphology") was found as an exceptional case in a male of common teal (*Anas crecca*) and shows up almost always in hybrids between the northern pintail (*A. acuta*) and common teal (*A. crecca*) (Figure 2); although, the parental species would not lead one to expect this. Variations of this pattern can be found repeatedly in other hybrids, especially when the mallard (*Anas platyrhynchos*), the Eurasian wigeon (*A. penelope*), or the northern shoveler (*A. clypeata*) are involved (comp. Harrison and Harrison 1971).

Corresponding ethological features were also reported, for example, in the case of the above-mentioned hybrid of *Mergus* and *Tadorna*. It partially exhibited non-intermediate behavioral traits which neither of the parents demonstrated. Within the Anatini it has been repeatedly reported that hybrids exhibit behavioral traits which are not present in either parent but are present in other species (Kaltenhäuser 1971). Also, combinations of known movement patterns have been observed that could not be detected in any other species (Lorenz 1958, 1960; von de Wall 1963). That this fact differs from hybrid to hybrid was shown in a detailed study by Sharpe and Johnsgard (1966) who investigated 22 $F_2$ hybrids of the mallard (*Anas platyrhynchos*) and the northern pintail (*A. acuta*). They came to the conclusion that "the $F_2$ hybrids did not exhibit a single action which was not typical of at least one parent; but one $F_3$ hybrid did, when, during social interaction, it showed a movement pattern called 'jerking up' (which is known to normally only follow copulation) immediately after the 'head-up, tail-up' movement" (Sharpe and Johnsgard 1966).

Figure 3. A short, hen-like bill is observed within different groups of anatids. From top to bottom: *Ceropsis novaehollandiae* (Cape Barren goose, Cereopsini); *Chloephaga melanoptera* (Andean goose, Tadornini); *Nettapus auritus* (cotton teal, Cairinini). Art work by Thomas Hilsberg.

Figure 4. Bills from *Anas clypeata* (northern shoveler) and *Malacorhynchus membranaceus* (pink-eared duck). The fromer is a European species while the latter is found in Australia. Usually, the form of the bill is interpreted as an example of convergence. Artwork by Marion Häberle.

Von de Wall (1963) researched mating movements of hybrids within the Anatini and made a number of interesting observations. Hybrids sometimes show movement patterns that cannot be observed in either parent but can be observed in other species of the group. Also, combinations of basic movement patterns, as well as coordinations, are observed that are not shown by the parents but are shown by other species.

### 10.2.2. Comparative Biology of the Anatidae

**10.2.2.1. Morphology.** The basic structure of the bill is typical in all representatives of the Anatidae, but the specific characteristics cannot always be utilized for the systematic placement of individual species. A short and comparatively thick bill could belong to a South American goose (*Chloephaga*), a Cape Barren goose (*Cereopsis*), or to a pygmy goose (*Nettapus*) (Figure 3). Although almost all mergansers have long, thin bills, the smew (*Mergellus albellus*) possesses a shorter, narrower bill, similar to the torrent ducks (*Merganetta*) or the common goldeneye (*Bucephala clangula*). A remarkable similarity is seen in the flat and elongated bills of the northern shoveler (*Anas clypeata*) and the pink-eared duck (*Malacorhynchus membranaceus*) (Figure 4).

A notable trait of some—often only distantly related—species is a longitudinal mark on the head. It can surround the eye, as in the case

Figure 5. Putative hybrid between domesticated duck and mallard (*Anas platyrhynchos*) exhibiting some sort of "bimaculated face pattern." The hybrid was observed in a flock of mallards with plumage ranging from pure white to wild type coloring. Spring 1982, Mainau Island, Lake Constance, Germany.

of the black-necked swan (*Cygnus melanocoryphus*), the torrent duck (*Merganetta armata*), the flying steamer duck (*Tachyeres patachonicus*), and the grey duck (*Anas superciliosa rogersi*). The black-bellied whistling duck (*Dendrocygna autumnalis*) and the white-winged scoter (*Melanitta fusca*) show only a ring around the eye. Harrison (1978) and Harrison and Harrison (1971) pointed out a pattern of particular interest in this case. The "bimaculated face pattern"—two bright fields on the side of the head are separated by a darker area—is especially well exhibited in the Baikal teal (*Anas formosa*). This pattern is somewhat different in the red-breasted goose (*Branta ruficollis*) and the harlequin duck (*Histrionicus*); frequently, only the part of the pattern which is near the bill is expressed, as, for example, in the bronze-winged duck (*Anas specularis*) and Barrow's goldeneye (*Bucephala islandica*). The females of the white-winged scoter (*Melanitta fusca*) and the surf scoter (*M. perspicillata*) also show this pattern, whereas it is absent in their males.

Recently, in the Bay of Constance and in the northwest arm of Lake Constance (Bodensee) around Mainau Island, individual mallards are observed with a notable white coloration. In all probability they are (frequently occurring) hybrids between the mallard and the domestic duck (*Anas platyrhynchos f. domesticus*). Some of the males clearly show the "bimaculated face pattern" (Figure 5). Previously this pattern had been known only as a result of cross-breeding between different species and was considered an atavism (an expression of a character of the ancestral form of the Anatidae) by Harrison (1945, 1953). It is particularly remarkable that there is not even a trace of the "bimaculated face pattern" in either the wild mallard or the conspecific domestic duck. On the other hand, it is fully or partially expressed in other species of the family. The occurrence of this trait does not depend on assumed relationships (Table 3).

**10.2.2.2. Aberrant Types.** The anatids have been described as a well-defined group. However, many representatives of the group are difficult to place, as far as their relationships within the family are concerned. These are called "aberrant types." On the other hand, they are good evidence for the systematic unity of the family. Each of the following species (see Figure 13) shows a combination of characters

| Spot in front of the eye | Spot behind the eye | Complete pattern |
|---|---|---|
| Dendrocygna viduata (white-faced whistling duck) | Branta canadensis (Canada goose) | Branta ruficollis (red-breasted goose) |
| Anser albifrons (greater white-fronted goose) | Bucephala albeola (bufflehead) | Anas sibilatrix (Chiloë wigeon) |
| Anser erythropus (lesser white-fronted goose) | Mergus cucullatus (hooded merganser) | Anas formosa (Baikal teal) |
| Anas specularis (bronze-winged duck) | Histronicus histronicus (harlequin duck) | |
| Anas discors (blue-winged teal) | Nettapus pulchellus* (green pygmy goose) | |
| Anas rhynchotis (Australian shoveler) | Malacorhynchus membranaceus* (pink-eared duck) | |
| Bucephala islandica (Barrow's golden-eye) | | |
| Bucephala clangula (common golden-eye) | | |

Table 3. The bimaculated face pattern is found in a variety of distantly related anatid species. * indicates that the pattern is only weakly developed.

of entirely different groups, which makes a systematic placement very difficult (Johnsgard 1978): 1) The white-backed duck (*Thalassornis leuconotus*) is generally placed within the stiff-tailed ducks (Oxyurini), but in its anatomy, voice, and feather proteins, it more closely resembles the whistling ducks (Dendrocygnini). 2) The Coscoroba swan (*Coscoroba coscoroba*) is often described as a swan with a number of goose-like features. According to its external appearance and the plumage of the young, however, it should belong to the whistling ducks. 3) The Cape Barren goose (*Cereopsis*) combines traits of the sheldgeese and shelducks (Tadornini; for example, the shape of the bill in *Chloephaga*), the swans and geese (Anserini), and the dabbling ducks (Anatini). 4) In spite of its dabbling duck-like appearance, the freckled duck (*Stictonetta*) shows affinities to the geese and swans in some of its anatomical features. 5) The steamer ducks (Tachyerini), which only occur in South America, are placed within the vicinity of the Tadornini according to their feather proteins but within the Anatini according to their skeletal features and trachea. Furthermore, the chemistry of the uropygial gland waxes indicate a close affinity to the greater eiders (*Somateria*) and dabbling ducks (*Anas*) and a greater taxonomic distance from *Tadorna* (Jacob 1977). 6) The spur wing (*Plectropterus*) is generally placed within the Cairinini but shows clear affinities to the Tadornini.

**10.2.2.3. Chemotaxonomy of Uropygial Gland Waxes.** The Anseriformes are well studied in regard to the chemistry of their uropygial gland waxes: 33 species from 12 tribes have been examined. The taxonomic evaluation of the alcohol/methylene-substituted fatty acids, however, is not suitable for construction of a phylogenetic tree. What it does allow, according to Edkins and Hansen (1972), Jacob and Glaser (1975), and Jacob (1977), is a clear distinction between the groups listed

| Group | Species | Tribe |
|-------|---------|-------|
| 1 | *Anser anser* (greylag goose)<br>*Anser caerulescens* (snow goose)<br>*Cygnus cygnus* (whooper swan)<br>*Cygnus columbianus* (whistling swan) | Anserini<br>Anserini<br>Anserini<br>Anserini |
| 2 | *Dendrocygnus viduata* (white-faced whistling duck)<br>*Cygnus melanocoryphus* (black-necked swan) | Dendrocygnini<br>Anserini |
| 3 | *Cygnus olor* (mute swan)<br>*Anser fabilis* (bean goose)<br>*Anser indicus* (bar-headed goose)<br>*Tadorna tadornoides* (Australian shelduck)<br>*Aythya ferina* (common porchard) | Anserini<br>Anserini<br>Anserini<br>Tadornini<br>Aythyini |
| 4 | *Stictonetta naevosa* (freckled duck)<br>*Cereopsis novaehollandiae* (Cape Barren goose)<br>*Branta leucopsis* (barnacle goose)<br>*Coscoroba coscoroba* (Coscoroba swan)<br>*Tadorna tadorna* (common shelduck)<br>*Tadorna ferruginea* (ruddy shelduck)<br>*Mergus serrator* (red-breasted merganser)<br>*Mergus albellus* (smew)<br>*Melanitta nigra* (black scoter)<br>*Cairina moschata* (Moscovy duck) | Stictonettini<br>Cereopsini<br>Anserini<br>Anserini<br>Tadornini<br>Tadornini<br>Mergini<br>Mergini<br>Mergini<br>Cairinini |
| 5 | *Aythya fuligula* (tufted duck)<br>*Somateria molissima* (common eider) | Aythyini<br>Mergini |
| 6 | *Tachyeres patagonicus* (long-winged steamer duck) | Tachyerini |
| 7 | *Anas strepera* (gadwall)<br>*Anas clypeata* (northern shoveler)<br>*Anas platyrhynchos* (mallard)<br>*Cygnus atratus* (black swan)<br>*Chenonetta jubata* (maned wood duck) | Anatini<br>Anatini<br>Anatini<br>Anserini<br>Cairinini |
| 8 | *Nettapus pulchellus* (green pygmy goose) | Cairinini |
| 9 | *Biziura lobata* (Australian musk duck) | Oxyurini |
| 10 | *Anseranas semipalmata* (magpie goose) | Anseranatini |

Table 4. Groupings of anatid species based on uropygial gland waxes. Wax composition is the same for the members of a group. Note that the composition is not correlated with tribal classification. After Scherer and Sontag (1986), who compiled data from various authors.

in Table 4. With their predominance of singly and doubly methylene-substituted fatty acids, the dabbling ducks (*Anas*) stand out and are separated from the maned wood duck (*Chenonetta*) and the black swan (*Cygnus atratus*).

Contrary to the view of Poltz and Jacob (1973, 1974a) and Jacob (1977) that uropygial gland waxes might be of taxonomic value among birds generally, the results presented in Table 4 seem to indicate that the uropygial gland waxes, at least of the Anseriformes, are less suited to clarifying systematic affinities. This is particularly obvious in the chemotaxonomic distance between the common porchard (*Aythya ferina*) and the tufted duck (*A. fuligula*), as well as between the black swan (*Cygnus atratus*) and the mute swan (*C. olor*). Both pairs hybridize easily and are very similar morphologically. The Anserini, as well as the Tadornini, do not form a unified group. Interestingly, the uropygial gland wax of *Anseranas* is completely out of line of the variation range within the Anatidae (as are the amino acid sequences of the hemoglobins).

Although an unequivocal taxonomy within the Anatidae cannot be based on the uropygial gland waxes, a comparison with other avian orders shows that the uropygial gland waxes can be used as a taxonomic character at the level from families to orders (and therefore probably is also useful for distinguishing basic types) (Poltz and Jacob 1974b, Jacob and Grimmer 1975, Jacob and Horschelmann 1982).

**10.2.2.4. Hennigian Analysis of Electrophoretic Patterns of Proteins.** The comparative electrophoretic investigation of proteins has methodical limitations. Since the behavior of a protein in an electric field depends on a number of factors, the taxonomic weight of electrophoretic patterns cannot always be determined accurately. The obvious approach, therefore, is to circumvent these limitations by applying parallel electrophoretic tests to several genetic products. This was first done for the Anatidae by Patton and Avise (1983), who did a comparative investigation of 82 "electromorphic characters" (representing 17-19 gene loci) of 26 species. Although a quantitative evaluation and cluster analysis of the electrophoretically determined similarities were conducted (see Brush and Witt 1983), using the data to construct a phylogenetic tree according to Hennig's criteria (Hennig 1966) leads to a much closer match with classical phylogenetic trees. Notably the Anserini are clearly separated from other anatids, and the Mergini do not represent a unified group, which is confirmed by the scattered interspecific hybridizations within the Mergini (see Figure 1). In general, individual pairs of species appear to be much more distant on the basis of their proteins than was previously assumed. An extreme case is the pair, gadwall (*Anas strepera*) and American wigeon (*A. americana*), which are as different from each other as swans are from geese.

**10.2.2.5. Hemoglobin Sequences.** The hemoglobins have been well studied (Perutz 1983). Numerous amino acid sequences are known, but it was only in the recent past that a team under Braunitzer clarified the primary structure of some avian hemoglobins. Generally it appears that the α-globin are much more homogenous than the β-globin. This is unusual since in other animals the opposite is the case. Braunitzer and Oberthür (1979), as well as Oberthür et al. (1980), saw an explanation for this in the avian-specific regulation of the $O_2$ affinity by inosite pentaphosphate on the β-globin molecule. This function would limit the

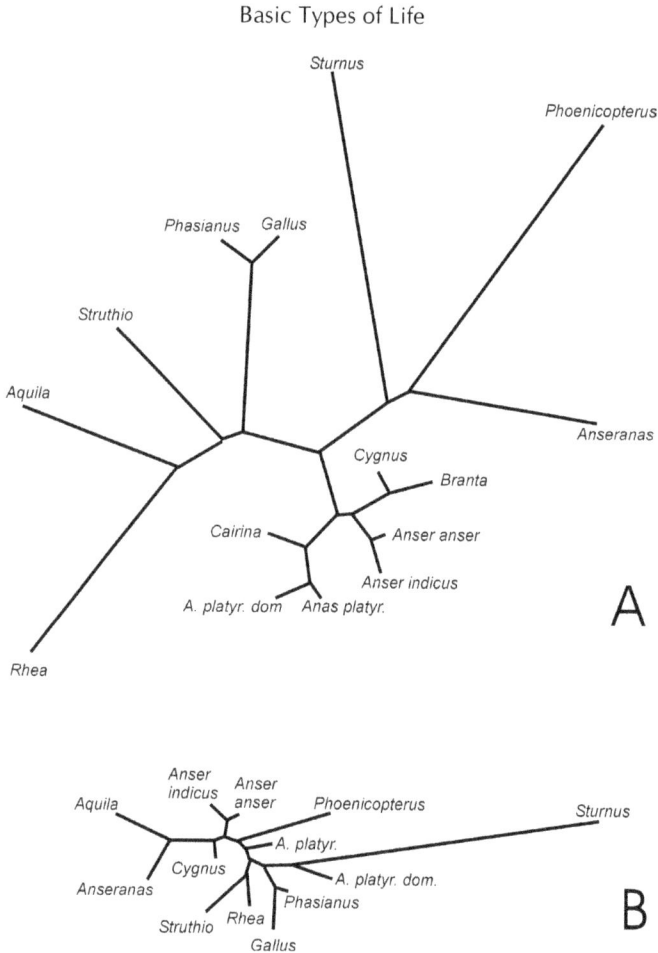

Figure 6. Similarity networks of some birds based on α-globin (A) and β-globin (B).

range of variability—the norm of mutative reactivity of the β-globin—considerably.

For a better presentation of the data, known sequences were used to calculate graphic similarity networks which do not necessarily represent the actual phylogenetic relationships (Figure 6). It is particularly obvious that the Anatidae with reference to their α-globin are separated from all other avian orders. With reference to both hemoglobins, the magpie goose (*Anseranas*) cannot be placed within the Anatidae. The Flamingos (Phoenicopteriformes) are usually considered to be a sister group of the Anseriformes, but this hypothesis is not confirmed by the hemoglobin sequences. The pheasant (*Phasianus colchicus*) and domestic chicken (*Gallus gallus*) show clear similarities to each other. Interestingly, these two species hybridize (Gray 1958; Stadie 1967; comp. Klemm 1993 concerning the basic type of the Phasianidae). According to classical systematics, the ostrich (*Struthio*) and the greater rhea (*Rhea*) are placed within the same order Struthioniformes (Wolters 1975-1982), which is confirmed by the β-globin but not by the α-globin. The starling (*Sturnus*), as the only representative of the passerine birds (Passeriformes), occupies an isolated position with reference to its β-globin. A comparison of both similarity networks shows that two proteins from the same species can indicate very different levels of similarity.

The notable position of the domestic duck (*Anas platyrhynchos* f. *domesticus*) deserves special attention. The closest relative of the domestic duck with reference to its α-globin is the wild form (which is to be expected), but according to its β-globin it does not even belong to the anatid family. This finding illustrates how rapidly multiple amino acid differences in isolated small populations can become fixed.

**10.2.2.6. DNA-Analysis[3].** So far, nine dabbling ducks (*Anas*) and four pochards (*Aythya*) have been investigated comparatively by analyzing their mitochondrial DNA (Kessler and Avise 1984) (the gene fragments resulting from splitting were characterized electrophoretically with restriction endonucleases). Brown et al. (1979) assumed that the mitochondrial DNA of mammals evolves 5 to 10 times faster than nuclear DNA and might therefore be particularly useful for a reliable delimitation of lower taxa from each other (comp. Barton and Jones 1983). Kessler and Avise (1984) designed a dendrogram on the basis of their analysis, which largely correlates with classical systematic concepts. It must, however, be noted that *Anas* and *Aythya* belong to the taxonomically problematic anatids.

A basic problem of systematics lies in the difficulty of evaluating and comparing quantitatively an adequately high number of characters of the organisms in question. DNA-DNA-hybridizations provide an average measure for sequence similarities of all single-copy genes of two genomes. This method has been used for numerous comparative studies, especially of birds (overview in Sibley and Ahlquist 1990); however, the results are not undisputed. The first problem to be considered is the exactness of measuring the differences in the average melting point temperature of the DNA-hybrids. Plus, for a reliable construction of the dendrogram, all paired distances of the investigated species have to be known. To determine those by DNA-DNA-hybridization consumes so much time and material that it cannot be done with a large number of species. The data that are known thus far, as well as other characters, can, however, be used as a point of reference to evaluate levels of similarity.

---

3    Today, a wealth of DNA sequence data is available. Recently, a comprehensive study of 30kb of DNA from 169 birds has been published (Hackett 2008, cited after Fehrer 2009). These data suggest that the Anhimidae and Anseranatidae are deeply separated from the Anatidae, which corresponds to the morphological and anatomical traits described in this paper as well as to the lack of hybrids between these three families.

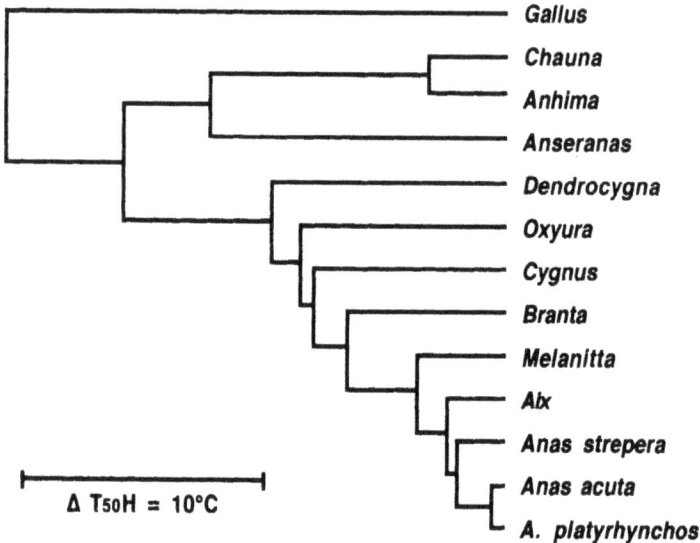

Figure 7. Dendrogram based on DNA-DNA hybridization. Adapted from Sibly & Alquist (1990). Note that *Anseranas* clusters towards Anhimidae. Anhimidae and Anseranatidae are well separated from Anatidae.

With reference to the Anseriformes, Madsen et al. (1988), as well as Sibley and Ahlquist (1990), provide some details (Figure 7). Contrary to Sibley and Ahlquist, Madsen et al. place the magpie goose (*Anseranas*) within the anatid clade. The other results are identical.

DNA-sequences of only a small piece of repetitive DNA have been investigated within several anatids (Madsen et al. 1992), but the results have not been evaluated systematically. An increase of the data base of anatid sequences is necessary.

**10.2.3. The Fossil Record of the Anatidae.**[4] It is generally assumed that the fossil record of the birds is of such a fragmentary nature that useful phylogenetic deductions are hardly possible. An extensive overview of avian paleontology (Olson 1985), however, shows that this is not true for all groups of birds. Museum collections contain a considerable number of fossil birds; many of them have not been investigated. In this section we will discuss the oldest anseriform fossils. Pliocene and Pleistocene fossils do not contribute significantly to resolving the question of the origin of the anatid family, because they represent forms that are always congeneric and often conspecific with recent species.

---

4    Clarke et al (2004) describe a partial skeleton of *Vegavis* (Anseriformes) from the first Cretaceous fossil within the extant bird radiation. They suggest that the families Anatidae and Galliformes are old and were already differentiated when duck-like and chicken-like birds lived together with dinosaurs.

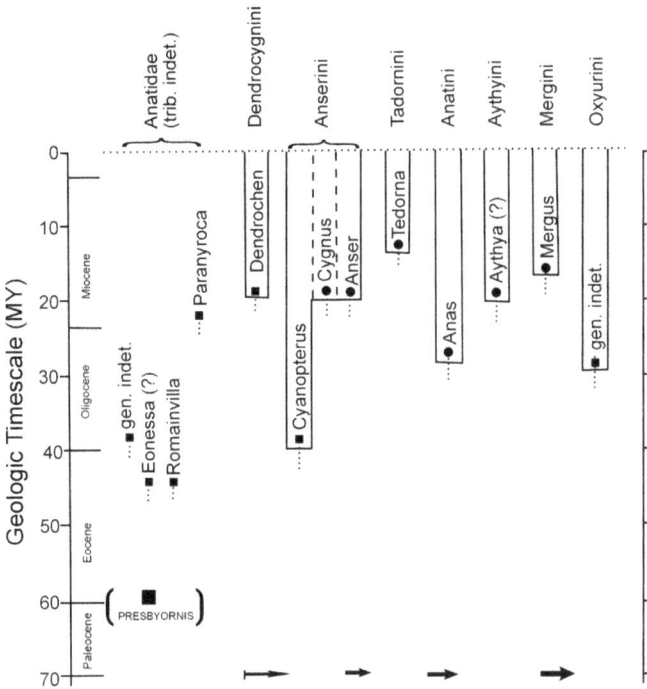

Figure 8. Early bird fossils belonging to Anatidae, compiled from Wetmore (1938), Howard (1964), Alvarez & Olson (1978), Olson & Feduccia (1980), and Olson (1985).

By 1964, approximately 170 fossil anatid species had been described (Howard 1964). This impressive number is actually an overestimate because species descriptions often had to be done on the basis of fragmentary material. Older studies have to be interpreted carefully, as far as taxonomic placements are concerned. In spite of this, some well-supported findings remain, providing a rough overview of the fossil history of the anatids. In Figure 8, the earliest fossils of anatid genera have been compiled from different sources.

Wetmore (1938) ascribed fragments of a right wing from the Upper Eocene as *Eonessa* to the Anseriformes, which according to Olson and Feduccia (1980) is not tenable. The oldest undisputed anatid fossil is *Romainvilla* from the Upper Eocene of France. Although clearly anatid, *Romainvilla* shows so many peculiarities that a separate subfamily Romainvillinae was created. The animal resembles dabbling ducks (Anatini) in its robust proportions but possesses a relatively long tarsometatarsus (leg), similar to that of whistling ducks (Dendrocygnini). Its coracoid indicates a connection to *Anseranas* (Anseranatidae) and the swans (Anserini) (Howard 1964). This combination of traits is reminiscent of living aberrant types. The oldest anatid fossils can hardly be placed into recent tribes, which is also generally true of avian fossils from the

Basic Types of Life

Figure 9. Reconstruction of *Presbyornis* redrawn from Olson & Feduccia (1980). Height is 40-50 cm.

Oligocene (Olson 1985). The oldest anatid fossil which could be placed in a recent tribe is *Cyanopterus* (Anserini) from the Lower Oligocene.

The findings known so far and compiled in Figure 8 create the impression that the Anatidae arose at the beginning of the Miocene in a considerable variety of forms and were widely distributed. In the Lower Miocene/Upper Oligocene they are represented in Africa (Anatini), Australia (Oxyurini), North America (Dendrocygnini), Siberia (Anatini) and Europe (Anserini, Anatini, Aythyini). Because highly specialized forms arose very early (e.g., *Mergus* in the Middle Miocene of Virginia: Alvarez and Olson 1978), Olson and Feduccia (1980) concluded that the differentiation of most anatid genera was concluded by the Middle Miocene. The radiation of the family had to have been "extremely rapid."

In 1926, Wetmore described a tarsometatarsus (leg) from the Lower Eocene (Green River Formation, Utah), named it *Presbyornis pervetus,* and placed it within its own family Presbyornithidae. Wetmore realized the close affinities to the Recurvirostridae (stilts) and therefore placed *Presbyornis* within the order Charadriiformes (which include the snipe, osytercatchers, plovers, curlews, and gulls). Later, many more fossils were found in the Green River Formation (McGrew and Feduccia 1973; Feduccia 1978), suggesting the catastrophic burial of a breeding colony. Feduccia and McGrew's (1975) interpretation of the new findings was an affinity of *Presbyornis* to the Phoenicopteridae (flamingos). The discovery of an almost complete skull led to a new revision (Olson and Feduccia 1980). *Presbyornis* was described as a mosaic form with charadriiform, phoenicopteriform and anseriform traits (Figure 9). The postcranial skeleton is undoubtedly charadriiform, but the skull and especially the bill shows typical duck-like features. Although the authors left *Presbyornis* within the order Charadriiformes, they regard this animal as an ancestor of the Anseriformes, a hypothesis most authors oppose (see below).

## 10.3. ... and Interpretations
### 10.3.1. The Anatidae as a Basic Type
#### 10.3.1.1 The Anatidae within the Classification of the Birds. The order Anseriformes is generally considered to be a primitive and ancient

group. However, there is no unity among the experts concerning its affinities to other avian orders. Generally, gallinaceous birds (Galliformes) are considered to be the closest relatives of the Anseriformes, due especially to the morphology of the skull (Bock 1969). Cracraft (1981), too, positioned the Anseriformes close to the Galliformes, but believes that this issue has not yet been settled. Others would like to combine the Anseriformes, via the flamingos, with the Ciconiiformes, *i.e.*, storks and herons (Sibley and Ahlquist 1972). Both positions are disputed by Olson and Feduccia (1980), who consider the

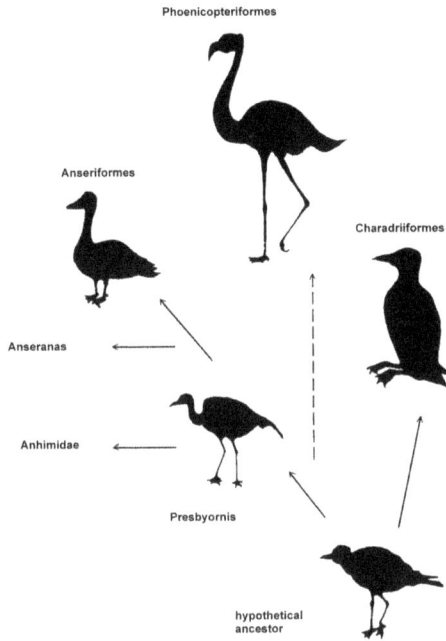

Figure 10. Putative phylogenetic relationship between flamingos, anatids, and charadriiform birds, interpreting *Presbyornis* as a missing link. Adapted from Feduccia (1977), artwork by Marion Häberle.

Anseriformes to be derived Charadriiformes. Since, according to these authors, the flamingos also descended from the ancestral Charadriiformes, they must be considered the sister group of the Anseriformes (Figure 10).

This last interpretation of affinities is based primarily on *Presbyornis*. If this bird was indeed the ancestor of the Anseriformes, they are undoubtedly closely related to the Charadriiformes (wading and shorebirds). Can *Presbyornis* indeed be considered a "missing link" or "proto-duck?" Cracraft (1981) explains in detail why he does not agree with this interpretation. He thinks Olson and Feduccia made the (frequently occurring) mistake of regarding the characters of fossils *ipso facto* as primitive. Olson (1982) countered that Cracraft disregarded the basic rules of phylogenetic systematics—the same reproof that Raikow (1981) directed towards Olson. This short excursion to an often emotional controversy, at least, shows that experts have not acknowledged *Presbyornis* to be the "proto-duck." Obviously it has not been possible to achieve unity of opinion by way of morphological analyses of the available fossil material. In such situations, other areas of comparative biology must corroborate the arguments. Molecular traits such as crystallin sequences (Stapel *et al.* 1984) and DNA-DNA hybridizations show that the Anseriformes could be a sister group of the Galliformes and Craciformes (curassows and mound-builders) (Sibley *et al.* 1988; Sibley and Ahlquist 1990). The team surrounding Sibley combines the three orders within the Superorder Gallomorphae, based on DNA-DNA-hybridization data and place the Gallomorphae into the infraclass "Eoaves," as opposed to the "Neoaves," which comprise the Charadriiformes, Phoenicopteriformes and Ciconiiformes. These results would exclude *Presbyornis* from being an ancestor of the ducks, since this animal undoubtedly belongs to the Charadriiformes, and, thus, to a different infraclass from the ducks. Even Olson and Feduccia themselves consider the skull of *Presbyornis* to show not only distinctly charadriiform traits, but also numerous differences from the Anatidae. Plus, according to Olson (1985), samples of a fossil humerus found in the Upper Cretaceous of New Jersey and described as *Telmatornis rex* Shuffeldt showed duck-like proportions. Its systematic position has not been determined, but if *Telmatornis* really is an anatid, then the much younger (Eocene) charadriiform *Presbyornis* cannot be the ancestor of the ducks.

Based on currently known data, it would be better to interpret the duck-like bill of *Presbyornis* and the "true" anatid bill as functional rather than phylogenetic similarities. Thus, *Presbyornis* would be an extinct charadriiform bird which may have occupied similar ecological niches as modern anatids or flamingos. It was probably not the "proto-duck."

**10.3.1.2. The Systematic Position of *Anseranas*.** The systematic position of the magpie goose (*Anseranas*) has been disputed for a long time. The order Anseriformes is divided into two or three families; but, among other things, there is disagreement about whether *Anseranas* should have family rank like the screamers (Anhimidae). Johnsgard (1961b), due to ethological observations, concludes that *Anseranas* should at least be placed into its own subfamily [within the Anatidae]. On the other hand, Woolfenden (1961) argued that osteology requires that it be segregated as a monotypic family. Based upon comparative investigations of the leg musculature of numerous anatids, Schulin (1981) agrees that *Anseranas* has to be taken out of the Anatidae. Otherwise, *Anseranas* is known to share a number of traits with the Anhimidae, including some morphological, anatomical, and ethological characters (Delacour 1954-1964 Volume1; Johnsgard 1961a, 1961b, 1978; Woolfenden 1961; Olson and Feduccia 1980). Peters (1976) concluded that "Overall, the peculiarities of the magpie goose separate more than connect it and the Anatidae." Contrary to Johnsgard (1978), Wolters (1975-1982) places the magpie goose into its own family. Biochemical and molecular biological findings point in the same direction (Scherer and Sontag 1986). Especially the amino acid sequence analysis of the α-globin, and the DNA-DNA hybridizations clearly delimit *Anseranas* from the Anatidae (comp. Figure 6 and 7). In summary, it appears to be appropriate to place *Anseranas,* at least, into its own family.

Although hybrids between *Anseranas* and the Anatidae have never been reported, this species was previously placed into the basic type Anatidae (Scherer and Hilsberg 1982). This view is now untenable because *Anseranas* is known to deviate strongly from the Anatidae in a wide spectrum of molecular, anatomical, morphological, and ethological characters.

Regrettably, to this date no molecular taxonomic studies have been done on any of the species within the family Anhimidae (screamers), so the question whether *Anseranas* indeed belongs to the Anhimidae must remain unanswered at this time. Should a closer affinity to the Anhimidae be confirmed, the duck-like bill would have to be regarded as a convergence with the Anatidae. Olson and Feduccia (1980), on the other hand, regard *Anseranas* and the Anhimidae as derived Anseriformes, according to which view the Anhimidae would have secondarily lost the anatid bill.

**10.3.1.3. The Importance of Hybrids.** With the exception of the magpie goose (*Anseranas*; see above), all 149 (according to the taxonomy of Johnsgard 1978) species of ducks, geese, and swans belong to the Anatidae and could hardly be placed into any other group. Anatomical, morphological, and especially genetic data demonstrate very close relationships among the individual species. The numerous hybrids call for the conclusion that not only are there external similarities, but the similarities are indeed the result of a phylogenetic and, thus, historical relationship. A common ancestral population could be postulated as the point of origin of the Anatidae. In the course of the differentiation of this population, speciation processes could have occurred which resulted in the formation of various biological species. As discussed elsewhere (Junker 1993), the term "biological species concept" has its rightful place within the field of speciation; although, in the antids it remains difficult to apply as numerous formations of hybrids in their natural environment clearly demonstrate. The taxonomic difficulty of defining a species is made evident by the fact that systematics experts disagree concerning numerous forms. Johnsgard (1978), for example, lists seven different species for the swans (*Cygnus*). Other authors list only five species; they regard *C. bewickii* as a subspecies of *C. columbianus*, and *C. buccinator* as a subspecies of *C. cygnus* (Klös and Klös 1975). Such variations of classification also occur frequently among the Anatini, for example, in the case of *Anas platyrhynchos* and *A. poecilorhyncha* (=*superciliosa*), which, on the Mariana Islands, form a hybrid population described as *A. oustaleti.* Therfore, one could justify—in opposition to the present thesis— uniting *A. poecilorhyncha* with *A. platyrhynchus.* These examples should

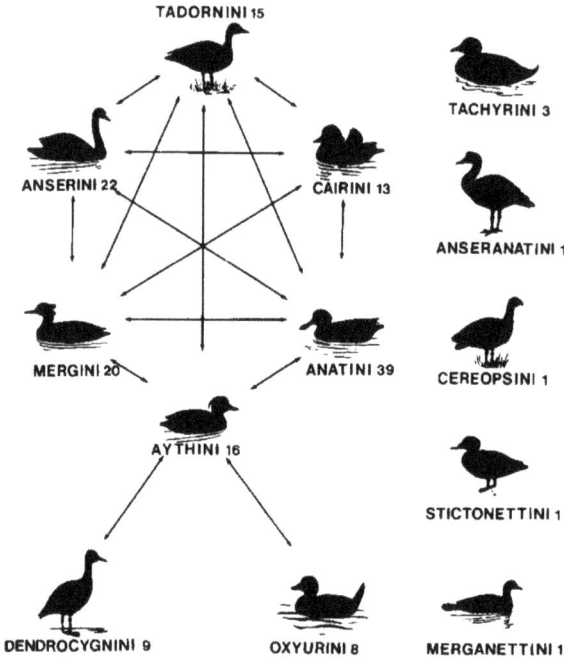

Figure 11. Hybridization network showing the known hybrid connections amoung the 13 tribes of anatids. Very often, genera within a tribe are connected by hybridization. Although only a few hundred hybrids have been observed, these data provide a reasonable basis for the hypothesis that Anatidae form a basic type. Number of species indicated following tribal name. Artwork by Thomas Hilsberg.

sufficiently demonstrate that it can be difficult to unequivocally define biological species within the Anatidae; although, this is not always the case. More difficult problems, however, arise when it comes to defining genera and tribes. This fact is emphasized not only by the hybridization data from Figure 1 but also by the numerous contradicting classifications of the Anatidae having considerable differences at the generic and tribal ranks (see Table 1).

On the other hand, there are hardly any problems when it comes to the question of whether a species belongs to the Anatidae, which is answered not only by anatomy and morphology but also by hybridization research (Johnsgard 1960). Only 26 out of 149 anatid species are not known to hybridize; all others are connected by the reported hybrids (Figure 11). The absence of hybrids for these 26 species does not mean that their hybridization is impossible but that further experiments are needed. Plus, in many cases these species represent aberrant types within the Anseriformes. Although no interspecific hybrids have been observed, they should not be excluded from the basic type (unlike the magpie goose) since their other characters suggest their affinity to the Anatidae.

Basically, the hypothesis that the family Anatidae (excluding the magpie goose and the screamers) is a basic type seems to be well-founded but is preliminary. Further hybridization data could lead to a revision.

**10.3.2.    Genetically Polyvalent    Ancestral Forms?**    Characters    of organisms can be described somewhat objectively, using biological methods. Their interpretation,    however, whether    within    cladistic or phylogenetic systems of

Figure 12. The bimaculated face pattern is found in a variety of anatid species. Top: *Branta ruficollis* (red-breasted goose), *Anas formosa* (Baikal teal); bottom: *Bucephala clangula* (common goldeneye), *Anas specularis* (bronze-winged duck). Artwork by Thomas Hilsberg.

classification, is frequently disputed. Especially the issue of the history of a group of organisms is often influenced by preconceptions. The great majority of biologists interpret the history of living beings under the premise of "primitive" (genetically less complex) ancestral forms.

The opposite premise is that the ancestral forms of basic types were genetically complex. By "complex" we mean a high potential of variability such as does not exist to the same extent in present-day forms; that is, to a certain extent a "pool" of possible characters were distributed to present-day biological species through the historical processes of speciation. (A hypothesis which is similar, but not identical to this one, is the concept of generalized ancestors.) Speciation basically means the loss of genetic potential and, as a result, specialization. The following discussion investigates whether the characters of anatids can be interpreted within the framework of complex (genetically polyvalent) ancestral forms, and it points out the problems that arise when an interpretation within the framework of "primitive" ancestors is sought.

**10.3.2.1. Morphological Traits.** The surprising similarities between hybrids from two species and a third species could be interpreted as a result of the same basic genetic potential in all three species. The expression of this potential has been partially suppressed until the chromosomes of two species were united, thus releasing the available (however suppressed) information. (The molecular genetic basis of such processes is as yet unknown.)

The specific plummage markings are generally not determined by function, although the pattern in its entirety could, for example, serve for mating purposes. Thus, it is not easy to explain similarities by convergence because no corresponding selective pressure can be identified. The similarities between individual species from very distinct tribes within the Anatidae can be better explained by assuming that their potential was already present in the ancestral forms but was exhibited in differing degrees in the descendants (comp. Harrison and Harrison 1971). This hypothesis is discussed below, using the example of the "bimaculated face pattern."

According to Harrison (1978), a spot in front of or behind the eye is a remnant of the original pattern (Figure 12), which is particularly well exhibited in the Baikal teal (*Anas formosa*). Therefore, this species is considered to be an original representative of the Anatidae (Harrison and Harrison 1971). The derived form would have resulted from the loss of a trait. The hybrid between domestic ducks and mallards described above (see Figure 5) shows that the genes responsible for the "bimaculated face pattern" are present in the mallard in a latent form. During evolution from an ancestral form, this genetic complex lost its function (probably not due to mutations in the structural genes but to changes in the regulatory genes). The splitting of the mallard into the domestic and wild duck may have led to such different alleles in their respective regulatory areas that the two genomes are no longer balanced in hybrids, allowing the ancestral pattern to be expressed. Either this was caused by small changes, or the domestic duck diverged significantly from the mallard during domestication (which is confirmed by molecular biological data: Godovac-Zimmermann and Braunitzer 1983). After all, in all other cases the "bimaculated face pattern" is exhibited only in hybrids between different species. One fact that does, however, argue for the first option, is the occurrence of the respective atavism in rare variations of the common teal (*Anas crecca*) that occur without being preceded by hybridization (Harrison 1945; Harrison and Harrison 1971).

The fact that the "bimaculated face pattern" in its complete, reduced, or altered form occurs independently of taxonomic position (see Table 3) confirms Harrison's (1953) assumption concerning the Anatidae. That is, "... one must, I think, postulate that in the course of evolution the genes of the original genotype of the family have been dispersed and are carried in the species and its races throughout the global range of the family...."

Within the Anatidae, numerous characters occur independently in various forms. Livezey (1986), upon an examination of 120 characters, suggests an optimal phylogenetic tree (a tree that postulates the least amount of convergences or reversals). Nevertheless, Livezey had to assume convergences or reversals for 47 out of these 120 characters.

However, the usual explanation of convergent evolution (Lack 1974) is plausible only if a sufficiently selective advantage for the respective trait can be identified, which is generally difficult. One of the few possible examples might be the shape of the bill (see Figure 3). Studies of African finches (*Pyrenestes*) in their natural environment show that the size of their bill can vary within a few years. This trait is influenced by only *one* genetic factor. The polymorphism that was observed probably came as a result of single mutations, followed by disruptive selection (Smith 1993). Investigations on the shape of bills are not available.

**10.3.2.2. Ethological Traits.** Important foundational facts on the ethology of ducks were provided by Konrad Lorenz. Lorenz (1960) summarized his findings on the structure and genesis of movements as follows:

> Many coordinations that have been described as special movements are just species-specific combinations of certain basic coordinations that can be found in almost all Anatini. When observed directly, these combinations appear to be very different from each other; however, a film analysis shows almost complete agreement between their elements. The most recent step in the evolution of the respective movements is most certainly their quantitative increase or reduction, sometimes their disappearance; the next recent step is the coupling and uncoupling of elementary movements that are hard to change. The absence of the latter is most certainly a consequence of a secondary loss which is confirmed by the behavior of hybrids.
>
> The linkage of hitherto independent elements of movement into a stable sequence is part of a phyletic process, the importance of which was recognized already by J. S. Huxley in 1914, who named it "ritualization". Although this process has only been studied within the comparative morphology of expressive movements, the results of its investigation will certainly provide important insights into the evolutionary process of inherited coordinations as such.
>
> Besides the coordinative blending of two heterogeneous impulse sequences into one movement, another evolutionary process plays an important role in ritualization. Through this process, the newly formed inherited coordination becomes independent from the stimuli that had initiated its original components. The newly formed coordination can, with regard to its stimulation, become completely autonomous, or it can become dependent upon another initiating factor.

Older ethological studies (Heinroth 1910, Lorenz 1941) also suggest that the presence of some inherited coordinations represents the original state of the group and that derived forms developed through the loss of certain elements of movement. It can therefore be assumed that, in the original representatives of the dabbling ducks, there was a "pool" of elementary movements, from which, through the loss, recombination, or variation of the original elements, behavioral patterns were formed. This assumption is supported also by behavioral patterns observed in hybrids that prove to correlate closely with those of third species. However, this hypothesis is presently limited to the Anatini, since other anatids have not been investigated.

**10.3.2.3. Molecular Traits.** One of the first studies of the molecular taxonomy of the birds on the basis of electrophoretic data (Sibley 1960, p. 211) already contains the following remarkable statement: "The conclusion seems inescapable that the Anatidae are a very closely related group and that the characters of male plumage and bill shape which have, in the past, been used as the basis for generic separation do not reflect accurately the degrees of relationship among the species." There is no doubt that the characters mentioned above are very useful for the distinction of different biological species (Mayr 1967). The differentiation of ancestral forms into biological species (for an overview see Junker 1993) can occur most rapidly through a change of regulatory gene regions that influence, for example, the color patterns of the plumage and, thus, the mating behavior. Once a reproductive barrier has been established (whether for geographical, behavioral, or genetic reasons), a strong selective pressure towards a differentiation of the external features is no longer there, unless it has to do with an adaptation to special ecological niches. External appearance and molecular characters can therefore show a wide gap. (This leaves us with the foundational, and, at this point, unanswerable question, of which characters should provide the basis for systematics.)

The uropygial gland waxes of the anatids are among those characters that can hardly be used for classification below the taxonomic level of the family. This could be interpreted as an indication of the presence of an enzymatic system within the ancestral anatids that provided the basis for the synthesis of the uropygial gland waxes. In the course of evolution, the concentrations of the individual enzymes were changed by mutations in regulatory regions, which resulted in a variety of uropygial gland wax chemistries in the descendants. The likelihood of this hypothesis was demonstrated by a study of the malonyl-CoA-decarboxylase of different anatids (Kim and Kolattukudy 1978). This enzyme occurs in all anatids examined and is very similarly structured. However, its concentration in the uropygial gland varies, which significantly influences the content of polymethyl-substituted fatty acids. The changes of the concentration

Figure 13. Aberrant types of anatids, exhibiting a mixture of characters form dfferent tribes. Aberrant types are difficult to classify within other tribes. Clockwise from top left = steamer duck (*Tachyeres*), Cape Barren goose (*Cereopsis novaehollandiae*), Coscoroba swan (*Coscoroba coscoroba*), Egyptian goose (*Alopochen aeyptiaca*).

of these enzymes seem to have happened independently in different anatids, thus making its taxonomic evaluation impossible. On the other hand, the ancestral forms of other basic types were equipped with other enzymatic systems and, therefore, produced different uropygial gland waxes (comp. Poltz and Jacob 1974b, Jacob and Grimmer 1975, Jacob and Hoerschelmann 1982).

The fact that a comparative interpretation of amino acid sequences presents certain problems is clearly shown by the molecular differentiation of domesticated ducks. In the course of domestication, five(!) positions of the β-globin of the domestic duck changed in comparison to its wild progenitor, so that, according to this measure, the domestic duck would not even belong to the Anatidae any more. The α-globin, on the other hand, changed in only two positions within the same time frame. The hemoglobin of the domestic goose shows only one difference from that of the greylag goose. Similar conditions can be expected for the ovalbumins and lysozymes of the Anseriformes and Galliformes. Such findings clearly demonstrate that massive microevolutionary changes of proteins can occur rather rapidly (*i.e.*, within a few thousand years). Also, in the light of these facts, a conversion of differences in amino acid sequences into millions of years (e.g., see Oberthür and Braunitzer 1984) becomes questionable. After all, nobody knows whether individual species, due to functional aspects of proteins or extended periods of extreme isolation in very small populations (bottle neck-effect of Ayala and Valentine

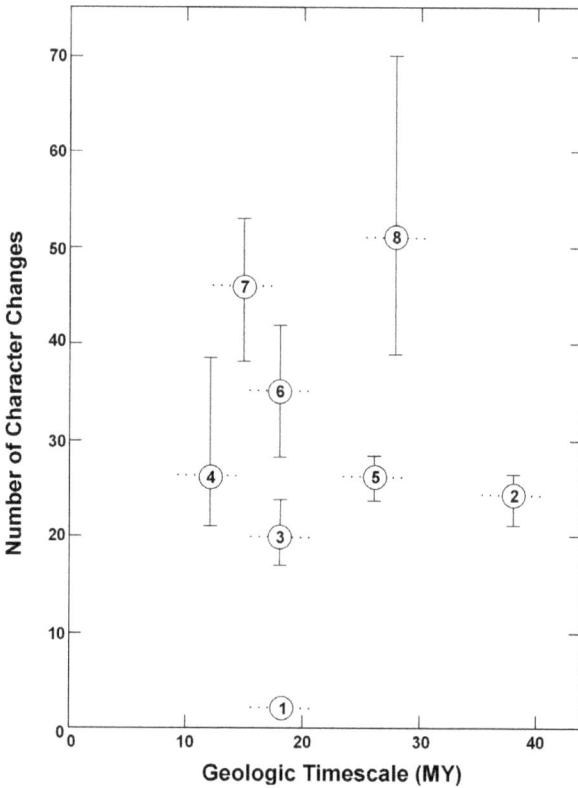

Figure 14. This graph shows character changes which were postulated by Livezey (1986) when constructing a phylogenetic tree using the minimum total number of character changes. The number of putative character changes of tribes is plotted versus the geologic time of first appearance in the fossil record. Tribes with high numbers of character changes are derived and, therefore, should be found in top strata (*i.e.*, left in the graph). Horizontal dotted lines indicate possible variation of stratigraphical appearence while vertical bars give deviations in number of character changes when different genera of a tribe are considered. Note that there is virtually no relation between stratigraphical position and hypothetical evolutionary stage. 1 = Dendorocygnini; 2 = Cygnini; 3 = Anserini; 4 = Tadornini; 5 = Anatini; 6 = Aythyini; 7 = Mergini; 8 = Oxyurini.

1979, pp.117-120), might not have experienced an accelerated fixation of mutations. If we assume that the separation of the mallard and the domestic duck started 3000-5000 years ago, then the β-globin would show an exchange rate of approximately 1 amino acid per 600-1000 years. However, in hemoglobins, an exchange rate of 1 amino acid in 6-10 million years is commonly assumed (Zuckerkandl and Pauling 1962, comp. Oberthür et al. 1982, 1983). Thus, in the course of domestication, the evolutionary speed of the β-globin would have increased by a factor of 10,000. This shows that a phylogenetic interpretation of amino acid sequences *per se*, especially within the framework of the "molecular clock", is questionable (comp. also Scherer 1990).

**10.3.2.4. Aberrant Types.** The numerous "aberrant types" within the Anatidae, such as the Cape Barren goose, the Coscoroba swan, the freckled duck and many others (Figure 13), illustrate the taxonomic problems that result from conflicting combinations of characters. It is impossible to include these species in a taxonomic system. Is coincidence, convergence, or narrowing of the genetic potential specific to the basic type during speciation responsible for mosaic-type combinations of traits in aberrant types on all biological levels? The hypothesis of coincidence is highly unlikely. Convergence as an explanation requires the presence of strong selective pressures, which are, however, unknown. This author, therefore, prefers the hypothesis that the aberrant anatids, as specialized end products of microevolutionary speciation processes, indicate an ancestral form with a complex variation potential. Similarities of an aberrant form with several other tribes can be explained by the assumption that the respective characters were already potentially present in the ancestral form. In Hennig's terminology they would be called plesiomorphies.

**10.3.2.5. Fossil ducks.** The sequence of individual historical events in the differentiation of the anatids cannot be deducted from the paleontological record, at least not thus far. Usually, a phylogenetic differentiation from a "primitive" to a "complex" or, in less subjective words, from an original to a derived state is assumed, but whether a character should be considered to be derived or primitive, is often disputed.

Livezey (1986), in an extensive study, attempted a phylogenetic analysis of all recent anatid genera, based on 120 individual morphological characters, and suggested a corresponding phylogenetic tree. His evaluations basically agree with those of older authors (e.g., Delacour 1954-1964, Johnsgard 1978). According to him, the original tribe would be the whistling ducks (Dendrocygnini), the most derived tribes the stiff-tailed ducks (Oxyurini), and the scoters, goldeneyes, and mergansers (Mergini). Livezey's study provides a basis for calculating the number of character changes for each genus or the average number of character changes for each tribe, if the assumed branching point between *Anseranas* and the Anatidae is chosen as a point of reference. When these data are compared with the fossil record (Figure 14), there is no correlation to traditional phylogenetic concepts. Highly specialized genera can be found in old strata, whereas the supposedly primitive Dendrocygnini arise comparatively late. Figure 14, at least, does not show any increase in the specialization of forms with younger strata.

If on the other hand the ancestral population of the anatids was not "primitive" but initially possessed a high potential of variation, then one would not expect a distinct correlation between the degree of specialization and the stratigraphic position. This is because the specific expression of characters can be considered to be due to coincidence of function.

# References

Alvarez, R. and S.L. Olson. 1978. A new merganser from the Miocene of Virginia (Aves: Anatidae). *Proceedings of the Biological Society of Washington* 91:522-532.

Ayala, F.J. and J.W. Valentine. 1979. *Evolving: The Theory and Process of Organic Evolution*. Benjamin Cummings, Menlo Park, CA.

Barton, N. and J.S. Jones. 1983. Mitochondrial DNA: New clues about evolution. *Nature* 306:317-318.

Bock, W.J. 1969. Origin and radiation of birds. *Annals of the New York Academy of Sciences* 167:147-155.

Braunitzer, G. and W. Oberthür. 1979. Die Primärstruktur des Hämoglobins der Graugans *(Anser anser)*: Die ungleiche Evolution der β-Ketten (Versuch einer biochemischen Analyse des Verhaltens). *Hoppe-Seyler's Zeitschrift für Physiologische Chemie* 360:679-683.

Brown, W.M., M. George and A.C. Wilson. 1979. Rapid evolution of animal mitochondrial DNA. *Proceedings of the National Academy of Sciences* 76:1967-1971.

Brush, A.H. and H.H. Witt. 1983. Intraordinal relationships of the Pelecaniformes and Cuculiformes: Electrophoresis of feather keratins. *Ibis* 125:181-199.

Clarke, J.A., C.P. Tambussi, J.I. Noriega, G.M. Erickson, and R.A. Ketcham. 2004. Definitive fossil evidence for the extant avian radiation in the Cretaceous. *Nature* 444:780.

Cracraft, J. 1981. Toward a phylogenetic classification of the Recent birds of the world (Class Aves). *Auk* 98:681-714.

Delacour, J. 1954-1964. *The Waterfowl of the World*. Country Life Limited, London.

Delacour, J and E. Mayr. 1945. The family Anatidae. *Wilson Bulletin* 57:4-54.

Dillon, L.S. 1978. *Evolution—Concepts and Consequences*. Second edtion. CV Mosby Company, Saint Louis.

Edkins, E. and I.A. Hansen. 1972. Wax esters secreted by the uropygial glands of some Australian water-fowl, including the magpie goose. *Comparative Biochemistry and Physiology (Part B)* 41:105-112.

Feduccia, A. 1977. Hypothetical stages in the evolution of modern ducks and flamingos. *Journal of Theoretical Biology* 67:715-721.

Feduccia, A. 1978. *Presbyornis* and the evolution of ducks and flamingos. *American Scientist* 66:298-304.

Feduccia, A and P.O. McGrew. 1975. A flamingo-like wader from the Eocene of Wyoming. *University of Wyoming Contributions to Geology* 13:49-62.

Fehrer, J. 2009. Eine neue Phylogenie der Vögel: Was sagen die Daten wirklich? *Studium Integrale Journal* 16, in press.

Fitch, W.M. and E. Margoliash. 1967. Construction of phylogenetic trees: A method based on mutation distances as estimated from cytochrome sequences is of general applicability. *Science* 155:279-285.

Gillham, E., J.M. Harrison, and J.G. Harrison. 1966. A study of certain *Aythya* hybrids. *The Wildfowl Trust Annual Report* 17:49-65.

Godovac-Zimmermann, J. and G. Braunitzer. 1983. The amino-acid sequence of northern mallard *(Anas platyrhynchos platyrhynchos)* hemoglobin. *Hoppe-Seyler's Zeitschrift für Physiologische Chemie* 364:665–674.

Gray, A.P. 1958. *Bird Hybrids: A Check-List with Bibliography.* Commonwealth Agricultural Bureaux, Farnham Royal, England.

Hackett, S.J. and 17 others. 2008. A phylogenomic study of birds reveals their evolutionary history. *Science* 320:1763-1767.

Harrison, J.M. 1945. Exhibition of two varieties of the teal. *Bulletin of the British Ornithologists' Club* 66:24.

Harrison, J.M. 1953. On the significance of variation of pattern in birds. *Bulletin of the British Ornithologists' Club* 73:37–40.

Harrison, J.M. 1959. Comments on a wigeon × northern shoveler hybrid. *Bulletin of the British Ornithologists' Club* 79:142.

Harrison, J.M. 1978. Hybridisation in waterfowl. In: Gooders, J., ed. *Birds of Ocean and Estuary.* Orbis Publishing, London, pp 248–253.

Harrison, J.M. and J.G. Harrison. 1963. A gadwall with a white neck ring and a review of plumage variants in wildfowl. *Bulletin of the British Ornithologists' Club* 83:101–108.

Harrison, J.M. and J.G. Harrison. 1965. A cinnamon teal × northern shoveler hybrid. *Bulletin of the British Ornithologists' Club* 85:107-110.

Harrison, J.M. and J.G. Harrison. 1966. Hybrid greylag × Canada goose suggesting influence of giant Canada geese in Britain. *British Birds* 59: 547-550.

Harrison, J.M. and J.G. Harrison. 1968. Wigeon ×Chiloe wigeon hybrid resembling American wigeon. *British Birds* 61:168-171.

Harrison, J.M. and J.G. Harrison. 1971. Notes on a further Pintail × Teal hybrid. *Bulletin of the British Ornithologists' Club* 91:28–32.

Heinroth, O. 1910. Beiträge zur Biologie, namentlich Ethologie und Psychologie der Anatiden. *Verhandlungen des V Internationalen Ornithologen-Kongresses*, pp. 589-702.

Hennig, W. 1966. *Phylogenetic Systematics.* University of Illinois Press, Urbana, IL.

Howard, H. 1964. Fossil anseriforms. In: Delacour, J. *The Waterfowl of the World* Volume 4. Country Life Limited, London.

Jacob, J. 1977. Die systematische Stellung der Dampfschiffente *(Tachyeres)* innerhalb der Ordnung Anseriformes. *Journal für Ornithologie* 118:52–59.

Jacob, J. and A. Glaser. 1975. Chemotaxonomy of Anseriformes. *Biochemical Systematics and Ecology* 2:215–200.

Jacob, J. and G. Grimmer. 1975. Composition of uropygial gland waxes in relation to the classification of some passerine birds. *Biochemical Systematics and Ecology* 3:267–271.

Jacob, J. and H. Hoerschelmann. 1982. Chemotaxonomische Untersuchungen zur Systematik der Röhrennasen (Procellariiformes). *Journal für Ornithologie* 123:63–84.

Johnsgard, P.A. 1960. Hybridisation in the Anatidae and its taxonomic implications. *Condor* 62:25–33.

Johnsgard, P.A. 1961a. The taxonomy of the Anatidae–A behavioural analysis. *Ibis* 103a:71–85.

Johnsgard, P.A. 1961b. The breeding biology of the magpie goose. *Wildfowl Trust, 12th Annual Report*, pp. 92–103.

Johnsgard, P.A. 1965. *Handbook of Waterfowl Behaviour*. Cornell University Press, Ithaca , NY.

Johnsgard, P.A. 1978. *Ducks, Geese and Swans of the World*. University of Nebraska Press, Lincoln, NE.

Junker, R. 1993. Prozesse der Artbildung. In: Scherer, S., ed. *Typen des Lebens*. Pascal, Berlin, pp. 31-45.

Kaltenhäuser, D. 1971. Über Evolutionsvorgänge in der Schwimmentenbalz. *Zeitschrift für Tierpsychologie* 29:481–540.

Kessler, L.G. and J.C. Avise. 1984. Systematic relationships among waterfowl (Anatidae) inferred from restriction endonuclease analysis of mitochondrial DNA. *Sytematic Zoology* 33:370–380.

Kim, Y.S. and P.E. Kolattukudy. 1978. Malonyl–CoA-decarboxylase from the uropygial gland of water fowl: Purification, properties, immunological comparison, and role in regulating the synthesis of multimethyl-branched fatty acids. *Archives of Biochemistry and Biophysics* 190:585–597.

Klemm, R. 1993. Die Hühnervögel (Galliformes): Taxonomische Aspekte unter besonderer Berücksichtigung artübergreifender Kreuzungen. In: Scherer, S., ed. *Typen des Lebens*. Pascal, Berlin, pp. 159-184.

Klös, H.G. and U. Klös. 1975. Die Gänsevögel: Wehrvögel und Entenvögel. In: *Grzimeks Tierleben* Volume 7. Kindler, Zurich, pp. 246–321.

Kolbe, H. 1984. *Die Entenvögel der Welt*. Neumann-Neudamm, Melsungen, Germany.

Lack, D. 1974. *Evolution Illustrated by Waterfowl*. Blackwell Scientific Publications, Oxford.

Lind, H and H. Poulsen. 1963. On the morphology and behaviour of a hybrid between goosander and shelduck (*Mergus merganser* and *Tadorna tadorna*). *Zeitschrift für Tierpsychologie* 20:558–569.

Livezey, B.C. 1986. A phylogenetic analysis of recent anseriform genera using morphological characters. *Auk* 103:737–754.

Lorenz, K. 1941. Vergleichende Bewegungsstudien an Anatiden. *Journal für Ornithologie* 89:194–293.

Lorenz, K. 1958. The evolution of behaviour. *Scientific American* 199:67–78.

Lorenz, K. 1960. Prinzipien der vergleichenden Verhaltensforschung. *Fortschritte der Zoologie* 12:265–294.

Madsen, C.S., K.P. McHugh, and S.R. Dekloet. 1988. A partial classification of waterfowl (Anatidae) based on single copy DNA. *Auk* 105:452–459.

Madsen, C.S., K.P. McHugh, and S.R. Dekloet. 1992. Characterization of a major tandemly repeated DNA sequence (RBMII) prevalent among many species of waterfowl (Anatidae). *Genome* 35:1037–1044.

Marsh, F. 1976. *Variation and Fixity in Nature.* Pacific Press Association, Mountain View, CA.

Mayr, E. 1967. *Artbegriff und Evolution.* Parey, Hamburg.

McGrew, P.O. and A. Feduccia. 1973. A preliminary report on a nesting colony of Eocene birds. In: Schell, E.M., ed. *Wyoming Geological Association 25th Conference Guidebook*, pp. 163-164.

Oberthür, W. 1980. *Primärstruktur der Hämoglobine von Graugans (Anser anser), Streifengans (Anser indicus) und Strauß (Struthio camelus).* Dissertation. Tübingen University, Tübingen, Germany.

Oberthür, W. and G. Braunitzer. 1984. Hämoglobine vom Gemeinen Star (*Sturnus vulgaris*, Passeriformes): Die Primärstruktur der α und β-Ketten. *Hoppe-Seyler's Zeitschrift für Physiologische Chemie* 365:159–173.

Oberthür, W., G. Braunitzer, and L. Würdinger. 1982. Das Hämoglobin der Streifengans (*Anser indicus*): Primärstruktur und Physiologie der Atmung, Systematik und Evolution. *Hoppe-Seyler's Zeitschrift für Physiologische Chemie* 363:581-590.

Oberthür, W., G. Braunitzer, R. Baumann, and P.G. Wright. 1983. Die Primärstruktur der α- und β-Ketten der Hauptkomponenten der Hämoglobine des Straußes (*Struthio camelus*) und des Nandus (*Rhea americana*) (Struthioformes). *Hoppe-Seyler's Zeitschrift für Physiologische Chemie* 364:119-154.

Olson, S.L. 1982. A critique of Cracraft's classification of birds. *Auk* 99:733-739.

Olson, S.L. 1985. The fossil record of birds. In: Farner, D.S., J.R. King, and K.C. Parkes, eds. *Avian Biology*, Volume 8, Academic Press, Orlando, FL, pp. 79-238.

Olson, S.L. and A. Feduccia. 1980. *Presbyornis* and the origin of the Anseriformes. *Smithsonian Contributions to Zoology* 323:1-23.

Patton, J.C. and J. Avise. 1983. An empirical evaluation of qualitative Hennigian analysis of protein electrophoretic data. *Journal of Molecular Biology* 19:244-254.

Perrins, C. 1961. The "lesser scaup" problem. *British Birds* 54:48-54.

Perutz, M.F. 1983. Species adaptation in a protein molecule. *Molecular Biology and Evolution* 1:1-28.

Peters, D.S. 1976. Evolutionstheorie und Systematik. *Journal für Ornithologie* 117:329-344.

Poltz, J. and J. Jacob. 1973. Bürzeldrüsensekrete von Webervögeln (Ploceidae). *Zeitschrift für Naturforschung C* 28:449-452.

Poltz, J. and J. Jacob. 1974a. Das Bürzeldrüsensekret vom Graureiher (*Ardea cinera*). *Journal für Ornithologie* 115:103-105.

Poltz, J. and J. Jacob. 1974b. Bürzeldrüsensekrete bei Ammern (Emberizidae), Finken (Fringillidae) und Webern (Ploceidae). *Journal für Ornithologie*

115:119-127.

Raikow, R.J. 1981. Old birds and new ideas: Progress and controversy in paleornithology. *Wilson Bulletin* 93:407-412.

Sage, B.L. 1966. Chilean pintail × red-crested pochard. *Bulletin of the British Ornithologists' Club* 86:50-54.

Scherer, S. 1990. The protein molecular clock: Time for a reevaluation. *Evolutionary Biology* 24:83-106.

Scherer, S. and T. Hilsberg. 1982. Hybridisierung und Verwandtschaftsgrade innerhalb der Anatidae—eine systematische und evolutionstheroretische Betrachtung. *Journal für Ornithologie* 123:357-380.

Scherer, S. and C. Sontag. 1986. Zur molekularen Taxonomie und Evolution der Anatidae. *Zeitschrift für zoologische Systematik und Evolutionsforschung* 24:1-19.

Schulin, R. 1981. *Vergleichende morphologische Untersuchungen zu Struktur und Funktion der Beinmuskulatur von Anatidae (Aves)*. Dissertation. Zurich University, Zurich.

Sharpe, R.S. and P. A. Johnsgard. 1966. Inheritance of behavioural characters in $F_2$ mallard × pintail (*Anas platyrhynchos* × *Anas acuta*). *Behaviour* 27:259-272.

Sibley, C.G. 1960. The electophoretic patterns of avian egg-white proteins as taxonomic characters. *Ibis* 102:215-259.

Sibley, C.G. and J. E. Ahlquist. 1972. A comparitive study of the egg-white proteins of non-passerine birds. *Bulletin of the Peabody Museum of Natural History* 39:1-267.

Sibley, C.G. and J. E. Ahlquist. 1990. *Phylogeny and Classification of Birds: A Study in Molecular Evolution*. Yale University Press, New Haven, CT.

Sibley, C.G., J.E. Ahlquist and B.L. Monroe. 1988. A classification of the living birds of the world based on DNA-DNA hybridization studies. *Auk* 105:409-423.

Smith, T.B. 1993. Disruptive selection and the genetic basis of bill size polymorphism in the African finch *Pyrenestes*. *Nature* 363:618-620.

Stadie, C. 1967. Verhaltensweisen von Gattungsbastarden *Phasianus colchicus* × *Gallus gallus* f. *domestica* im Vergleich mit denen der Ausgangsarten. *Verhandlungen der Deutschen Zoologischen Gesellschaft* 60:493-510.

Stapel, O.S, J.A.M. Leunissen, M. Versteeg, J. Wattel, and W.W. de Jong. 1984. Ratites as oldest offshoot of avian stem—Evidence from "α-crystallin A sequences. *Nature* 311:257-259.

Steinebrunner, B. 1991. Der Sturz der Vernunft. Bringen Methodenkritik und Theorienpluralismus mehr Erkenntnis? In: Scherer, S., ed. *Die Suche nach Eden*. Hänssler, Neuhausen, Germany, pp. 13-44.

Voous, K.H. 1955. Hybrids of scaup duck and tufted duck (*Aythya marila* × *Aythya fuligula*). *Ardea* 43:284–286.

von de Wall, W. 1963. Bewegungsstudien an Anatinen. *Journal für Ornithologie* 104:1-43.

Weller, M.W. 1957. Growth, weights and plumages of the redhead, *Anthya americana. Wilson Bulletin* 69:4-38.

Wetmore, A. 1926. Fossil birds from the Green River deposits of eastern Utah. *Annals of the Carnegie Museum* 16:391-402.

Wetmore, A. 1938. A fossil duck from the Eocene of Utah. *Journal of Paleontology* 12:280-283.

Wolters, H.E. 1975-1982. *Die Vogelarten der Erde.* Parey, Berlin.

Woolfenden, G.E. 1961. Postcranial osteology of the water fowl. *Bulletin of the Florida State Museum* 6:1-129.

Zuckerkandl, W.E. and L. Pauling. 1962. Molecular disease, evolution and genic heterogeneity. In: Kasha, M. and B. Pullman, eds. *Horizons in Biochemistry.* Academic Press, New York, pp. 189-225.

# 11. The Gallinaceous Birds (Galliformes): Interspecific Hybridization and Classification

ROLAND KLEMM

## Abstract

The analysis of interspecific hybrids is of taxonomic importance because hybridization and embryonic development are open to experimental validation. As a result, the basic type is conceived as a systematic category above the level of the species. The gallinaceous birds are a suitable group for basic type studies since many interspecific hybrids have been reported in the literature.

The taxonomy of the gallinaceous birds lists 250-285 species in 70-94 genera. Within the order Galliformes, three groups can clearly be distinguished: mound-builders (Megapodiidae) with 12 species, curassows (Cracidae) with 43 species, and pheasants (Phasianidae) with 203 species in 15 subfamilies. These groups are introduced and described briefly.

A total of 158 successful hybridizations were taken into consideration. Hybrids with representatives of other orders have not been observed. The percentage of known hybridizations within the various families / subfamilies differs considerably. Among the Megapodiidae, no hybrids have been reported. Hybridizations between representatives of the Cracidae and the Phasianidae could not be verified beyond reasonable doubt. In 11 out of 15 subfamilies of the Phasianidae, hybridizations have been observed; all 11 subfamilies are connected either directly or indirectly by hybrids. Twenty-eight percent of all hybridizations crossed the borders between subfamilies; 60% are intergeneric. Thirty-four hybridizations (9 of them transgeneric) resulted in fertile offspring.

Based on the available data, the gallinaceous birds currently include three basic types. Being morphologically and ethologically similar to one another, these basic types correspond to the families Megapodiidae (mound-builders), Cracidae (curassows), and Phasianidae (pheasants).

## 11.1. Introduction

From a taxonomic point of view, the class of birds (Aves) is considered well known. However, we are far from a commonly accepted system of relationships within the group. Among other things, sufficient criteria for combining larger taxa such as orders and families are missing. The fact that at least 17 different phylogenetic trees, based on various

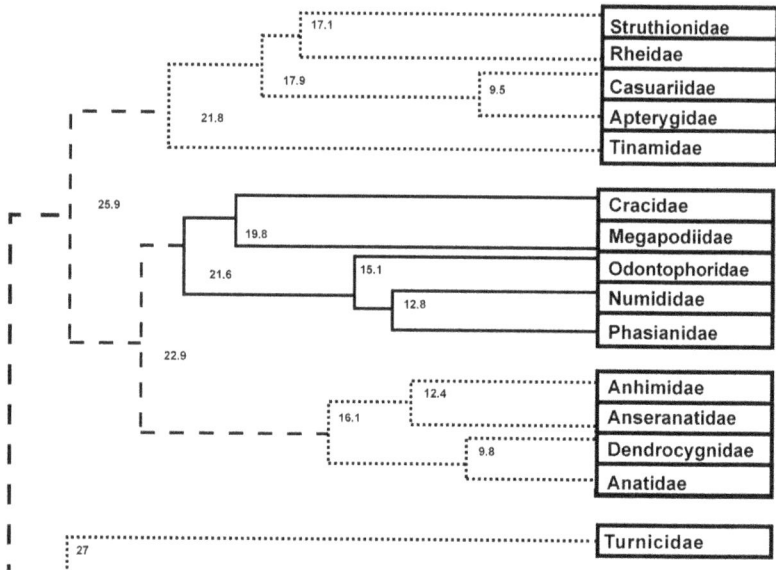

Figure 1. Pertinent portion of a phylogenetic tree (UPGMA), proposed by Sibley and Ahlquist (1990) and based on the distances of DNA-DNA hybridizations.

methods and assumptions, have been published for the gallinaceous birds (Randi et al. 1991) illustrates the problem for this major group.

According to Jacob (1989), only a comparative investigation and evaluation of as many characters as possible will lead to a better understanding of affinities among the different species. However, the importance of biochemical and biomolecular data, for example DNA-DNA hybridization or protein polymorphism, in the study of relationships is often exaggerated.

As the foundational systematic category, the biological species is the basis for evaluating hybridization data in the Galliformes presented in this study. Since, however, a species definition that is satisfying "in every regard" does not exist (Lange 1985), different taxonomic studies vary when it comes to assigning species status. In most cases, Mayr's (1967) definition of the biological species is utilized. According to this definition, "a species comprises groups of populations with geographic and ecologic coherence, and in which immediate neighbors interbreed wherever they come in contact with each other, or are potentially able to do so, even if geographical or ecological barriers prevent such immediate contact."

The more or less frequently occurring hybrids between biospecies demonstrate the difficulties of defining the species but, on the other hand, provide an opportunity to gain taxonomic insights from the analysis of interspecific hybridizations.

The ability of individuals from different species to cross-breed is a criterion that can be validated experimentally and can be used for the study of kinship connections. It is the foundation for the basic type concept as a systematic category above the level of the species (Scherer 1993, this volume).

The gallinaceous birds as a major taxon are well suited for the study of basic types since numerous hybrids have been described in the group. The reasons for this are the use of certain species for decorative purposes (e.g., pheasants), as well as frequent cross-breeding between domestic species like the chicken, turkey, guineafowl, and Japanese quail.

## 11.2. Systematic Divisions and Characters of the Gallinaceous Birds (Galliformes; synonym: Phasianiformes)

Recent systems of the gallinaceous birds list 250-285 species in 70-94 genera (order Galliformes, superorder Gallomorphae: Sibley and Ahlquist 1990). The gallinaceous birds represent a rather uniform group, and it is difficult to derive them from other groups. Distances derived from DNA-DNA hybridizations performed by Sibley and Ahlquist (1990, Figure 1) also point to this fact. Currently, two extinct subfamilies assigned to the Cracidae are known only from fossils. Among these is the oldest known gallinaceous bird *Palaeophasianus meleagroides*. Fifty-three fossil species are included in modern families and genera (Brodkorb 1964). The fossil record does not allow an unequivocal derivation of the Galliformes from other groups.

Until the beginning of the 20th century the tinamous (Tinamiformes: 46 species) and buttonquails (Turniciformes: 17 species) were regarded as gallinaceous birds but have been separated from the Galliformes in recent systems (see Figure 1). Heinroth and Heinroth (1966) commented on this delimitation as follows: "All gallinaceous birds are characterized by a special development of the juvenile primaries and secondaries, as well as a peculiar change of the wings of the juvenile animals. The fact that they differ from all other birds proves the close affinity of the gallinaceous birds *sensu stricto*, the curassows, and the mound-builders, as well as their distance from the buttonquails and tinamous."

Some authors still place the hoatzin (*Opisthocomus hoazin*), the only representative of the order Opisthocomiformes, into the order Galliformes. For this reason, its peculiarities and systematic affinities will be discussed below (see also Table 2).

The most important anatomical, morphological, and ethological characteristics of the gallinaceous birds are summarized in Table 1. Exceptions and peculiarities will be pointed out below.

Some species of the group differ considerably in their size. There is quite a difference between the smallest representative of the order, the Chinese painted quail with a body weight of 45 g, and the peafowl, the

| morphological - anatomical | ethological |
|---|---|
| • short, strong, lightly curved bills<br>• large, strong legs, anisodactyle feet (position of toes I towards the back, II-IV towards the front)<br>• legs covered with two rows of horny scales<br>• rounded wings held close to the body, eutaxic (5th primary is present, tail feather very large)<br>• skull with slit palate (as in cranes, gulls/waders, woodpeckers)<br>• weak vomer, palate without lamina interna<br>• nostrils holorhinous<br>• large notarium (fused shoulder vertebrae)<br>• 6 cervical vertebrae<br>• ossified tracheal rings<br>• large crop and gizzard, long caecum | • specialized terrestrial birds<br>• chicks are precocial and feed selves, some reach the ability to fly on the first day of their life<br>• take dust baths<br>• are unable to swim<br>• the majority feed on seeds and plants<br>• seek food by scraping<br>• short-range flyers<br>• strong polygyny |

Table 1. Characteristics of the gallinaceous birds. After Ziswiler (1976), Glutz von Blotzheim *et al.* (1973), Ilicev & Flint (1989), and Sibley & Ahlquist (1990).

domestic turkey and the capercaillie, which weigh 6-8 kg. All geographic areas of the earth, with the exception of Antarctica, Polynesia, and the Galápagos Islands, are inhabited by gallinaceous birds. Many species showing a wide variety of morphological traits dwell in the tropical and subtropical rain forests of East and Southeast Asia.

Table 2 shows the differences in the taxonomic systems of the gallinaceous birds below the level of order, based on six sources. According to Wolters (1975-1982), three distinct groups are recognized: mound-builders (Megapodiidae) with 12 species, curassows (Cracidae) with 43 species, and the largest family, the pheasants (Phasianidae) with 203 species. Various authors differ over the degree of relationship between the Megapodiidae and the Cracidae, as well as between both these groups and the Phasianidae. Sibley and Ahlquist (1990) place mound-builders and curassows into a separate order, indicating their close affinity (see Figure 1). Johnsgard (1973) combines both groups into one superfamily (Cracoidea). An anatomical peculiarity of the species in both groups is the position of the hind toe on the same level as the

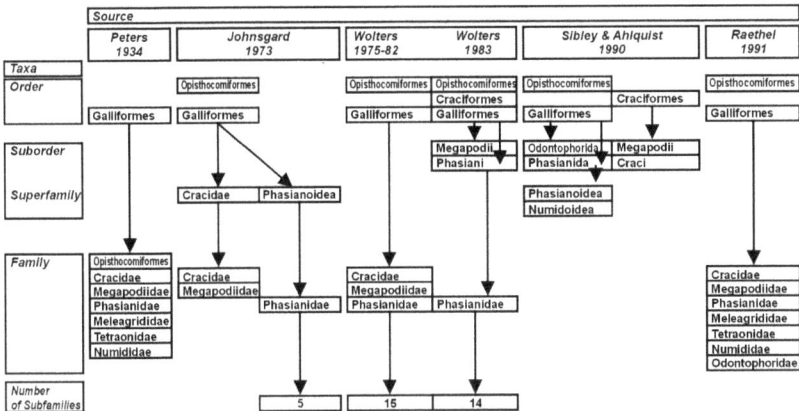

| Source | Peters 1934 | Johnsgard 1973 | Wolters 1975-82 | Wolters 1983 | Sibley & Ahlquist 1990 | Raethel 1991 |
|---|---|---|---|---|---|---|
| **Taxa** | | | | | | |
| Order | Galliformes | Opisthocomiformes / Galliformes | Opisthocomiformes / Galliformes | Opisthocomiformes / Craciformes / Galliformes | Opisthocomiformes / Galliformes / Craciformes | Opisthocomiformes / Galliformes |
| Suborder | | | | Megapodii / Phasiani | Odontophorida / Phasianida / Megapodii / Craci | |
| Superfamily | | Cracidae / Phasianoidea | | | Phasianoidea / Numidoidea | |
| Family | Opisthocomiformes / Cracidae / Megapodiidae / Phasianidae / Meleagrididae / Tetraonidae / Numididae | Cracidae / Megapodiidae / Phasianidae | Cracidae / Megapodiidae / Phasianidae | Cracidae / Megapodiidae / Phasianidae / Phasianidae | | Cracidae / Megapodiidae / Phasianidae / Meleagrididae / Tetraonidae / Numididae / Odontophoridae |
| Number of Subfamilies | | 5 | 15 | 14 | | |

Table 2. Systematic arrangement of the gallinaceous birds according to different authors.

front toes. In the Phasianidae, this toe is positioned higher up and is not as strong. Wolters (1983) segregates the Cracidae as a separate order (Craciformes) and divides the mound-builders as suborder Megapodii from the pheasants (suborder Phasiani). Vuilleumier (1965), in a summary article on systematic studies, concludes that the Cracidae are more closely related to the Phasianidae than was assumed in the past. The differences between the mound-builders and curassows on the one side and the pheasants on the other are essentially due to their reproductive behavior (see below).

Furthermore, divisions of the pheasants (family Phasianidae) in Table 2 show essential differences. According to some newer investigations, there seems to exist a closer relationship among the representatives of this family. For example, Johnsgard (1973) and Wolters (1975-1982; 1983) divide the genera only to the level of subfamilies. However, Sibley and Ahlquist (1990) and Raethel (1991) assume a greater taxonomic distance for the New World quails (Odontophoridae) and the guinea fowl (Numididae).

An extensive historical overview of taxonomic studies of gallinaceous birds is given by Sibley and Ahlquist (1990). The basis for the taxonomic system chosen in this paper is Wolters 1975-1982 (see Table 2).

**11.2.1. Opisthocomiformes – Hoatzins (1 species).** An analysis by Barnikol (1952) shows that, since 1837, the hoatzin (*Opisthocomus hoazin*; Figure 2) has been placed within the Galliformes (29 times), Musophagiformes (12 times), Columbiformes (9 times), Ralliformes (9 times), and Cuculiformes (6 times). Electrophoretic examinations of the protein patterns of the eggs by Sibley and Ahlquist (1972) point to a close relationship to the cuckoo birds, which is also assumed by Wolters (1983). A closer analysis indeed shows hardly any similarities with the

Figure 2. Hoatzin (*Opisthocomus hoazin*) in the Bronx Zoo, New York. Photo by A. Johann

gallinaceous birds. The differences are the moulting process, as well as the ability of the juvenile birds to climb the twigs of trees by means of claws located at the front edge of the wings. They escape enemies by letting themselves fall into the water, and they are able to swim under water. The pheasant-like, brown, approximately 24-inch (60 cm) hoatzin, which lives in northern South America, cannot be placed unequivocally into any system. Parker (1891) described the hoatzin as "One of the most mixed forms in the whole range of ornithology." The reason is a combination of so-called primitive characters (structure of the wings, formation of the feathers, length of the caecum) and "specialized" characters (highly developed crop, reduced wing muscles, configuration of the syrinx). Limited by the defective flight ability, the populations live completely isolated in numerous small river valleys. Studies done in Venezuela (Grajahl and Strahl 1991) show that the hoatzin is the only bird that, by means of symbiotic bacteria, is able to digest its food (leaves, blossoms, and fruits of freshwater mangroves) in the front part of the intestine.

The claws on its wings, which resemble those of *Archaeopteryx*, have in the past lead to the conclusion that the hoatzin is a living fossil. Presently, due to the discovery of its specialized feeding habits and the resulting anatomical changes, the claws of the hoatzin are considered to be a secondary development. The numerous special characters allow for the conclusion that the hoatzin represents its own basic type. Hybrids with other species are not known.

**11.2.2. Megapodiidae – Mound-builders (12 species in 7 genera).** The systematic placement of these 12 species, which live in Australia and the Indo-Australian archipelago, has already been mentioned. The Latin name of this family is derived from their long toes and strong claws. The most significant difference between the Megapodiidae and other families is their habit of incubating their eggs by means of sand warmed from the sun or decaying plants. Grzimek (1980) describes the breeding habit of these shy, inconspicuously colored terrestrial birds. He concludes that the breeding habit cannot be regarded as primitive since it differs from the breeding habit of the reptiles in many ways. The typical moulting process of these species enables their chicks to fly beginning the first day of their lives. Johnsgard (1988) assumes a "megapode-like ancestor" at the root of a hypothetical phylogenetic tree of the Perdicinae

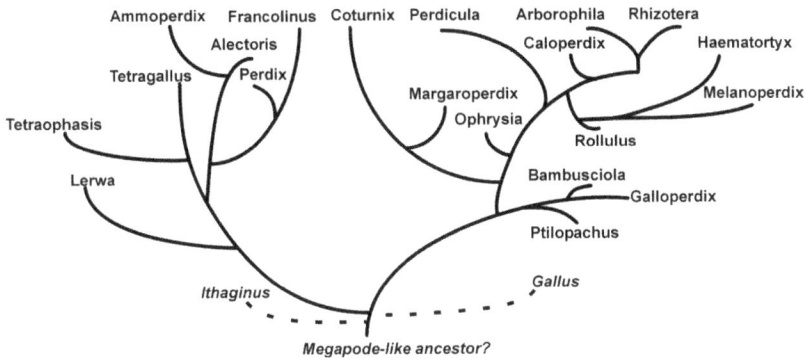

Figure 3. Assumed systematic relationships within the Phasianidae, subfamily Perdicinae (Old World quails) (after Johnsgard 1988).

(a subfamily of the Phasianidae; see Figure 3).

**11.2.3. Cracidae – Curassows and Their Kin (43 species in 10 genera).** The species of this family live in Central and South America (with the highest species density in the northwest) and are divided into two groups: the curassows (larger species up to 40 inches (100 cm) in four genera, see Figure 4) and the chachalacas or guans (smaller species in six genera). Extensive studies of the systematic relationships within the Cracidae were done by Vuilleumier (1965) and Vaurie (1968). In these studies both morphological and ethological characters were taken into consideration. In contrast to other gallinaceous birds, these animals are exceptionally well adapted to an arboreal life. Outside of the Cracidae, the only other galliforms that nest in trees are the Congo peafowl and tragopans, and only the peacock-pheasants have clutch sizes of as few as two to four eggs (the egg shells of the Cracidae are rough and white). With the peacock-pheasants and the true peafowl, they also share a mode of tail moulting that deviates from that in all other gallinaceous birds (Stresemann 1965). Other specific characters are the remarkable wax-like membrane on the base of the bill, the absence of aftershafts on the feathers, and the presence of a penis in males.

Fossil Cracidae were already mentioned; most fossils were found in North America, so a north-south dispersal process can be assumed. Brodkorb (1964), however, also lists records from the Upper Eocene of France. According to Vuilleumier (1965), even the oldest fossils differ little from recent species. The current distribution of this family in America and the degree of its morphological divergence and isolation allow the assumption that the colonization of South America was accomplished by recent species.

**11.2.4. Phasianidae – Pheasants and Their Kin (203 species in 70 genera).** In spite of many common characters, this family is divided into several subfamilies in which the genera display a wide variety of

Figure 4. Bare-faced curassow (*Crax fasciolata*) in the Walsrode Bird Park. Photo by R. Klemm

Figure 5. The domestic guineafowl (*Numida meleagris* f. *domesticus*) is used for breeding and economic purposes; its body shape corresponds to that of the wild form (helmeted guineafowl). Photo by R. Klemm.

morphological and ethological traits. The systematic relationships, however, have not been sufficiently examined. The phylogeny is still largely uncertain. The New World quails, grouse, and guineafowl have a relatively isolated position. The division of the family into the Perdicinae, Odontophorinae and Phasianinae is not unequivocal and, since the study performed by Ogilvie-Grant (1893), has been done differently by different taxonomists (Johnsgard 1988). Below we will briefly characterize the 15 subfamilies of Wolters (1975-1982) and discuss their systematic relationships.

**11.2.4.1. Numidinae – Guineafowl (6 species in 4 genera).** The social guineafowl are common birds of Africa with a short, slightly bent beak and high, usually spurless legs, rounded wings, short tail, as well as a bare head and upper neck. They can run well and stay in trees at night. Most species show the well-known light pearl dotting. The helmeted guineafowl (*Numida meleagris*), as forerunner of the domesticated form (Figure 5), comprises 23 subspecies, which populate the entire African continent south of the Sahara desert (von Boettcher 1954). The two species of the genus *Agelastes* are regarded as generalized species – as "original guineafowl" (Wissel et al. 1966.)

**11.2.4.2. Pavoninae – Peafowl (3 species in 2 genera).** The two species of the genus *Pavo* are among the largest gallinaceous birds in the Old World; they inhabit Southeast Asia. A well-known trait is the peacock's train, which consists of the much elongated covert feathers of the tail. During the mating period as the peacock is strutting, he displays his train by spreading it in the form of a semicircular, raised fan. The legs have long spurs, similar to those of the turkey.

The Congo peafowl (*Afropavo congensis*) was discovered as late as 1936; it inhabits the jungles of central Africa and differs from the Asian peacocks in numerous peculiarities. Its systematic position is often debated. According to Verheyen (1965), this species was called an "aberrant francolin," "a more primitive generalized peafowl," as well

as a link between Numidinae and
Perdicinae. There are conspicuous
similarities between *Afropavo* and
*Penelope* (a genus of the family
Cracidae), as far as their breeding
biology is concerned. Both are
monogamous and lay up to a
maximum of 3 eggs in arboreal nests
(Verheyen 1965). The breeding
periods are equally long. Randi *et*
*al.* (1991) constructed phylogenetic
trees based on the analysis of
protein polymorphisms (21 loci),
which show a closer relationship

Figure 6. Rooster of the ring-necked pheasant, a race of the common pheasant, *(Phasianus colchicus)*; 33 different races of this species inhabit Eurasia from the Atlantic to the Pacific.

between *Pavo* and *Numida* than between *Pavo* and the remaining
Phasianidae. Examinations of the karyotype and the immunology also
point to this close relationship (Verheyen 1965).

**11.2.4.3. Meleagridinae – Turkeys (2 species in 1 genus).** The
turkeys inhabit North and Central America and are large, long-legged
gallinaceous birds with a metallic shiny plumage. During mating time
the rooster raises its tail and displays it in form of a fan. The hens are
much smaller and less colorful. The head and neck are bare with warts
and skin protrusions. The turkeys are polygamous and stay in trees at
night. Domestic turkeys originated from the species *Meleagris gallopavo*,
which includes 7 subspecies.

**11.2.4.4. Argusianinae – Peacock-pheasants (8 species in 3 genera).**
The peacock-pheasants live in pairs in the jungles of Southeast Asia. The
rather inconspicuous plumage is decorated with metallic shiny feathers.
The nest normally contains only 2 eggs. These peacock-sized pheasants
display a strongly specialized mating behavior and plumage peculiarities
— the secondaries, for example, are longer than the primaries. Besides
that, they are the only gallinaceous birds without a uropygial gland.

**11.2.4.5. Phasianinae – Pheasants (21 species in 8 genera).** Raethel
(1991) proposed one possible division of these 21 species, which are
characterized by a splendid plumage of the roosters (except in the cheer
pheasant). He distinguishes common pheasants (*Phasianus*), long-tailed
pheasants (*Syrmaticus, Calophasis, Graphephasianus*), ruffed pheasants
(*Chrysolophus*), gallopheasants (*Lophura*), cheer pheasants (*Catreus*)
and eared-pheasants (*Crossoptilon*). Common pheasants (*Phasianus*
*colchicus*) are characterized by a long tail consisting of 18 feathers.
The head is adorned by ear tufts; the bare face lobes ("wattles") expand
during mating time by being filled with blood. The single species (Figure
6) includes 33 races (divided into five groups by Delacour 1977), which
inhabit Eurasia from the Atlantic to the Pacific and, thus, represent the

Figure 7. The red junglefowl (*Gallus gallus*) lives in Indo-Asia and is considered to be the ancestral form of our various breeds of domestic chicken. Photo by R. Klemm.

most widespread wild gallinaceous species.

The long-tailed pheasants inhabit mountain forests in east and southeast Asia. The tail is unusually long (in most species) and consists of 16-20 feathers, the middle pair always being the longest. The roosters of the ruffed pheasants are magnificently colored birds with a head bonnet consisting of shiny feathers and a neck collar, which can be spread into a fan during the mating time. They inhabit the mountains of Central and Western China. The nine species of gallopheasant are divided into six genera by Raethel (1991); Wolters (1975-1982) unites all of them into a single genus (*Lophura*). The best-known representative is the silver pheasant (22 races in two groups). All species live in east and southeast Asia. The most specialized representative of this subfamily is Bulwer's wattled pheasant from Borneo with 32 tail feathers (a unique number among the birds). The more or less sexually monomorphic eared-pheasants are inhabitants of high mountain regions in eastern Asia. The cheer pheasant; also monomorphic and deviating in many respects from the other pheasants, inhabit the Himalayan Mountains.

**11.2.4.6. Lophophorinae – Monals (3 species in 1 genus).** The monals are large birds, which due to their strong short legs appear clumsy. They inhabit high mountain regions in Asia. Their beak is rather long and slightly bent. In the roosters, the plumage of the upper side has a strong metallic shine; the hens are colored brown.

**11.2.4.7. Pucrasiinae – Koklass Pheasants (1 species).** The 10 subspecies of the Koklass pheasant inhabit the Himalayan Mountains. They are moderate-sized gallinaceous birds. Their plumage (sexual dimorphism in the plumage color) consists primarily of lancet-shaped feathers. The roosters wear a vertical bonnet and hairy crests on both sides of the head which are raised during the mating period.

**11.2.4.8. Ithagininae – Blood Pheasants (1 species).** The 14 subspecies of the blood pheasant live in the high mountain regions of central Asia. They resemble partridges. Their beak is short, stout, and bent; the plumage coloring shows sexual dimorphism.

**11.2.4.9. Gallinae – Junglefowl (4 species in 1 genus).** The junglefowl (*Gallus*) live in Indo-Asia and are characterized by a comb, neck lobes, and an almost bare face. The colorful roosters possess a long sharp spur on each leg. The lineage of the domestic chicken with a remarkable number (approximately 100) of morphologically very different breeds is traced back to the red junglefowl (Figure 7) (Herre and Röhrs 1990). Since all four species of this genus can be crossbred with each other (see below), it has frequently been debated whether our domestic chicken originated with the participation of genetic material from the other three species (Hertwig 1936a; Engelmann 1984). Ghigi (1922, according to Hertwig 1936a), for example, traces the dotted pattern of the feathers of domestic breeds and the short lancet feathers (Indian fighting chicken) back to the green junglefowl.

**11.2.4.10. Tragopaninae – Tragopans (5 species in 1 genus).** As in the case of the blood pheasants (Ithagininae), the tragopans, too, closely resemble partridges (see Figure 3). The roosters of these species, which inhabit mountain regions in Asia, are remarkably colored and possess fleshy lobes on the sides of the vertex, which can be raised during the mating season. As with the Cracidae, tragopans build their nests in trees.

**11.2.4.11. Galloperdicinae – Spurfowl (3 species in 1 genus).** The three species of spurfowl (*Galloperdix*) live in tropical forests of the Indian subcontinent. They display sexual dimorphisim in the plumage color. The roosters possess one to three spurs. According to Johnsgard (1988), this genus helps to bridge the gap between Perdicinae and Phasianinae. A study of skeletal traits (Crowe and Crowe 1985) points to a close relationship to *Francolinus* (a genus of the Perdicinae), in contrast to their position in Figure 3.

**11.2.4.12. Ptilopachinae – Stone Partridges (1 species).** The five subspecies of the stone partridge (*Ptilopachus petrosus*) live south of the Sahara. Their distribution raises some questions of zoogeography and microevolution because the nearest relatives (the bamboo partridges [*Bambusciola*] and spurfowl [*Galloperdix*]; see Figure 3) are representatives of the oriental fauna. Both sexes are equally colored grayish-brown and live in predominantly rocky terrain.

**11.2.4.13. Perdicinae – Partridges, Francolins, and Old World Quails (98 species in 27 genera).** With 98 species, the Old World partridges form the largest subfamily of the Phasianidae. Johnsgard (1988) includes *Galloperdix* and *Ptilopachus* in this group and distinguishes 21 genera, whose systematic relationships are presented in Figure 3. They are usually dull-colored, small to middle-sized birds. Most species posses a spur. The Perdicinae include monotypic genera (for example the crested partridge), genera with relatively few species (for example partridges, rock partridges and snowcocks [similar to grouse]), and genera with many species (bush quails and francolins). The closely related francolins (according to Johnsgard all 41 species in 1 genus; according to Wolters in 6 genera but with 24 of the species belonging to the genus *Pternistis*) predominate in sub-Saharan and tropical Africa where numerous species are variously distributed. Five species live in the prairies of western and Southeast Asia. Representatives of the genus *Coturnix* are the only true migratory birds among the gallinaceous birds. Quails (*Coturnix coturnix*) can fly over long distances only with considerable effort. For example, flocks that fly across the Black Sea and arrive at the coast of Turkey are so weakened that they can be caught by hand (Ilicev and Flint 1989). Similar observations are reported when they cross the Red Sea and arrive at the Sinai Peninsula. This is remniscent of the Old Testament reports (Exodus 16; Numbers 11) of swarms of quail, which served as food for the Israelites in the desert.

**11.2.4.14. Odontophorinae – New World Quails (31 species in 9 genera).** These gallinaceous birds live exclusively in the New World (from southern Canada to northern Uruguay). They resemble the Old World quails, partridges and francolins. Despite the external similarity, the taxonomic delimitation is not completely clear. According to Johnsgard (1988), there is a greater taxonomic distance between the New World quails and the Old World partridges than between the latter and the Old World pheasants. Results of DNA-DNA hybridization (Sibley and Ahlquist 1990) also show a closer relationship between the Perdicinae and the guineafowl and the turkey than between the Perdicinae and the New World quails. Protein polymorphisms indicate the reverse for at least two genera and illustrate the current confusion regarding the systematic relationships within the Perdicinae, as well as between Perdicinae and Odontophorinae.

The New World quail differ from other gallinaceous birds due to a tooth-like structure on the sheath of the upper beak. In addition, they possess a hook-like downward-bent tip of the beak, as well as spurless legs with a long rear toe. Many species are characterized by plumes and crests. New World quails are monogamous and form flocks in the winter; none of the species are known for long migrations.

**11.2.4.15. Tetraoninae – Grouse (16 species in 9 genera).** The grouse are inhabitants of northern tundras and forests; North American species are adapted for life on the prairie. In Europe's native forests, the capercaillie and the black grouse, as typical representatives of this subfamily, have become very rare. The partridge- to turkey-sized species of this subfamily in certain ways differ considerably from the remaining gallinaceous birds. They are characterized by mostly feathered legs, feathered nose holes and short, downward-bent beaks. The digestive system with its long caecum is adapted to the consumption of conifer needles and buds. The roosters of all species are known for their impressive mating behavior. Inflatable airbags and hormone-controlled, colored erectile bodies on the skin ("combs and wattles") reinforce the imposing behavior of the roosters of some species. The species with specialized courtship behavior are polygamous, lacking pair bonds, whereas, hazel grouse and ptarmigans are monogamous. The grouse survive freezing

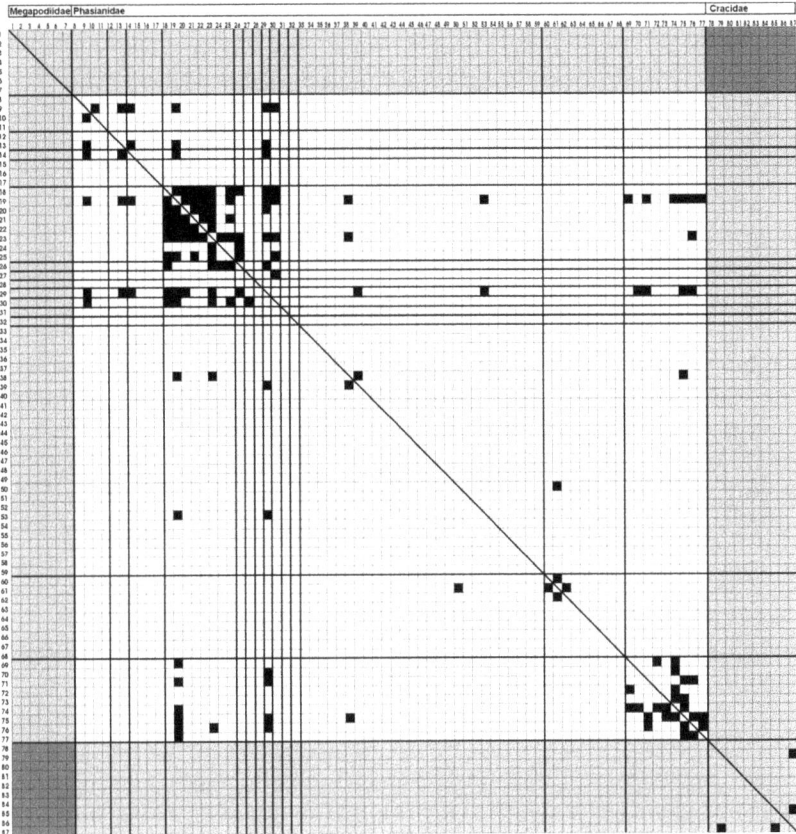

Fig.ure 8. Hybridization matrix of intergeneric hybrids in gallinaceous birds. The numbers correspond to the numbers of genera listed in the appendix.

winter nights and blizzards in snow burrows. The Tetraoninae possess a lesser resistance to poultry diseases than most other gallinaceous birds.

## 11.3. Interspecific hybrids Among the Gallinaceous Birds

### 11.3.1. Evaluation of Results from the Scientific Literature. In this evaluation, all available data indicating successful hybrids between species listed by Wolters (1975-1982; see appendices) have been included. The taxonomic assignment of species versus subspecies may differ from that by other authors. In this paper, subspecies are not listed separately. To the best of our knowledge, crosses between subspecies produce fertile hybrids (exceptions see below). Raethel (1991) provides detailed results for the individual subspecies.

An overview of the distribution and frequency of the known intergeneric hybrids is given in the hybridization matrix (Figure 8), which is laid out symmetrically about the diagonal; *i.e.*, each hybridization is represented by two dots. Each genus is assigned a unique number between 1 and 87 and included in the matrix in Figure 8 and in the first column of the appendices. The patterns and lines in the matrix represent the classification chosen in this paper, and the systematic distance between hybridizing partners is represented by the distance of the dots from the diagonal line. Interspecific hybridizations are listed in the last column of Appendix B and document the matrix.

Due to the study performed by Gray (1958), an evaluation of the older literature is available. In the last 30 years very little has been published in the primary literature of interspecific hybridizations (although some interest in connection with cytogenetic investigations has developed). Only 15 hybridizations can be added which were not mentioned by Gray (see appendices). One problem in the evaluation is the accuracy or testability of the results, especially those from the 19th century. Since much of the primary literature was not available to this author, consideration was given to Gray's critical remarks concerning these results. Questionable hybridizations — namely those which have not been confirmed unequivocally — have not been included in this evaluation. Some of the questionable cases are discussed in the text.

Data concerning the fertility of hybrids, as far as they were available in the literature, were included in the evaluation (see Table 4 and Appendix A). Since this study is qualitative in character, the author did not attempt an extensive presentation and discussion of the quantitative results of interspecific hybridizations (fertilization success, embryonic mortality rate, hybrid fertility). Some of these data will either be dealt with in the discussion of the individual subfamilies of the Phasianidae, or respective literature will be quoted.

**11.3.2. Overview of Hybridization Results.** The probability of hybridization between the different species pairs is not equal. The fact that different subfamilies have a different numbers of species limits the degree to which they can be compared (as in Table 3). The percentage of observed hybridizations differs in the various families/subfamilies. The 158 verified hybridizations represent 0.5% of all possible hybridizations. According to Scherer and Hilsberg (1982), this same measure is almost 2% in the ducks. When this measure is compared among the subfamilies, those subfamilies with a comparable number of species show values ranging from 23% (pheasants) to 14% (grouse) to 2% (American quail). In the junglefowl and turkeys, the measure is 100%, *i.e.*, all species of these subfamilies have hybridized successfully with each other. On the other hand, the measure is only 0.3% in the partridges, a subfamily with many species. In the Cracidae, too, only 1% of all possible hybridizations have been observed. Due to the differing numbers of species and the important role of certain species (ring-necked pheasant, domestic chicken) the measures between different subfamilies can hardly be compared. The main reason for these differences is the absence of methodical hybridization experiments. Thus certain subfamilies are underrepresented.

Twenty-eight percent of all hybridizations cross the lines between subfamilies; a total of 60% are intergeneric; (within the ducks the measure of intergeneric hybridizations is 52%: Scherer and Hilsberg 1982). In all 11 subfamilies of the Phasianidae in which hybrids are known, at least one hybrid crossed the line between subfamilies. For the study of basic types this is an important fact (see discussion).

In many cases it cannot be determined for sure whether the hybrids are fertile or sterile. In general, it is frequently true that intergeneric hybrids are fertile (compare Table 4 and Appendix A). A total of 34 hybridizations (24.7%) resulted in fertile hybrids; 9 hybridizations(5.7%) are intergeneric. In reality, the percentage of fertile hybrids is probably higher; in many cases the necessary experiments are lacking.

**11.3.3. Megapodiidae – Mound-builders.** No hybridizations among the 12 species of this family or between any of them and any other gallinaceous birds have been observed. Keartland (1901, according to Gray 1958) does report three hybrids from a cross between Latham's brush turkey (*Alectura lathami*) and the domestic chicken (*Gallus gallus*). However, because this information is highly questionable it has not been considered relevant.

**11.3.4. Cracidae – Curassows and Their Kin.** The intergeneric and interspecific hybrids of this family are listed in the appendices. The intergeneric hybrids are visually represented in Figure 8. All of them have been observed in captivity. Taibel (1961) mentions fertile hybrids of *Crax alberti* and *Mitu mitu*. He observed fertile offspring, both among the $F_2$, as well as among backcrosses. Gray (1958) lists a total of five hybrids between Cracidae and Phasianidae and mentions that the speckled chachalaca (*Ortalis motmot* subspecies *guttata*) and the domestic chicken (*Gallus gallus*) "are said to cross easily." The other hybrids are: *Crax* sp. and domestic chicken; *Crax* sp. and domestic turkey; *Penelope* sp. and domestic chicken. However, all these statements are questionable since no documentation is available.

**11.3.5. Phasianidae – Pheasants and Their Kin.** Hybrids have been documented in 11 of the 15 subfamilies. Those 11 subfamilies represent approximately 94% of all the species.

**11.3.5.1. Numidinae – Guineafowl.** The hybrids that were observed connect the Numidinae with five other subfamilies. Within the guinea fowl, hybrids are known only between helmeted (*Numida meleagris*) and vulturine guineafowl (*Acryllium vulturinum*), and they have been described as expressing intermediate characters states. Hybrids between domestic chicken rooster and the domestic guineafowl hen (Figures 9

Figure 9. Portrait of an intergeneric hybrid of the domestic rooster ("White Leghorn" breed) and the domestic guineafowl hen. Photo by R. Klemm.

Figure 10. Intergeneric hybrids of the domestic rooster (*Gallus gallus* f. *domesticus*) and the domestic guineafowl hen (*Numida meleagris* f. *domesticus*) at age of 16 weeks. Photo by R. Klemm.

and 10) look more like guineafowl. They either do not have a crest or they only show a rudiment of a crest. The horns and lobes which are typical of the helmeted guineafowl (see Figure 5), are absent. The color of the plumage depends largely upon the breed of the hybridization partner. Thus, the Color Inhibitor I (a hereditary gene influencing the color of offspring) of the breed white leghorn caused the white plumage coloration of the hybrids in Figures 9 and 10. Damme (1992b) quotes a fertilization rate of 8.6% (domestic guineafowl rooster and domestic chicken hen) and 28.7% (domestic chicken rooster and guineafowl hen) and, thus, shows distinct positional effects (differences in reciprocal hybridizations), which is also true of the hatching rate. Hybrids between subspecies of the helmeted and the plumed guineafowl (*Gattera plumifora*) were described by von Boetticher (1954). The phenotype of hybrids between the West African helmeted guineafowl (*Numida meleagris galeata*) and the Northeast African helmeted guineafowl (*Numida meleagris meleagris*) resembles the Southwest African helmeted guineafowl (*Numida meleagris papillosa*), indicating a close affinity between those three subspecies.

**11.3.5.2. Pavoninae – Peafowl.** Besides fertile hybrids within the genus, successful crossbreeding has been done between the domestic peafowl (*Pavo cristatus*) and representatives of four subfamilies. Hybrids have not been documented for the Congo peafowl. Its systematic position was already pointed out. Deserving mention here is a statement by Verheyen (1965) that a hybrid between domestic guineafowl and domestic peafowl showed strong phenotypic similarities with the Congo peafowl (see discussion).

**11.3.5.3. Meleagridinae – Turkeys.** Via the domestic turkey, this subfamily is connected to five other subfamilies through hybridization. Through a questionable cross with the capercaillie, the domestic turkey is possibly connected also with the Tetraonidae. Reciprocal crossbreeding between capercaillie (T), and domestic chicken (H) and capercaillie (T) and common pheasant (F) result in clear positional effects in the

| Family/subfamily | Number of species | observed hybrids | | | | Ratio |
|---|---|---|---|---|---|---|
| | | total | within the subfamily | between genera | between different subfamiles | T:P × 100 |
| | 1 | 2 | 3 | 4 | 5 | 6 |
| Megapodiidae | 12 | 0 | | | | |
| Phasianidae | | | | | | |
| Numidinae | 6 | 6 | 1 | 1 | 5 | 7.0 |
| Pavoninae | 3 | 5 | 1 | 0 | 4 | 33.0 |
| Meleagridinae | 2 | 5 | 1 | - | 4 | 100.0 |
| Argusianinae | 8 | 0 | | | | |
| Phasianinae | 21 | 74 | 48 | 27 | 27 | 23.0 |
| Lophophorinae | 3 | 5 | 0 | - | 5 | 0.0 |
| Pucrasiinae | 1 | 1 | - | - | 1 | - |
| Ithagininae | 1 | 0 | | | | |
| Gallinae | 4 | 21 | 6 | - | 15 | 100.0 |
| Tragopaninae | 5 | 11 | 5 | - | 6 | 50.0 |
| Galloperdicinae | 3 | 0 | | | | |
| Ptilopachinae | 1 | 0 | | | | |
| Perdicinae | 98 | 23 | 16 | 2 | 7 | 0.3 |
| Odontophorinae | 31 | 11 | 10 | 3 | 1 | 2.0 |
| Tetraoninae | 16 | 30 | 17 | 14 | 13 | 14.0 |
| Cracidae | 43 | 9 | 9 | 4 | - | 1.0 |
| Sum | 258 | 158 | 114 | 51 | 44 | |

Table 3. Quantitative analysis of the crossbreeding results of gallinaceous birds. It has to be noted that the hybrids between representatives from different subfamilies in columns 2 and 5 are listed twice. This is not true for the sums of the last line. * within the subfamily, T: number of hybrids observed within the subfamily; P: number of potentially possible hybrids per subfamily.

fertilization rates after artificial insemination (Makos and Smyth 1970: H × T = 58.6%; T × H = 4.8%; F × T = 65.1%; and T × F = 37.6 %). Causes for this effect are being studied.

**11.3.5.4. Argusianinae – Peacock-pheasants.** As far as the peacock-pheasants are concerned, no hybrids have thus far been observed, either within the subfamily or with representatives of other subfamilies. Bronzini (1946) mentions an attempted cross between a male great argus (*Argusianus argus*) and the hen of the crested argus (*Rheinardia ocellata*). No copulation was observed, and all eggs were unfertilized.

**11.3.5.5. Phasianinae – Pheasants.** Among the 21 species, of this subfamily, 23% of the possible hybrids have been described. Close relationships within the Pasianidae are supported by 27 documented hybrids between members of this subfamily and represtatives of eight other subfamilies.The largest number of hybrids occurred between the common pheasant (*Phasianus colchicus*) and the silver pheasant (*Lophura nycthemera*) (see appendix).

With 42% of the hybrids being fertile in 48 crosses within the subfamily, there appears to be little genetic isolation among pheasant

| Species | fertile hybrids observed with species: | |
| | within the genus | of other genera |
|---|---|---|
| 32 gold pheasant | 32 | 34', 36' 48' |
| 33 Lady Amherst's pheasant | 32 | 34' |
| 34 common pheasant | | 32', 33', 46', 48' |
| 35 copper pheasant | | 36 |
| 36 Reeve's pheasant | | 32, 34', 35' |
| 38 Elliot's pheasant | 39 | |
| 39 Mikado pheasant | 38 | |
| 41 crested fire back | 48 | |
| 42 Siamese fire back | 48 | |
| 45 Edward's pheasant | 46,47,48 | |
| 46 imperial pheasant | 45,47,48 | |
| 47 Swinhoe's pheasant | 45,46,48 | |
| 48 silver pheasant | 41', 42', 45, 46, 47 | 32', 34', 51' |
| 49 white eared-pheasant | 50,51 | |
| 50 blue eared-pheasant | 49,51 | |
| 51 brown eared-pheasant | 9,50 | 48' |

Table 4. Fertile hybrids in the Subfamily Phasianinae. (see Gray [1958], Raethel [1991], Felix [1964] and Hertwig [1936b]). Numbers refer to the number in the appendix. ' denotes hybrids where only the male is fertile.

species. A summary is shown in Table 4. Due to the constellation of homogametic sexual chromosomes of male birds, often only the male hybrids are fertile. Hertwig (1936b) discusses questions regarding different degrees of sterility. When the silver pheasant is paired with three other species, fertile hybrids of both sexes are generated; in another five crosses, only the roosters emerge as fertile hybrids. The golden pheasant and Lady Amherst's pheasant have bred so frequently that it is almost impossible today to get "pure-breed" pairs of Lady Amherst's pheasant (Raethel 1991). A phenotypic description of hybrids of eared-pheasants is given by Felix (1964). Gray lists two cases of three-species hybrids: 1. A hybrid rooster of Mikado pheasant (*Syrmaticus midako*) × Elliot's pheasant (*S. ellioti*) crossed with the hen of the Reeves's pheasant (*S. reevesii*); 2. A rooster of Edward's pheasant (*Lophura edwardsi*) crossed with a hybrid hen of silver pheasant × imperial pheasant (*L. imperialis*).

Results of fertilization and hatching rates in reciprocal crossbreedings between the common pheasant and domestic chicken were summarized by Damme (1992a). As in the case of domestic chicken rooster × domestic turkey and domestic chicken rooster × domestic guineafowl, the cross of domestic chicken rooster × pheasant hen possessed a much higher rate of fertility. Figures 11 and 12 illustrate these hybrids. Stadie (1968) gives a comparative description of morphological and ethological traits of the hybrid progeny. Sarvella (1970) describes crosses between the pheasant rooster and the Japanese quail with a fertilization rate of

Figure 11. Intergeneric hybrid from the pairing of a common pheasant and a domestic chicken ("White Rock" breed). Photo by B. Chelmonska.

Figure 12. Intergeneric hybrid from the pairing of a pheasant rooster (*Phasianus colchicus*) with the domestic hen (*Gallus gallus* f. dom.) ("White Rock" breed). Photo by K. Damme.

58%. In one case eight animals out of 162 fertilized eggs reached the adult stage.

Eight hybrids between Phasianinae and Tetraoninae point to the close relationship between the grouse and the pheasants. Boag *et al.* (1971), who characterized known hybrids between the pheasant and capercaillie, may be mentioned as well.

**11.3.5.6. Lophophorinae – Monals.** Five hybrids involving the Himalayan monal (*Lophonhorus impejanus*) show a connection of this subfamily to the Phasianinae and the Tragopaninae.

**11.3.5.7. Pucrasiinae – Koklass Pheasants.** Hybrids among three of the ten subspecies of the Koklass pheasants (*Pucrasia macroplopha*) have been described (Gray 1958). The successful transspecific pairing of a hen of *Pucrasia macrolopha* with a rooster of Temminck's tragopan (*Tragopan temmonckii*) produced a sterile hybrid.

**11.3.5.8. Ithagininae – Blood Pheasants.** To date, no hybrids with this species are known. As suggested in Figure 3, the blood pheasant is closely related to the Perdicinae. This is confirmed by common traits in the tail moulting.

**11.3.5.9. Gallinae – Junglefowl.** Among the four species of junglefowl fertile crossability exists (the details are given by Gray 1958). Little is known about the phenotype of the hybrids (e.g., Hein 1954). With 18 successful crosses, the domestic chicken is the most frequent crossbreeding partner within the Galliformes besides the common and silver pheasant. Apart from the above-mentioned debatable connections to the mound-builders and curassows, the domestic chicken has been crossbred with representatives of seven subfamilies (qualitative aspects of crossbreeding with species of above-mentioned subfamilies have already been discussed). A summary of the crosses between the

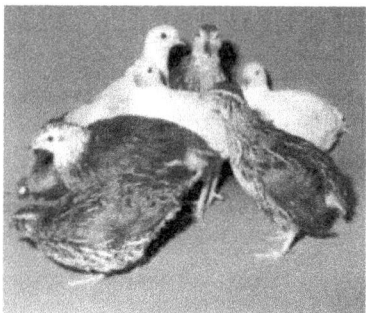

Figure 13. Hybrid progeny of the pairing of the domestic rooster (*Gallus gallus* f. *domesticus*) × Japanese quail hen (*Coturnix japonica*); light chicks from the White Rock rooster, dark chicks from the Rhode Island Red rooster; in the foreground 2 Japanese quails. Photo by K. Damme

domestic chicken and Japanese quail is given by Damme (1990; 1991). The fertilization rates were between 21% and 29%. With an average body mass of 600 g after 20 weeks, the hybrids (Figure 13) expressed a coefficient of variation of 25%, compared to approximately 10% in the original species. This was also observed in pairings of the pheasant and the domestic chicken (see Figure 14) and points to high genetic variability in interspecific crosses.

**11.3.5.10.    Tragopaninae – Tragopans.** In three of the five species, a total of six hybridizations have been observed across the boundaries of the subfamily. In this subfamily, 50% of all possible combinations have been observed as actual crosses.

**11.3.5.11. Galloperdicinae – Spurfowl.** In spite of the close relationship between the species *Galloperdix spadicea* and *G. lunulata*, no hybridizations have been described thus far (Johnsgard 1988). Hybrids with other gallinaceous birds are also not known.

**11.3.5.12. Ptilopachinae – Stone Partridges.** Hybrids of this species, which is closely related to the partridges, are not known. This may be due to the fact that stone partridges are only rarely held in captivity.

**11.3.5.13. Perdicinae – Partridges, Francolins, and Old World Quails.** The few hybrids reported are usually observed between closely related species (see appendices). The low frequency in comparison with the pheasants is not surprising since only a few species of this subfamily are kept as ornamental birds. Crossbreeding connections exist with four other subfamilies. In seven out of eight cases, the partridge or the Japanese quail is involved. Of special interest is the hybrid between the sand partridge (*Ammoperdix heyi*) and the California quail (*Callipepla californica*) because this is the only hybrid among the New World quails which crosses the boundaries of that subfamily. This report (Seth-Smith 1906), however, is viewed with caution as it lacks a detailed description.

Qualitative aspects of chicken-quail hybrids have already been discussed in the section on the Gallinae. Baumgartner *et al.* (1983) crossbred domestic turkeys and guineafowl with Japanese quails and could not find any embryos in the 547 resulting eggs. The germinal disks showed strong symptoms of decomposition.

**11.3.5.14. Odontophorinae – New World Quails.** Within this subfamily, three species are involved in a total of 10 hybrids. No hybrids of *Odontophorus*, the genus with the most species, are known. The only hybrid which crossed the boundaries of the subfamily ( b e t w e e n *Ammoperdix heyi* and *Callipepla californica*) has already been m e n t i o n e d .

Figure 14. Same-aged hybrids of the pairing of the common pheasant rooster (*Phasianus colchicus*) × domestic hen (*Gallus gallus f. domesticus*). Photo by B. Chelmonska.

Consequently, there is no certain proof of a crossbreeding connection between the New World quails and other subfamilies of the Galliformes. Crossbreeding experiments with the Cracidae (which are exclusively New World) would be very enlightening. This author suspects that successful hybrids with representatives of the partridges (*e.g., Coturnix japonica*) are possible. This could be proven experimentally since several species of New World quails are held in captivity.

**11.3.5.15. Tetraoninae – Grouse.** Among the subfamilies with many species and aside from the Phasianinae, the Tetraoninae have the largest number of hybrids, and most of those are intergeneric. There are crossbreeding connections to four other subfamilies. Well-known examples are hybrids (the German "Rackelhühner") between the black grouse rooster (*Lyrurus tetrix*) and capercaillie hen (*Tetrao urogallus*), which also occur in nature (there is a limited reproductive capability between the male hybrids and the capercaillie hen). The reciprocal cross of the capercaillie rooster and black grouse hen is documented only in captivity. Details can be found in the works of Höglund and Porkert (1989) and Klaus *et al.* (1989). A possible cause for the hybridizations within the Tetraoninae is the fact that the roosters are not able to distinguish the hens of the various species (which are very similar) if they are given the same invitation to mate. Unfortunately, the observed hybridizations are not well documented, either morphologically or ethologically. An interesting observation was published by Rensel and White (1988) on the cross of the greater sage grouse (*Centrocercus urophasianus*) and the blue grouse (*Dendragapus obscurus*). It was pointed out that the female hybrids resemble a third species, the sharp-tailed grouse (*Tympanuchus phasianellus*) in the color and pattern of the plumage. The data suggest that there is a surprisingly low degree of genetic separation among the species of this subfamily.

| Taxon | | | Trait | | |
| --- | --- | --- | --- | --- | --- |
| Genus | Family | Subfamily | tail moulting | location of nest | uropygial glands |
| | Phasianidae | | | | |
| *Afropavo* | | Pavoninae | ? | b | ? |
| *Meleagris* | | Meleagridinae | b | a | b |
| *Argusianus* | | Argusianinae | c | a | d |
| *Phasianus* | | Phasianinae | b | a | c |
| *Gallus* | | Gallinae | b | a | a |
| *Tragopanus* | | Tragopaninae | a | b | ? |
| *Perdix* | | Perdicinae | a | a | b |
| *Coturnix* | | Perdicinae | a | a | a |
| | Cracidae | | | | |
| *Crax* | | | c | b | ? |

Table 5. Comparison of common selected traits among some genera of the gallinaceous birds. Left column: a: from the center to the outside (centrifugal), b: from the outside to the center (centripetal), c: begins with the feather that lies between the inner and the outer one; middle column: a: ground breeders, b: tree breeders; right column: a: alcan-2,3-diols (erythro-and threo-forms), b: alcan-2,3-diols (only erythro-form), c: only erythro-octadecan-2,3-diol (data from Jacobs 1989); d: no uropygial gland present.

## 11.4. Discussion

The hybrids mentioned here allow for a preliminary delimitation of basic types within the gallinaceous birds. First, it can be pointed out that there are no hybrids that cross the boundaries of the gallinaceous birds (Table 2). Gray (1958) mentions three speculative hybrids (e.g., domestic chicken × mallard duck [*Anas platyrhychos*]), which, however, have neither been verified nor reproduced, and cannot, therefore, be taken into account. Furthermore, it was pointed out that the hoatzin, as a single representative of the Opisthocomiformes (hybrids have never been observed), deviates strongly from the gallinaceous birds and very likely represents a separate basic type.

As far as the position of the mound-builders is concerned, it was already determined that their breeding behavior cannot be regarded as ancestral. An analysis of the secretion of uropygial glands led Jacob (1989) to reach a simliar conclusion about the chemical structure of the secretions. In the Galliformes, this secretion is characterized by the existence of alcan-2,3-diols, which can occur in two stereoisomeric forms. The possession of both configurations is interpreted as ancestral (Table 5). In *Leipoa* (a genus of the mound-builders), however, only erythro-diols were found. These observations do not support the assumption of Johnsgard (1988) to place a "Megapode-like ancestor" at the root of the phylogenetic tree of the Perdicinae. Despite the morphological similarity to the Phasianidae, the lack of undisputable hybrids and exceptional breeding behavior suggest that the mound-builders are a separate basic type.

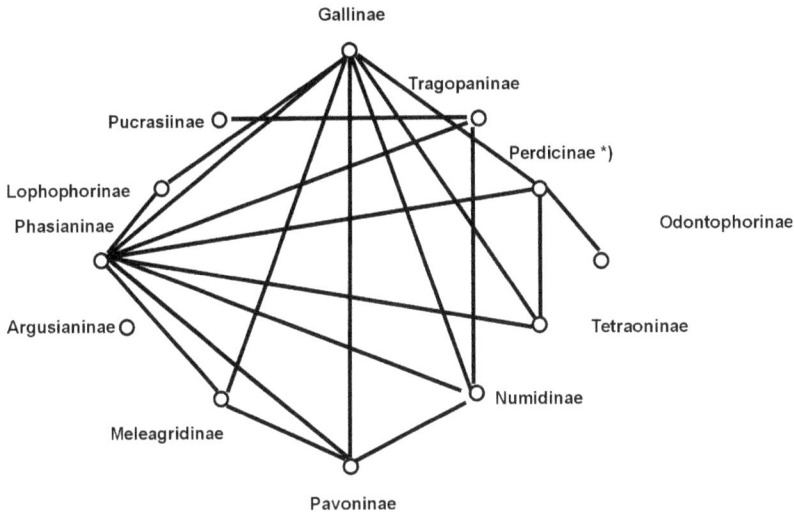

Figure 15. Crossbreeding connections among the subfamilies of the pheasants (Phasianidae). To emphasize their inclusion in this basic type, the Ithagininae, Ptilopachinae, and Galloperdicinae are included in the Perdicinae following Johnsgard (1988) as indicated by "*".

In accordance with Vuilleumier (1965), it can be assumed that the curassows are a phylogenetically young family. Based on the observation that hybridizations occur in captivity but were never observed in nature, Vuilleumier (1965) concludes that genetic isolation mechanisms are absent. Thus, the species are closely related and experienced a process of speciation by means of ethologic and ecologic isolation. The absence of an unequivocal hybrid between the Cracidae and the Phasianidae does not allow for a conclusive statement regarding a basic type delimitation. The most recent results of Sibley and Ahlquist (1990; see Figure 1) point to a greater taxonomic distance between these two families. The proof of a successful cross (e.g., between the speckled chachalaca and domestic chicken) would be very significant for determining the basic type boundaries. It should at this point be assumed that the curassows represent a separate basic type.

The close relationships within the family of the Phasianidae, which in the present study is divided into 15 subfamilies, has already been pointed out. Eleven of the subfamilies are connected either directly by hybrids or indirectly by at least one case of hybridization with a mutual species (Figure 15) (as far as the New World quails are concerned, the connection is based on one uncertain hybridization and is, therefore, rather weak). Results of Sibley and Ahlquist (1990) suggest a greater taxonomical distance for the New World quails. According to these authors (Figure 1), the degree of deviation is surprisingly high. Perhaps afer a geographical islation event, parallel microevolutionary development led to convergences. On the other hand, comparative karyotype studies of Stock and Bunch (1982) suggest a generally closer relationship of the New World quails, grouse, turkeys, partridges, and pheasants. Examination of protein polymorphisms also shows a close relationship of *Coturnix japonica* with *Callipepla californica* and *Coturnix japonica* with *Colinus virginianus* (Randi et al. 1991). In biochemically-based taxonomic methods, the partridges appear to be non-homogeneous (e.g., Randi et al. 1991). Phylogenetic trees based on an analysis of 21 loci of 12 species of the Phasianidae, on one hand, show a close relationship between *Perdix* and *Phasianus, Meleagris* and the grouse and, on the other hand, between the New World quails and both *Alectoris* and *Coturnix*.

On the basis of the present study of hybridization, it can be postulated that the representatives of the family Phasianidae belong to one basic type. This includes the representatives of the four subfamilies without hybrids on the basis of their similarity (see Appendix A). Apart from the peacock-pheasants, the five species of the three remaining subfamilies are generally placed within the Perdicinae (see Figure 3; as was also done in Figure 15). Nevertheless, an experimental examination of the hybridization potential of the peacock-pheasants would be very helpful.

It is remarkable that—in contrast to the New World quails and the curassows—the turkeys (which also inhabit only the New World) and the grouse (the only subfamily which inhabits both the Old World and the New World) show a significantly higher tendency to hybridize.

The statements about the degree of fertility of the hybrids substantiate the close relationship of species within the genera. With the exception of the hybrid between Reeves' pheasant (*Syrmaticus reevesii*) and the copper pheasant (*Graphephasianus soemmerringi*), none of the other 24 hybrids with fully fertile descendants cross the generic boundaries. Examples are *Gallus, Alectoris,* and *Callipepla* (see appendices).

In the case of eight other transgeneric hybrids, a restricted fertility was observed. The hybrid between the hen of the capercaillie and the rooster of the Black Grouse has already mentioned. In seven hybrids within the subfamily Phasianinae (Table 4), only the male hybrids are fertile, pointing to the taxonomic distance of the genera within this subfamily. Regarding the gallopheasants, the present analysis (see the fertile hybrids in Table 4) supports the view of Wolters to unite all nine species into one genus (*Lophura*). Other authors (e.g., Raethel 1991) divide these species into six genera.

An analysis of the taxonomic literature shows that it is difficult to construct a phylogenetic tree for the order Galliformes even if the number of characters is maximized (including the analysis of interspecific hybridizations in the present study). The comparison of morphological, ethological, and biochemical criteria frequently points to different distances between the same taxonomic units. Examples are listed in Table 5 where different genera are compared with each other. According to Randi et al. (1991), there is only a limited correspondence in the results of systematic relationships between biomolecular and genetic analyses on one hand and the traditional morphologically-based classifications on the other.

Individual reports on similarities of hybrids with a third species (see the sections on the Pavoninae and Tetraoninae) indicate the phenotypic expression of character states that existed in the original gene pool. Hybridizations which cross the boundaries of genera/subfamilies (e.g., the similarity between the hybrid of a domestic peafowl and a domestic guineafowl with the Congo peafowl, *Afropavo congensis*) are particularly enlightening. Regarding the phenotypic appearance of characteristics that are not detectable in the parental species, Darwin (1868) made the following observation: he reports the crossbreeding of domestic chicken varieties (a dark-green Spanish rooster with a White Silkie hen) resulting in a hybrid which resembled the wild red junglefowl (*Gallus gallus*). A good documentation of the morphological and ethological characteristics of hybrids, therefore, is important for further examinations of the systematic relationships. It seems possible that despite the phenotypic variation of the species, genera, and subfamilies, the variation of original characteristics is a sufficient explanation for the variability within the Phasianidae.

To clarify questions regarding the delimitation of basic types, the gallinaceous birds are especially suited for experimental crossbreeding because crossbreeding partners can be easily obtained (domestic chicken, Japanese quail, domestic turkey, domestic guineafowl, common pheasant, etc.), and methical questions of the sperm extraction and insemination are described in the literature on animal husbandry. For the important hybrids in the Cracidae, Argusianinae and Odontophorinae, suitable individuals can be raised in captivity and selected as partners for the above-mentioned species. It should, however, be noted that, in crossbreeding experiments, a sporadic parthenogenetic development of the eggs is possible. In some breeds of domestic turkey, this happens regularly (Olsen 1974). In birds, this always leads to diploid male parthenogens with a 2AXX-constitution. The experiments of Olsen (1974) show that the parthenogenetic development is subject to genetic control. In crossbreeding experiments, it is therefore recommended to test the eggs of the female animals for such a development before the pairing. Cytogenetic tests are recommended as well.

The present study confirms the initial claim that the gallinaceous birds represent a relatively close systematic unit, including a preliminary number of *three* basic types. These morphologically and ethologically related basic types correspond to the families Megapodiidae (mound-builders), Cracidae (curassows) and Phasianidae (pheasants), according to the system of Wolters (1975-1982).

## Acknowledgements

I would like to thank Prof. Dr. Siegfried Scherer (Weihenstephan) for initiating this study, as well as Mr. A. Johann (Rheine), Dr. K. Damme (Grub), and Prof. B. Chelmonska (Wroclaw), for providing the graphic illustrations. I also would like to thank Dr. Köhler (Leipzig) for providing literature references, as well as Dr. F. Zimbelmann (Weingarten) and Dr. H. Kutzelnigg (Ratingen) for their critical examination of the manuscript. Last but not least, I would like to thank my wife for her support with the writing of this contribution.

# References

Barnikol, A. 1952. Vergleichende anatomische, taxonomische und phylogenetische Studien am Kopf der Opisthocomiformes, Musophagidae, Galli, Columbae und Cuculi. *Zoologische Jahrbuche* 81:488-526.

Baumgartner, J., V. Illes, J. Saulic, and J. Muravska. 1983. Attempts to obtain intergeneric hybrids Turkey × Quail and Guinea Fowl × Quail. In: *Proceedings of the 5th International Symposium on Actual Problems of Avian Genetics.* Slovak Society for Agriculture, Piestany Czechoslovakia, pp. 170-173.

Beer, J.V. 1992. Quality gamebird chicks. *International Hatchery Practice* 6(3):1417.

Boag, D.A, A. Watson, and N. Bousfield. 1971. Characteristics of Pheasant × Capercaillie hybrids. *Scottish Birds* 6(6):313-316.

Boback, W and D. Müller-Schwarze. 1968. *Das Birkhuhn* [Die Neue Brehm Bucherei 397]. A. Ziemsen, Wittenberg.

Brodkorb, P. 1964. Catalogue of fossil birds, Part 2. *Bulletin of the Florida State Museum* 8:195-335.

Bronzini, E. 1946. Dieci anni di attivita avicola del Giardino Zoologico di Roma. *La Rivista Italiana di Ornitologia* 2(Ser. 16):324.

Crowe, T.M. and A.A. Crowe. 1985. The genus *Francolinus* as a model for avian evolution and biogeography in Africa. In: Schuchmann, K.L. ed. *African Vertebrates: Systematics, Phylogeny and Evolution.* Zoologisches Forschungsinstitut und Museum Alexander König, Bonn, Germany, pp. 207-240.

Damme, K. 1990. Huhn-Wachtel-Kreuzungen. *Deutsche Geflügelwirtschaft und Schweineproduktion* 43:1260-1263.

Damme, K. 1991. Gattungskreuzung zwischen dem Haushahn (*Gallus domesticus*) und der Wachtel (*Coturnix coturnix japonica*). *Archiv für Geflügelkunde* 55:127-129.

Damme, K. 1992a. *Fasan-Huhn-Kreuzungen.* Geflügelbörse. 17:13.

Damme, K. 1992b. *Huhn-Perlhuhn-Kreuzungen.* Geflügelbörse. 18: 6.

Damme, K. 1992c. Intergeneric crosses among various galliform species. *Proceedings, 19th World's Poultry Conference* 3:347.

Darwin, C. 1868. *The Variation of Animals and Plants under Domestication.* John Murray, London.

Delacour, J. 1977. *The Pheasants of the World,* Second Edition. World Pheasant Association and Spur Publications, Hindhead, UK.

Eller, G. 1971. *Die Japanische Wachtel: Eine Monographie.* Dissertation. University of Giessen, Giessen, Germany

Engelmann, C. 1984. *Leben und Verhalten unseres Hausgeflügels.* Neumann-Neudamm, Melsungen, Germany.

Felix, J. 1964. *Ohrfasanen* [Die Neue Brehm Bucherei 339]. A. Ziemsen, Wittenberg.

Glutz von Blotzheim, U.N., K.M. Bauer, and E. Bezzel. 1973. *Handbuch der Vögel Mitteleuropas, Band 5: Galliformes.* Aula, Wiesbaden, Germany.

Grajal, A and S. Strahl. 1991. Hoatzin als "Wiederkäuer". *Naturwissenschaftliche Rundschau* 44.

Gray, A.P. 1958. *Bird Hybrids: A Check-List with Bibliography*. Commonwealth Agricultural Bureaux, Farnham Royal, England.

Grzimek, B. 1980. *Grzimeks Tierleben* Band 7: Vögel 1. Kindler, Zürich.

Hein, L. 1954. *Unser Haushuhn*. [Die Neue Brehm Bucherei 127]. A. Ziemsen, Wittenberg.

Heinroth, O and M. Heinroth. 1966. *Die Vögel Mitteleuropas*, Volume 3, Urania, Leipzig.

Herre, W. and M. Röhrs. 1990. *Haustiere zoologisch gesehen*. Fischer, Stuttgart, Germany.

Hertwig, P. 1936a. Rassen und Artbildung bei Haushühnern. *Archiv für Geflügelkunde* 10:144-150.

Hertwig, P. 1936b. Artbastarde bei Tieren. In: von Baur, E. and M. Hartmann, eds. *Handbuch der Vererbungswissenschaften*, Volume 2, Borntraeger, Berlin.

Höglund, N.H. and J. Porkert. 1989. Experimentelle Kreuzungen zwischen Auer und Birkhuhn (*Tetrao urogallus* et *Tetrao tetrix*). *Zeitschrift für Jagdwissenschaft* 35:221-234.

Ilicev, V.D. and V.E. Flint. 1989. *Handbuch der Vögel der Sowjetunion, Band 4: Galliformes, Gruiformes*. Aula, Wiesbaden, Germany.

Jacob, J. 1989. Bewertung und Ergebnisse chemischer Methoden für systematische Fragen in der Zoologie. *Zoologische Beiträge (N.F.)* 32:391-424.

Johnsgard, P.A. 1973. *Grouse and Quails of North America*. University of Nebraska Press, Lincoln, NE.

Johnsgard, P.A. 1988. *The Quails, Partridges and Francolins of the World*. Oxford University Press, Oxford, UK.

Klaus, S., A.V. Andreev, H.H. Bergmann, F. Moller, J. Porkert, and J. Wiesner. 1989. *Die Auerhühner* [Die Neue Brehm Bucherei 86]. A. Ziemsen, Wittenberg.

Lange, E. 1985. *Mechanismen der Evolution* [Die Neue Brehm Bucherei 433]. A. Ziemsen, Wittenberg.

Makos, J.J. and J.R. Smyth. 1970. A study of fertility following intergeneric crosses among certain gallinacaeous birds. *Poultry Science* 42:23-28.

Mayr, E. 1967. *Artbegriff und Evolution*. Paul Parey, Hamburg, Germany.

Ogilvie-Grant, W.R. 1893. Catalogue of the game birds in the collection of the British Museum. In: *Catalogue of Birds*, Volume 22, British Museum, London.

Olsen, M.W. 1974. Frequency and cytological aspects of diploid parthenogenesis in turkey eggs. *Theoretical and Applied Genetics* 44:216-221.

Parker, W.K. 1891. On the morphology of a reptilian bird, *Opisthocornus cristatus*. *Transactions of the Zoological Society of London* 13: 43-85, pls. 7-10.

Peters, J.L. 1934. *Checklist of Birds of the World*, Volume 2. Harvard University Press, Cambridge, MA.

Raethel, H.S. 1991. *Hühnervögel der Welt*. Natur Verlag, Augsburg, Germany.

Randi, E., G. Fusco, R. Lorenzini, and T.M. Crowe. 1991. Phylogenetic relationships and rates of allozyme evolution within the Phasianidae. *Biochemical Systematics and Ecology* 19:213-221.

Rensel, J.S. and O.M. White. 1988. First description of hybrid Blue × Sage Grouse. *The Condor* 90:716-717.

Sarvella, P. 1970. Raising a new hybrid: Pheasant × Japanese Quail. *Poultry Science* 50:298-300.

Scherer, S. 1993. Basic Types of Life. In: Scherer, S. ed. *Typen des Lebens*. Pascal, Berlin, pp. 11-30.

Scherer, S. and T. Hilsberg. 1982. Hybridisierung und Verwandtschaftsgrade innerhalb der Anatidae: Eine systematische und evolutionstheoretische Betrachtung. *Journal für Ornithologie* 123:357-380.

Seth-Smith. 1906. Bird notes from the Zoological Gardens. *Avicultural Magazine, New Series* 4:285-286.

Sibley, C.G. and J. E. Ahlquist. 1972. A comparitive study of the egg white proteins of non-passerine birds. *Bulletin of the Peabody Museum of Natural History* 39:1-276.

Sibley, C.G. and J. E. Ahlquist. 1990. *Phylogeny and Classification of Birds: A Study in Molecular Evolution*. Yale University Press, New Haven, CT.

Stadie, C. 1968. Verhaltensweisen von Gattungsbastarden (*Phasianus colchicus × Gallus g. f. dom.*) im Vergleich mit deren Ausgangsarten. *Zoologischer Anzeiger* 31 Supplement Volume.

Stock, A.D. and T.D. Bunch. 1982. The evolutionary implications of chromosome banding pattern homologies in the bird order Galliformes. *Cytogenetics and Cell Genetics* 34:136-148.

Stresemann, E. 1965. Die Mauser der Hühnervögel. *Journal für Ornithologie* 106:58-64.

Taibel, A.M. 1961. Esperimenti ibridologici tra specie di generi distinti: *Mitu e Crax*. Nota 1 and 2. *Archivio Zoologico Italiano* 46:181-226, 291-324.

Vaurie, C. 1968. Taxonomy of the Cracidae. *Bulletin of the American Museum of Natural History* 138(4):133-259

Verheyen, W.N. 1965. *Der Kongopfau* [Die Neue Brehm Bucherei 351]. A. Ziemsen, Wittenberg.

von Boetticher, H. 1954. *Die Perlhühner* [Die Neue Brehm Bucherei 130]. A. Ziemsen, Wittenberg.

Vuilleumier, F. 1965. Relationships and evolution within the Cracidae (Aves, Galliformes). *Bulletin of the Museum of Comparative Zoology* 134:1-27.

Wissel, C., M. Stefani, and H.S. Raethel. 1966. *Fasanen und andere Hühnervögel*. Neumann-Neudamm, Melsungen, Germany.

Wolters, H.E. 1975-1982. *Die Vogelarten der Erde*. Parey, Hamburg, Germany.

Wolters, H.E. 1983. *Die Vögel Europas im System der Vögel*. BIOTROPIC,

Ziswiler, V. 1976. Spezielle Zoologie Wirbeltiere, Volume 1. Georg Thieme, Stuttgart, Germany.

## Appendix A.

List of hybrids of the order Galliformes (species and taxonomy after Wolters 1975-1982), with reference to the number of observed interspecific hybrids.

*) 1st number: total number of hybrids, 2nd number: number of hybrids across the boundaries of subfamilies.

**) See Table 4 for details on the fertile hybrids of the subfamily Phasianinae.

| ID # | | Species | Common Name | observed species crosses* | | fertile hybrids with # |
|---|---|---|---|---|---|---|
| Genus | Species | | | | | |
| **Family** | | **Megapodiidae** | **mound-builders** | | | |
| | | no hybrids reported | | | | |
| **Family** | | **Phasianidae** | **pheasants and kin** | | | |
| **Subfamily** | | **Numidinae** | **guineafowl** | | | |
| 9 | 15. | Numida meleagris | helmeted guineafowl | 6 | 5 | |
| 10 | 16. | Acryllium vulturinum | vulturine guineafowl | 1 | | |
| **Subfamily** | | **Pavoninae** | **peafowl** | | | |
| 13 | 20. | Pavo cristatus | Indian peafowl | 5 | 4 | 21 |
| | 21. | P. muticus | green peafowl | 1 | | 20 |
| **Subfamily** | | **Meleagridinae** | **turkeys** | | | |
| 14 | 22. | Meleagris ocellata | ocellated turkey | 1 | | 23 |
| | 23. | M. gallopavo | wild turkey | 5 | 4 | 22 |
| **Subfamily** | | **Argusianinae** | **peacock-pheasants** | | | |
| | | no hybrids reported | | | | |
| **Subfamily** | | **Phasianinae** | **pheasants** | | | ** |
| 18 | 32. | Chrysolophus pictus | golden pheasant | 8 | 1 | |
| | 33. | C. amherstiae | Lady Amherst's pheasant | 8 | 2 | |
| 19 | 34. | Phasianus colchicus | common pheasant | 22 | 14 | |
| 20 | 35. | Grapheph. soemmerringii | copper pheasant | 7 | 1 | |
| 21 | 36. | Syrmaticus reevesii | Reeves' pheasant | 9 | | |
| 22 | 38. | Calophasis ellioti | Elliot's pheasant | 6 | | |
| | 39. | C. mikado | Mikado pheasant | 3 | | |
| 23 | 41. | Lophura ignita | crested fireback | 3 | | |
| | 42. | L. diardi | Siamese fireback | 5 | 1 | |
| | 43. | L. erythophthalma | crestless fireback | 1 | | |
| | 45. | L. edwardsi | Edwards's pheasant | 3 | | |
| | 46. | L. imperalis | imperial pheasant | 3 | | |
| | 47. | L. swinhoii | Swinhoe's pheasant | 11 | 1 | |
| | 48. | L. nycthemera | silver pheasant | 18 | 4 | |
| 24 | 49. | Crossoptilon crossoptilon | white eared-pheasant | 2 | | |
| | 50. | C. auritum | blue eared-pheasant | 2 | | |
| | 51. | C. mantchuricum | brown eared-pheasant | 4 | 1 | |
| 25 | 52. | Catreus wallichii | cheer pheasant | 6 | 2 | |
| **Subfamily** | | **Lophophorinae** | **monals** | | | |
| 26 | 53. | Lophophorus impejanus | Himalayan monal | 5 | 5 | |
| **Subfamily** | | **Pucrasiinae** | **Koklass pheasants** | | | |
| 27 | 56. | Pucrasia macrolopha | Koklass pheasant | 1 | 1 | |
| **Subfamily** | | **Ithagininae** | **blood pheasants** | | | |
| | | no hybrids reported | | | | |
| **Subfamily** | | **Gallinae** | **junglefowl** | | | |
| 29 | 58. | Gallus varius | green junglefowl | 3 | | 59,60,61 |

# Basic Types of Life

|  | No. | Species | Common name |  |  |  |
|---|---|---|---|---|---|---|
|  | 59. | *G. gallus (f. domestica)* | red junglefowl/chicken | 18 | 15 | 58,60,61 |
|  | 60. | *G. sonneratii* | grey junglefowl | 3 |  | 58,59,61 |
|  | 61. | *G. lafayettii* | Ceylon junglefowl | 3 |  | 58,59,60 |
| **Subfamily** |  | **Tragopaninae** | **tragopans** |  |  |  |
| 30 | 63. | *Tragopan satyra* | satyr tragopan | 4 | 1 | 65 |
|  | 64. | *T. blythii* | Blyth's tragopan | 2 |  |  |
|  | 65. | *T. temminckii* | Temminck's tragopan | 5 | 2 | 63 |
|  | 66. | *T. caboti* | Cabot's tragopan | 5 | 3 |  |
| **Subfamily** |  | **Galloperdicinae** | **spurfowl** |  |  |  |
|  |  | no hybrids reported |  |  |  |  |
| **Subfamily** |  | **Ptilopachinae** | **stone partridges** |  |  |  |
|  |  | no hybrids reported |  |  |  |  |
| **Subfamily** |  | **Perdicinae** | **partridges/francolins** |  |  |  |
| 38 | 94. | *Perdix perdix* | grey partridge | 5 | 3 |  |
| 39 | 97. | *Alectoris melanocephala* | Arabian partridge | 1 |  | 100 |
|  | 100. | *A. chukar* | chukar | 3 |  | 91,101 |
|  | 101. | *A. graeca* | rock partridge | 5 | 1 | 100,102 |
|  | 102. | *A. rufa* | red-legged partridge | 3 |  | 101 |
|  | 103. | *A. barbara* | Barbary partridge | 1 |  |  |
| 41 | 108. | *Francolinus pictus* | painted francolin | 1 |  |  |
|  | 109. | *F. francolinus* | black francolin | 1 |  |  |
| 42 | 110. | *Pternistis leucoscepus* | yellow-necked francolin | 1 |  |  |
|  | 111. | *P. rufopictus* | grey-breasted francolin | 1 |  |  |
|  | 112. | *P. swainsonii* | Swainson's francolin | 2 |  |  |
|  | 113. | *P. afer* | red-necked francolin | 1 |  |  |
|  | 114. | *P. erckelii* | Erckel's francolin | 1 |  |  |
|  | 127. | *P. natalensis* | Natal francolin | 1 |  |  |
|  | 131. | *P. bicalcaratus* | double-spurred francolin | 1 |  |  |
| 50 | 153. | *Ammoperdix heyi* | sand partridge | 1 | 1 |  |
| 53 | 156. | *Coturnix coturnix* | common quail | 1 |  | 157 |
|  | 157. | *C. japonica* | Japanese quail | 3 | 2 | 156 |
|  | 158. | *C. novaezelandiae* | New Zealand quail | 1 |  |  |
|  | 159. | *C. coromandelica* | rain quail | 2 |  |  |
|  | 160. | *C. delegorguei* | harlequin quail | 1 |  |  |
| **Subfamily** |  | **Odotophorinae** | **New World quails** |  |  |  |
| 60 | 169. | *Colinus virginianus* | northern bobwhite | 3 |  |  |
| 61 | 173. | *Callipepla squamata* | scaled quail | 4 |  | 175,176 |
|  | 174. | *C. douglasii* | elegant quail | 3 |  | 175 |
|  | 175. | *C. gambelii* | Gambel's quail | 4 |  | 173,174 |
|  | 176. | *C. californica* | California quail | 6 | 1 | 173 |
| 62 | 177. | *Oreortyx pictus* | mountain quail | 1 |  |  |
| **Subfamily** |  | **Tetraoninae** | **grouse** |  |  |  |
| 69 | 200. | *Tympanuchus phasianellus* | sharp-tailed grouse | 3 |  | 201 |
|  | 201. | *T. cupido* | greater prairie chicken | 2 | 1 | 200 |
| 70 | 202. | *Bonasa umbellus* | ruffed grouse | 2 | 1 |  |
| 71 | 203. | *Tetrastes bonasia* | hazel grouse | 5 | 2 |  |
| 72 | 205. | *Centrocercus urophasianus* | greater sage grouse | 2 |  |  |
| 73 | 206. | *Falcipennis canadensis* | spruce grouse | 2 |  |  |
| 74 | 208. | *Dendragapus obscurus* | blue grouse | 6 | 1 |  |
| 75 | 210. | *Lagopus mutus* | rock ptarmigan | 4 | 1 |  |
|  | 211. | *L. lagopus* | willow grouse | 9 | 3 |  |
| 76 | 212. | *Lyrurus tetrix* | Eurasian black grouse | 7 | 3 | 215 |
| 77 | 214. | *Tetrao parvirostris* | black-billed capercaillie | 1 |  | 215 |
|  | 215. | *T. urogallus* | western capercaillie | 4 | 1 | 212,214 |
| **Family** |  | **Cracidae** | **currasows and kin** |  |  |  |
| 79 | 229. | *Penelope marail* | Marail guan | 1 |  |  |
|  | 237. | *P. jacucaca* | white-browed guan | 2 |  |  |

| | | | | |
|---|---|---|---|---|
| | 239. | *P. pileata* | white-crested guan 3 | |
| 85 | 249. | *Mitu mitu* | Alagoas curassow 2 | 254 |
| 87 | 253. | *Crax rubra* | great curassow 4 | 258 |
| | 254. | *C. alberti* | blue-billed curassow 3 | 249,258 |
| | 258. | *C. fasciolata* | bare-faced curassow 3 | 253,254 |

## Appendix B.

Documentation of hybrids summarized in Appendix A and Figure 8. Sources: Gray (1958); Beer (1992) 100/102; Boback & Müller-Schwarze (1968) 48/212; Eller (1971) 34/157, 157/158; Felix (1964) 49-51; Glutz von Blotzheim *et al.* (1973) 100/101; Johnsgard (1988) 110/111, 112/113, 112/127; Rensel & White (1988) 200/205, 205/208; Vuillemier (1965) 237/253, 249/254; 253/254.

| ID Number Genus | Species | Successful crosses species # |
|---|---|---|
| Numidae | | |
| 9 | 15 | 16,20,23,34,59,66 |
| 10 | 16 | 15 |
| Pavonineae | | |
| 13 | 20 | 15,21,23,34,59 |
| | 21 | 20 |
| Meleagridinae | | |
| 14 | 22 | 23 |
| | 23 | 15,20,22,34,59 |
| Phasianinae | | |
| 18 | 32 | 33,34,35,36,47,48,52,59 |
| | 33 | 32,34,36,38,47,48,53,66 |
| 19 | 34 | 15,20,23,32,33,35,36,38,47,48, 52 ,59,66,94,157,201,203,208,210, 211,212,215 |
| 20 | 35 | 32,34,36,38,42,48,59 |
| 21 | 36 | 32,33,34,35,38,39,47,48,52 |
| 22 | 38 | 33,34,35,36,39,48 |
| | 39 | 36,38,47 |
| 23 | 41 | 42,47,48 |
| | 42 | 35,41,47,48,59 |
| | 43 | 48 |
| | 45 | 46,47,48 |
| | 46 | 45,47,48 |
| | 47 | 32,33,34,36,39,41,42,45,46,48,63 |
| | 48 | 32,33,34,35,36,38,41,42,43,45,46, 47,51,52,53,59,94,212 |
| 24 | 49 | 50,51 |
| | 50 | 49,51 |
| | 51 | 48,49,50,53 |
| 25 | 52 | 32,34,36,48,53,65 |
| Lophophorinae | | |
| 26 | 53 | 33,48,51,52,59 |
| Pucrasiinae | | |
| 27 | 56 | 65 |
| Gallinae | | |
| 29 | 58 | 59,60,61 |
| | 59 | 15,20,23,32,34,35,42,48,53,58,60, 61,101,157,202,203,211,212 |
| | 60 | 58.59.61 |
| | 61 | 58.59.60 |
| Tragopaninae | | |
| 30 | 63 | 47,64,65,66 |
| | 64 | 63,65 |
| | 65 | 52,56,63,64,66 |
| | 66 | 15,33,34,63,65 |
| Perdicinae | | |
| 38 | 94 | 34,48,101,102,211 |
| 39 | 97 | 100 |
| | 100 | 97,101,102 |
| | 101 | 59,94,100,102,103 |
| | 102 | 94,100,101 |
| | 103 | 101 |
| 41 | 108 | 109 |
| | 109 | 108 |
| 42 | 110 | 111 |
| | 111 | 110 |
| | 112 | 113,127 |
| | 113 | 112 |
| | 114 | 131 |
| | 127 | 112 |
| | 131 | 114 |
| 50 | 153 | 176 |
| 53 | 156 | 157 |
| | 157 | 34,59,156 |
| | 158 | 159 |
| | 159 | 158,160 |
| | 160 | 159 |
| Odontophoninae | | |
| 60 | 169 | 173,175,176 |
| 61 | 173 | 169,174,175,176 |
| | 174 | 173,175,176 |
| | 175 | 169,173,174,176 |
| | 176 | 153,169,173,174,175,177 |
| 62 | 177 | 176 |
| Tetraoninae | | |
| 69 | 200 | 201,205,208 |
| | 201 | 34,200 |
| 70 | 202 | 59,208 |
| 71 | 203 | 34,59,210,211,212 |
| 72 | 205 | 200,208 |
| 73 | 206 | 208,211 |
| 74 | 208 | 34,200,202,205,206,211 |
| 75 | 210 | 34,203,211,212 |
| | 211 | 34,59,94,203,206,208,210,212,215 |
| 76 | 212 | 34,48,59,203,210,211,215 |
| 77 | 214 | 215 |
| | 215 | 34,211,212,214 |
| Cracidae | | |
| 79 | 229 | 239 |
| | 237 | 239,253 |
| | 239 | 229,237,253 |
| 85 | 249 | 254,258 |
| 87 | 253 | 237,239,254,258 |
| | 254 | 249,253,258 |
| | 258 | 253,254,259 |

# 12. Basic Types of Diurnal Birds of Prey (Falconiformes)

FRIEDER ZIMBELMANN

**Abstract**

It may be possible to separate the diurnal birds of prey (Falconiformes) into a small number of basic types. These correspond to the families or subfamilies of the established system. Intergeneric hybridizations and the discovery that the greater anatomical differences occur only between families point in this direction. Variations of plumage color, shape and size of wings, tail, beak, and talons, as well as differing behavioral patterns, that have led to subdivisions into genera and species are within the range of variation of these basic types. Sufficient data are available to propose the following provisional basic types: the sea-eagles (Buteoninae, in part), the true eagles (Buteoninae, in part), the Old World vultures (Aegypiinae), the hawks and buzzards (Accipitrinae and Buteoninae, in part), the falcons (Falconidae), the New World vultures (Cathartidae), the secretary bird (Sagittariidae), and the osprey (Pandionidae). Already at their first occurrence in the Eocene and Oligocene, the five extant families can be identified. Transient forms are hitherto unknown. The family Neocathartidae became extinct in the Eocene. The geographical distribution of fossil New World vultures, Old World vultures, and the secretary bird is wider than the present distribution of these birds. Thus, their present distribution is merely a relict.

## 12.1. Introduction

The order Falconiformes comprises the predatory birds (also known as "birds of prey" or "raptors") that hunt during the day-time. It represents a systematic unit that is well-defined and clearly separated from other systematic groups. It is characterized by a bent upper beak with hook-shaped tip and prehensile feet with powerful sharp talons. With the exception of the palm-nut vulture (*Gypohierax angolensis*), all birds of prey feed on animals or carrion. Based on anatomical and ethological criteria, five groups (families) can be distinguished (Table 1).

The New World vultures (*Cathartidae*) with seven recent species can be found exclusively in North and South American (fossil species also in Europe). They are characterized by a perforated nose partition, weak rear toes, colored skin protrusions on the head, a distinctive sense of smell, and the absence of a syrinx.

The hawk-like birds (Accipitridae) include hawks, buzzards, kites, harriers, sea-eagles, eagles, and Old World vultures. A total of over 200

Order Falconiformes (diurnal birds of prey)
Family Cathartidae (New World vultures)
      G   *Cathartes, Coragyps, Sarcoramphus, Gymnogyps, Vultur*

Family Accipitridae (hawk-like birds)
    SF  1.  Elaninae
      G   *Elanus, Chelictinia, Macheiramphus, Gampsonyx, Elanoides*
    SF  2.  Perninae
      G   *Aviceda, Henicopernis, Pernis, Leptodon, Chondrohierax*
    SF  3.  Milvinae
      G   *Harpagus, Ictinia, Rostrhamus, Helicolestes, Haliastur, Milnus, Lophoictinia, Hamirostra*
    SF  4.  Accipitrinae
      G   *Accipiter, Melierax, Urotriorchis, Erythrotriorchis, Heterospiza*
    SF  5.  Buteoninae
      G   *Buteo, Geranoaetus, Parabuteo, Buteogallus, Busarellus, Leucopternis, Kaupifalco, Butastur, Harpyhaliaetus, Morphnus, Harpia, Pithecophaga, Harpyopsis, Oroaetus, Spizastur, Spizaetus, Lophoaetus, Stephanoaetus, Polemaetus, Cassinaetus, Hieraaetus, Aquila, Haliaeetus, Ichthyophaga*
    SF  6.  Aegypiinae (Old World vultures)
      G   *Gypohierax, Neophron, Gypaetus, Necrosyrtes, Torgos, Sarcogyps, Aegypius, Trigonoceps, Gyps, Pseudogyps*
    SF  7.  Circinae
      G   *Circus, Polyboroides, Geranospiza*
    SF  8.  Circaetinae
      G   *Terathopius, Circaetus, Spilornis*

Family Falconidae (falcon-like birds)
    SF  1.  Herpetotherinae
      G   *Herpetotheres, Micrastur*
    SF  2.  Polyborinae
      G   *Polyborus, Daptrius, Milvago, Phalcoboenus*
    SF  3.  Polihieracinae
      G   *Polihierax, Spiziapteryx, Microhierax, Neohierax*
    SF  4.  Falconinae
      G   *Falco, Nesierax, Jeracidea*

Family Pandionidae (ospreys)
      G   *Pandion*

Family Sagittariidae (secretary birds)
      G   *Sagittarius*

Table 1. Overview of the taxonomic position of the genera and families of the falconiform raptors (after Ziswiler 1976, Grzimek 1968, Feduccia 1984). G: genus, SF: subfamily.

recent species have been described (a large variety of Old World vultures also inhabited the New World as late as the Pleistocene). The wings of the hawk-like birds are wide and rounded, their upper beak is powerful and long and does not possess any tooth protrusion.

The falcon-like birds (Falconidae) include approximately 60 species. Their particular characteristics are narrow, pointed wings and a relatively short and round upper beak with a tooth-like protrusion ("falcon tooth").

The only recent species of the family Pandionidae is the osprey. It is a highly specialized fish hunter with a reversible outer toe and nostrils that can be closed. As a typical cosmopolitan, it can be found almost world-wide.

The secretary birds (Sagittariidae), too, are represented by just one extant species. The distribution of the recent species is limited to Africa; fossil species also lived in Europe. As a long-legged running bird, the secretary bird hunts reptiles in grasslands.

The systematic division below the family level is disputed (Brown 1979, p. 7). Depending on the author, different characteristics are taken into account and given different importance. For example, Moll (in Grzimek 1968) places the osprey as the subfamily Pandioninae into the hawk-like birds (Accipitridae), but Ziswiler (1976) and Weick (1980) regard the osprey as an independent family (Pandionidae). Another example is the group of the sea-eagles, whose eight recent species are united to one single species by Fischer (1982) and Brown and Amadon (1968) but are divided among three genera by Swann (1930-1945).

In the New World vultures and the falcon-like birds there is sufficient consistency in the characteristics, so that one can speak of the "vulture type" or "falcon type." Species are determined above all by morphological criteria.

In contrast, the family of hawk-like birds (Accipitridae) consists of groups that partially deviate from each other in their behavior and morphology. Both buzzards and hawks, for example, are united into the family of hawk-like birds. The European representatives of these two genera illustrate their differences: the common buzzard (*Buteo buteo*) primarily feeds on small mammals, frequently practices a hovering flight, and is characterized by long wings and a short swoop; instead, the northern goshawk (*Accipiter gentilis*) catches its prey almost exclusively in the air where it proves to be a fast and agile hunter (which never hovers), and its characteristics of short wings and long swoop are very useful for this purpose.

In contrast, the non-European representatives of the two genera demonstrate the great variety of this group in its behavior and morphology. The prey of the North American red-tailed hawk (*Buteo jamaicensis*), for example, consists not only of small mammals but also of snakes, lizards, birds, frogs, fish, and insects. This species inhabits the taiga, forests, deserts, prairies, and tropical rain forests (Grzimek 1968, p. 362). The African pale chanting goshawk (*Melierax musicus*) catches mostly lizards, insects, frogs, and snakes (Grzimek 1968, p. 358). Both hawks and buzzards exhibit a large variety of different hues and patterns, besides variations of nearly white and nearly black.

An objective systematic classification is frustrated by the lack of clear anatomical differences and especially by the wide spectrum of morphological characteristics and behavioral patterns. However, a significant systematic criterion is the ability to hybridize. Species that can be crossbred are characterized by a close conformity of their genotypes. Due to this fact, the attempt to establish "basic types," which are defined by documented cases of crossbreeding, seems worthwhile. All individuals that are directly or indirectly connected through crossbreeding belong to the same basic type (Marsh 1976; Scherer 1993, this volume).

According to this definition, the buzzards and the hawks belong to the same basic type because they can be crossbred. They demonstrate not only a close genetic relationship but also the high variability in morphology and behavior, which can sometimes occur within one and the same basic type. The phenetic differences between the buzzards and the hawks are obvious but do not justify a systematic separation.

In the following section the attempt is made to classify the Falconiformes utilizing the basic type concept.

## 12.2. Falconiform Basic Types

**12.2.1. Buzzards/Hawks.** As an important component of this basic type, the genus *Buteo* (buzzards) includes 27 relatively unspecialized species, which on the basis of external traits are hard to distinguish from each other. In particular, *Buteo* is characterized by great variability in body size, coloring, and pattern, not only within one species, but also within one geographic race or even local population (Brown 1979). Crossbreeding has been documented between the common buzzard (*B. buteo*) and the rough-legged hawk (*B. lagopus*) and between the common buzzard and the red-tailed hawk (*B. jamaicensis*) (Gray 1958). The conformity in morphology, anatomy, and behavior allows for the conclusion that all species of the genus *Buteo* belong to one and the same basic type.

This type further includes the genus *Parabuteo* (Harris' hawks) with its only species *P. unicinctus*, as shown by a hybrid between the Harris' hawk and the red-tailed hawk (*B. jamaicensis*) (Cade 1986).

The genus *Accipiter* (bird hawks) with 47 species, including the indigenous German species northern goshawk (*A. gentilis*) and Eurasian sparrowhawk (*A. nisus*), also belongs to this basic type. The species of *Accipiter* conform strongly to each other in morphology and behavior (e.g., the long swoop, long talons, and the habit of surprising their prey). The currently known interspecific hybrids are: *A. brevipes* (Levant sparrowhawk) × *A. nisus* (Eurasian sparrowhawk), *A cooperi* (Cooper's hawk) × *A. gentilis*, and *A. fasciatus* (brown goshawk) × *A. novaehollandiae* (gray goshawk) (Gray 1958). There is only one currently documented intergeneric hybrid: *A. gentilis* × *Buteo buteo*. Even so, when taken with the above information about *Accipiter*, one can extrapolate and assume that hybrids among all species of *Accipter* and *Buteo* are possible.

This basic type then is rather extensive and includes more than 70 species. The considerable differences between representatives of *Buteo* and *Accipiter* indicate that among the Falconiformes only a few basic types, with a large number of species each, can be expected – similar to the songbirds (Fehrer 1993, this volume), the ducks (Scherer and Hilsberg 1982), and the gallinaceous birds (Klemm 1993, this volume).

Figure 1. White-tailed sea-eagle (*Haliaeetus albicilla*)

**12.2.2. Sea-Eagles.** The eight recent species of the almost world-wide genus *Haliaeetus* (sea-eagles; Figure 1) are characterized by a high degree of conformity in anatomy and moderate degree in the behavior. They differ most notably in size and coloring (Fischer 1982). Likewise, there are differences in the spectrum of prey items each species chooses. Here, however, all species tend toward opportunism and can be regarded as generalists (Zimbelmann 1989). Although no hybrids have so far been described (and due to the geographical isolation are not to be expected), the high degree of conformity within the genus with continuous transitions in morphological characteristics (color patterns, proportions) and behavior seems to indicate that the formation of such hybrids should be possible. Therefore, one can assume the existence of a basic type at the level of the genus (to which all eight recent sea-eagle species would belong).

The genera *Ichthyophaga* (fish eagles), *Haliastur*, *Milvus* (black and red kites; Figure 2), *Ictinia*, and *Busarellus* (black-collared hawks) possibly also belong to the basic type of the sea-eagles. Apart from some similar traits in behavior and morphology, this is confirmed by a shared anatomical characteristic, the merged phalanges of the interior toes (Figure 3) (Olson 1982). However, thus far no transgeneric hybrids have been described.

**12.2.3. True Eagles.** What was said about the sea-eagles is essentially also applicable to the nine recent species of the genus *Aquila* (true eagles): conformity of traits in anatomy and behavior; differences in size and coloring. The food spectrum is even more uniform (Fischer 1976).

No natural hybrids among representatives of the true eagles have been documented, and, due to the geographical distribution of the various species, are not to be expected. However, a crossbreeding experiment

Figure 2. Red kite (*Milvus milvus*)

Figure 3. Left: skeleton of the foot of a typical diurnal raptor. The phalanges of the interior toe are not merged. Right: Foot bones of the genus *Ichthyophaga*. The phalanges of the interior toe are merged; the suture is still visible (arrow). (After Jollie 1977, modified.)

in captivity between the golden eagle (*Aquila chrysaetos*) and the eastern imperial eagle (*A. heliaca*) resulted in a hybrid (Claus Fentzloff, personal communication), so potential hybrids among all species of this genus can be postulated. Apart from the genus *Aquila,* the basic type "eagle" may include even further genera. Similarities in morphology and behavior have been observed, for example, in the genus *Hieraaetus* (African hawk-eagles) and *Spitzaetus* (hawk-eagles). It is also significant that there are no anatomical characteristics that delimit the genus *Aquila* from these other genera.

### 12.2.4. Old World Vultures.

The Old World vultures (subfamily Aegypiinae) also belong to the Accipitridae. This group whose recent species are limited to the Old World is represented by a large number of fossil species that were widespread also in the New World (Feduccia 1984). An especially large number of fossil species has been found in Pleistocene sediments of North America.

Among the recent species, hybrids have been formed between the Eurasian griffon (*Gyps fulvus*) and the monk vulture (*Aegypius monachus*), between the Eurasian griffon and Rüppell's griffon (*Gyps rueppellii*; Figure 4) and between Eurasian griffon and the white-backed vulture (*Pseudogyps africanus*) (Gray 1958, Cade 1986). The existence of transgeneric hybrids shows that possibly the entire subfamily of the Old World vultures (10 genera, 16 species) belong to a basic type. This is further confirmed by the fact that the Old World vultures are a very uniform group, whose characteristics are closely connected to their feeding habits. Since most of these characteristics are losses of certain traits compared to the remaining Accipitridae, there is even the possibility of an affiliation with another basic type of the Accipitridae.

The most important differences between the Old World vultures and the remaining subfamilies of the Accipitridae are the plumage, the form of the beak and the claws, as well as the behavior.

The head and the neck of the vultures are often only sparsely feathered—an adaptation to penetrating the body cavities of other animals; a normal plumage would be a disadvantage because the feathers would stick together. The secondary and primary feathers, as well as the tail feathers, are long and wide as an adaptation to gliding. Some species possess an additional two tail feathers (14 instead of 12), probably also as an adaptation to gliding. Additional tail feathers have also been observed in other species of birds (Zimbelmann 1989).

Figure 4. Rüppell's griffon (*Gyps rueppelli*)

The upper beak of the Old World vultures is usually elevated and has powerful cutting surfaces for the dividing of animal carcasses. The beak of birds is generally regarded as a very variable characteristic (Emslie 1988). The blunt claws of the Old World vultures are not suitable for seizing and killing live prey.

These characteristics possibly include both degenerative traits as well as microevolutionary adaptations (a "normal" and "elevated" upper beak as well as a "normal" and "elevated" number of tail feathers exists, for example, also within the genus *Haliaeetus*) that accompany the specialization of feeding on carrion.

**12.2.5. New World Vultures.** Within the Falconiformes, the New World vultures (*Cathartidae*) as a family are odd. The superficial similarity with the Old World vultures reflects adaptation to a similar way of life. The New World vultures, the recent species of which can be found only in the western hemisphere, also include fossil species from Europe. Together with the Old World vultures, they underwent an enormous radiation during the Pleistocene in North America, resulting in many species and individuals with transcontinental distributions (Feduccia 1984). Whether the extinct giant vultures (subfamily Teratornithinae), which included truly gigantic flying forms with a wingspan of up to 7.2 m, also belong to the basic type of the New World vultures must remain speculative but is supported by their close similarity to typical cathartids. In the recent New World vultures, hybrids have occurred between the turkey vulture (*Cathartes aura*) and the black vulture (*Coragyps atratus*) (Gray 1958). In this case, too, the existence of a hybrid indicates that the basic type is located at a transgeneric level. The entire family of the *Cathartidae* most likely belongs to one and the same basic type.

**12.2.6. Falcons.** The family of the falcon-like birds (Falconidae) includes approximately 60 recent species. Within this group, 35 species

Figure 5. Characteristics typical of falcons: rounded, relatively short beak with a "falcon tooth" (triangular projection along each lower edge of the upper beak, allowing the falcon to easily sever the neck bones in killing its prey). Drawing by R. Sanders

are notably simlar and are, thus, all united into one genus, the "true falcons" (*Falco*). The species of the genus *Falco* correspond very well to the "falcon type" (relatively short, round beak with a "falcon tooth:" see Figure 5; wings long, narrow, and pointed; oval nostrils; no nest building). The interspecific boundaries are generally difficult to determine. Individual species are often distinguished only on the basis of size, coloring, or behavioral differences.

Within the genus *Falco*, there are numerous hybrids, so that one can safely assume a basic type on at least the generic level. Although natural hybrids are rare, the frequently practiced artificial insemination experiments (to produce offspring of endangered species for falconry) have recently resulted in unexpected successes at crossbreeding. Of particular importance for the practice of falconry since its beginnings is the cosmopolitan peregrine falcon (*Falco peregrinus*), which catches its prey in flight and primarily feeds on birds up to the size of a pigeon. Thus, the peregrine falcon has been a prime object of crossbreeding experiments. The result is that it has been crossbred with practically all other species of the genus *Falco* with which such experiments have been performed. The details are listed in Table 2.

Since the gyrfalcon, too, is used for hunting, all hybrids known thus far – with the exception of the last two species pairs in Table 2 – are significant only for the purpose of falconry. The motive for the above-mentioned crossbreeding experiments is probably the endeavor to combine the characteristics of two or more species.

The peregrine falcon is a typical cosmopolitan with almost world-wide distribution. Its distribution and the fact that it can directly or indirectly crossbreed with many other species of the genus *Falco* suggests that the peregrine falcon comes fairly close to a hypothetical ancestral form of the falcons. It is interesting that the color patterns and tones of the individual species of the genus *Falco* (gray and brown tones, almost white, almost black or reddish-brown parts of the plumage, etc.) are also fully or partially present in the subspecies of the peregrine falcon. Thus, the coloring of the orange-breasted falcon (*F. deiroleucus*), a native of America, is almost exactly the same as that of *Falco peregrinus peregrinator*, a subspecies of the peregrine falcon found in India, Ceylon, and China. Also, the wide spectrum of hunting behavior confirms the

| a) | Barbary falcon (*Falco pelegrinoides*) |
|----|----------------------------------------|
|    | Saker falcon (*Falco cherrug*) |
|    | gyrfalcon (*Falco rusticolus*) |
|    | prairie falcon (*Falco mexicanus*) |
|    | merlin ("pigeon hawk", *Falco columbarius*) |
|    | lanner falcon (*Falco biarmicus*) |
|    | orange-breasted falcon (*Falco deiroleucus*) |
|    | Bat falcon (Falco rufigularis) |
|    | American kestrel ("sparrow hawk", (*Falco sparverius*) |
| b) | gyrfalcon (*F. rusticolus*) × merlin (*F. columbarius*) |
|    | gyrfalcon × Saker falcon (*F. cherrug*) |
|    | gyrfalcon × prairie falcon (*F. mexicanus*) |
|    | common kestrel (*F. tinnunculus*) × merlin |
|    | common kestrel × lesser kestrel (*F. naumanni*) |

Table 2. Hybrids of the genus *Falco* (Cade 1986; Haffer 1989): a) species with which the peregrine falcon ("duck hawk", *F. peregrinus*) has been able to form successful crosses; b) further hybrids within the genus.

position of the peregrine falcon as an ancestral species. Although it is almost exclusively an airborne hunter, it is very adaptable, and can inhabit completely different biomes, such as grasslands, open woodland, coastland, tundra, and occasionally even large cities (Fischer 1973; Newton 1979). As a rule, the remaining species of the genus *Falco*, on the other hand, each depend on a specific habitat.

The native species of Europe (peregrine falcon, Eurasian hobby [*Falco subbuteo*], common kestrel [*F. tinnuculus*], red-footed falcon [*F. vespertinus*]) clearly demonstrate a microevolutionary splitting with the tendency to avoid competition (ecological separation), as well as a differentiation in coloring and partially also in size. The peregrine falcon and the Eurasian hobby, for example, differ only in their size but not in their coloration or their behavior. Both are airborne hunters and feed almost exclusively on birds. While the larger peregrine falcon usually hunts thrush- to pigeon-sized birds, the smaller Eurasian hobby predominantly hunts songbirds of sparrow size (Fischer 1973, Fiuczynski 1987). There is, in this case, a close connection between the size and the prey spectrum. Although this is very likely a causal connection, it must remain open whether a changed prey spectrum led to the size change, or whether the size change happened first and either required or allowed a different prey spectrum.

In contrast, the common kestrel, red-footed falcon, and Eurasian hobby are similar in size, but clearly differ in coloration and prey spectrum. The common kestrel, in whose plumage brown and gray tones dominate, is a typical ground hunter and prefers to hunt field mice; the red-footed falcon, in whose plumage dark-gray and chestnut-red tones prevail, specializes in hunting insects. The Eurasian hobby with grayish-blue, white, and black color shades goes after small birds (Mauersberger 1971, Glutz von Blotzheim 1971). The differences in size, coloration, prey spectrum, and behavior (in part) allow for a sympatric distribution of these four species in central and eastern Europe.

The remaining genera of the Falconidae can be grouped as falconets, forest falcons, and caracaras. The falconets (*Spiziapteryx, Polihierax, Microhierax*) resemble the true falcons in their appearance, way of life, and behavior and correspond to the above-mentioned "falcon type." Although no hybrids have been observed thus far, one can assume membership in the basic type of the true falcons based on results within the hawk-like birds. The caracaras and forest falcons, on the other hand, are characterized by somewhat different traits. The forest falcons (*Micrastur, Herpetotheres* [laughing falcon]) possess short, rounded wings and a long swoop, which gives them high mobility in their forest habitat. The caracaras (*Dapirius, Phalcoboenus, Polyborus, Milvago*) are characterized by a slightly bent beak without a tooth, as well as slightly bent, partly blunt claws (Grzimek 1968) – traits that can be found also in other carrion-feeders. A slightly bent beak and slightly bent, blunt claws can be regarded as degenerative traits; rounded wings and long swoop are merely the result of an extension or reduction of the length of the feathers. Therefore, an affiliation of the forest falcons and caracaras with the falcon basic type is conceivable. Beak and claws, as well as the feathers of the wing and tail, are quite variable in size and form; thus, considerable differences can exist within the same genus or even the same species (Zimbelmann 1992).

## 12.3. Discussion

### 12.3.1. Fossil History of the Falconiformes.
If one compares the distribution of the recent genera and families of Falconiformes with the distribution of fossils ascribed to this order, one notices that in almost every case the present-day distrubution represents only a relic of an originally much larger area of distribution. The secretary bird, which today can only be found in Africa, inhabited large parts of Europe during the Miocene and Oligocene (Olson 1985). The wider distribution of fossil species is not only true for the falconiform raptors but is also true for most of the other orders of birds (Brodkorb 1964, Müller 1985). Examples include the ostriches (Struthionidae; recent species only in Africa; from the Miocene to the Pleistocene also in Europe and Asia) and the mound-builders (Megapodiidae; recent species only in the Australian faunal region; during the Eocene and Oligocene also in Europe). Olson (1988), therefore, considers all groups of birds that today can only be found in the southern hemisphere as relics of an originally much larger

distribution.

The New World and Old World vultures—which enjoyed a wide distribution during the Pleistocene ice age (with New World vultures in the Old World, and Old World vultures in the New World)—represent an especially interesting case. These gigantic populations of Pleistocene vultures possibly fed on large herds of animals, which were forced to migrate due to climactic changes during the ice age and were decimated because of catastrophe or lack of food. The disappearance of numerous species of mammals of the North American continent (among them elephants, horses, tapirs, antelopes, camels, giant sloths,

Figure 6. Fossil record of the five families of the Falconiformes.

and the mastodon) coincides with the extinction of most of the vulture species, which fed on their carcasses (Feduccia 1984). Emslie (1987) demonstrated that the Californian condor (*Gymnogyps californianus*), which in prehistoric times inhabited the entire North American continent, disappeared at the same time as the large mammals upon which they feed. A remarkable observation of the same author in this context is that the Californian condor experienced a new range expansion towards the end of the 18th century, when livestock brought to North America by Europeans quickly spread and multiplied across North America.

Meanwhile, the cause of the disappearance of the New World vultures in Europe remains uncertain. Possibly, after retreating glaciers worsened their own circumstances, they were unable to resist the pressure of competition of the Old World vultures.

The fossil distribution of all five falconiform families essentially begins in the Eocene and Oligocene (Figure 6).

The extinct group of the Neocathartidae is limited to the Eocene. Only one New World vulture of questionable systematic position is known from as early as the Upper Paleocene.

A comparison of falconiform fossils with the recent forms shows that an unambiguous classification of the fossil material within the five above-mentioned families is possible. Fossil falcons, vultures, *etc.* can be identified as such without any difficulty. Thus far, transitional forms between the individual groups are not known.

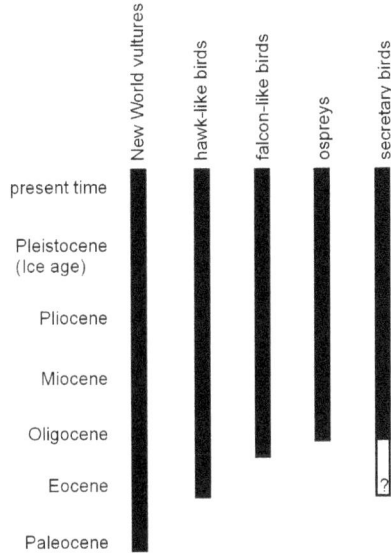

Although the oldest fossil secretary bird, *Amphiserpentarius*, was originally assigned to the Ciconiidae (storks), today its affiliation with the Sagittariidae (secretary birds) is an established fact. However, it does not qualify as the ancestor of the genus *Sagittarius* (Olson 1985), since it already possessed some highly specialized characteristics.

Fossil Pandionidae, too, can unequivocally be identified as such. The species *Pandion homalopteron* from the Miocene of North America corresponds in size to the recent species *P. haliaetus* and differs only minimally from the latter as far as anatomical traits are concerned (Olson 1985).

The oldest fossil falcon was found in Eocene-Oligocene sediments in France (Olson 1985) and the oldest hawk-like raptor, *Palaeocircus cuvieri*, in Eocene sediments in France and England (Brodkorb 1964). As to their family affiliations, there are no doubts. On the other hand, the fossil Falconiformes that have been found thus far do not allow for any final conclusions on the systematic relationships *among* the families. "On the basis of the fossil material we learn almost nothing at all about the evolution of the raptors... The systematic relationships among the different families of diurnal birds of prey are still largely unknown; the same is true for the relationships of this order to other bird groups... Also, the usual classification of the diurnal birds of prey in the vicinity of the ducks, geese and swans (order Anseriformes) cannot be justified" (Feduccia 1984).

For a while, the fossil species *Neocathartes grallator* from the Eocene of North America was regarded as a connecting link between ratites and diurnal raptors. The "running vulture," as it was once named, superficially resembled a stork. As a result, it was first interpreted as a New World vulture with reduced wings and long legs, which supposedly lived like a ratite. Subsequently, however, it has been determined that *Neocathartes* belongs to its own order, which is not related to any other group of birds (Feduccia 1984). Its position as a transitional form between storks and New World vultures is not entertained.

The groups that, according to the fossil record, are the older ones cannot be regarded as ancestral because they include highly specialized forms. For example, the turkey vulture (*Cathartes aura*), a representative of the New World vultures, detects its food with the sense of smell. Also, the Madagascar harrier (*Polyboroides radiatus*), a representative of the hawk-like birds, possesses an intertarsal joint that can be turned to the front as well as to the back.

**12.3.2. Genetically Polyvalent Ancestral Forms.** The hybrids between buzzards and hawks, as well as numerous other transgeneric hybrids, demonstrate that species of diurnal birds of prey clearly differing in their outer appearance and behavior are characterized by a conformity of their fundamental genotypic makeup. It can thus be assumed that present-day (as well as fossil) species can be traced back to a small number of basic types, whose genomes possessed a considerable plasticity towards the formation of morphological characteristics and behavioral patterns. Migrations and isolation of individual populations then resulted in the formation of specific characteristics. This process depended on the respective environment (habitat, food, competition, *etc.*), as well as on the scope of the gene pool of the respective population. Although the original plasticity of the genome of the ancestral form was reduced, it can still be estimated in those cases where crossbreeding is (still) possible. Examples are: the differences in morphology and behavior between the genera *Buteo* (buzzards; wide wings, wide tail, distinct gliders) and *Accipiter* (hawks; short rounded wings, long tail, agile hunters which surprise their prey), as well as considerable size differences between hybridizing species of the genus *Falco* (the total length of the gyrfalcon, *F. rusticolus*, for example, is on average 55 cm, and that of the merlin, *F. columbarius*, only 29 cm). Nevertheless, buzzards and hawks, on one hand, and the above-mentioned species of falcons, on the other hand, can all be traced back in each case to one and the same basic type. One can assume that the information for the expression of all the characteristics that later emerged in the different populations existed in the gene pool of this ancestral form.

The proposed division of the Falconiformes into a few basic types must be regarded as provisional since the currently known hybrids do not allow for a conclusive classification. The geographical isolation of potential crossbreeding partners makes hybridizations in nature unlikely or impossible. The strong aggressive behavior of birds of prey might be the cause for little success in future crossbreeding experiments, even with animals held in captivity. However, the recent technology of artificial insemination may provide new options to test the basic type concept. Artificial insemination experiments conducted primarily with falcons give reason for optimism.

Where no hybrids are known, one must depend on anatomical, morphological and ethological characteristics for classification of the birds of prey. According to current knowledge, only a few basic types (probably less than 10) exist within the Falconiformes at approximately the taxonomic level of families or subfamilies.

For the splitting of these basic types into the present-day species, extended time periods are not necessary because phenotypic changes can take place within decades (Mayr 1967; Zink and Remsen 1986). In the house sparrow (*Passer domesticus*), significant phenotypic changes were observed in less than 100 generations after introduction to the American continent (Johnston and Selander 1964). Significant differences in tones and colors as well as in size and body proportions emerged in this case.

# References

Brodkorb, P. 1964. Catalogue of fossil birds, Part 2: Anseriformes through Galliformes. *Bulletin of the Florida State Museum, Biological Sciences* 8(3):195-335.

Brown, L. 1979. *Die Greifvögel*. Paul Parey, Berlin.

Brown, L. and D. Amadon. 1968. *Eagles, Hawks and Falcons of the World*. Volume 1. Country Life Books, Feltham, UK.

Cade, T.J. 1986. Propagating diurnal raptors in captivity: A review. *International Zoological Yearbook* 24/25:1-20.

Emslie, S. 1987. Age and diet of fossil California condors in Grand Canyon, Arizona. *Science* 237:768-770.

Emslie, S. 1988. The fossil history and phylogenetic relationships of condors in the New World. *Journal of Vertebrate Paleontology* 8:212-228.

Feduccia, A. 1984. *Es begann am Jura-Meer*. Gerstenberg, Hildesheim, Germany.

Fehrer, J. 1993. Interspecies-Kreuzungen bei carduliden Finken und Prachtfinken. In: Scherer, S., ed. *Typen des Lebens*. Pascal, Berlin, pp. 197-215.

Fischer, W. 1973. Der Wanderfalke. *Die Neue Brehm-Bücherei* 380:1-152

Fischer, W. 1976. Stein-, Kaffern- und Keilschwanzadler. *Die Neue Brehm-Bücherei* 500:1-220.

Fischer, W. 1982. Die Seeadler. *Die Neue Brehm-Bücherei* 221:1-192..

Fiuczynski, D. 1987. Der Baumfalke. *Die Neue Brehm-Bücherei* 575:1-208.

Glutz von Blotzheim, U., ed. 1971. *Handbuch der Vögel Mitteleuropas, Band 4: Falconiformes*. Akademische Verlagsgesellschaft, Frankfurt, Germany.

Gray, A.P. 1958. *Bird hybrids: A Check-List with Bibliography*. Commonwealth Agricultural Bureaux, Farnham Royal, England.

Grzimek, B. ed. 1968. *Grzimeks Tierleben*. Volume VII. Kindler, Zürich

Haffer, J. 1989. Parapatrische Vogelarten der paläarktischen Region. *Journal für Ornithologie* 130:475-512.

Johnston, R.F. and R.K. Selander. 1964. House sparrows: Rapid evolution of races in North America. *Science* 144:548-550.

Jollie, M. 1977. A contribution to the morphology and phylogeny of the Falconiformes (Part 4). *Evolutionary Theory* 3:1-142.

Klemm, R. 1993. Die Hühnervögel (Galliformes): Taxonomische Aspekte unter besonderer Berücksichtigung artübergreifender Kreuzungen. In: Scherer, S., ed. *Typen des Lebens*. Pascal, Berlin, pp. 159-184.

Marsh, F.L. 1976. *Variation and Fixity in Nature*. Pacific Press, Mountain View.

Mauersberger, G. 1971. *Urania Tierreich, Band 2: Vögel*. Urania, Leipzig.

Mayr, E. 1967. *Artbegriff und Evolution*. Parey, Hamburg.

Müller, A.H. 1985. *Lehrbuch der Paläozoologie, Band III, Teil 2: Reptilien und Vögel*, Second edition. Gustav Fischer, Jena.

Newton, I. 1979. *Population Ecology of Raptors*. T. and A.D. Poyser, Berkhamsted, Hertfordshire, England.

Olson, S.L. 1982. The distribution of fused phalanges of the inner toe in Accipitridae. *Bulletin of the British Ornithologists' Club* 102:8-12.

Olson, S.L. 1985. The fossil record of birds. In: Farner, D.S., J.R. King and K.C. Parkes, eds. *Avian Biology,* Volume VIII. Academic Press, London, pp. 79-238.

Olson, S.L. 1988. Aspects of global avifaunal dynamics during the Cenozoic. In: Ouellet, H. ed. *Acta XIX Congressus Internationalis Ornithologici,* Volume 2. University of Ottawa Press, Ottawa, Canada, pp. 2023-2029.

Scherer, S. 1993. Basic types of life. In: Scherer, S., ed. *Typen des Lebens,* Pascal, Berlin, pp. 11-30.

Scherer, S and T. Hilsberg. 1982. Hybridisierung und Verwandtschaftsgrade innerhalb der Anatidae: Eine evolutionstheoretische und systematische Betrachtung. *Journal für Ornithologie* 123:357-380.

Swann, H.K. 1930-1945. *A Monograph of Birds of Prey.* Wheldon and Wesley, London.

Weick, F. 1980. *Die Greifvögel der Welt.* Parey, Berlin.

Zimbelmann, F. 1989. *Beitrag zur Phylogenetik und Revision der Seeadler* (Haliaeetus *Savigny 1809*). Dissertationen der Mathematisch-Naturwissenschaftlichen Fakultät. Heidelberg.

Zimbelmann, F. 1992. Ein vergleichend morphologischer Beitrag zur Phylogenetik der Seeadler (Haliaeetus Savigny 1809). *Mitteilungen aus dem Zoologischen Museum in Berlin.* 68, Supplement Annalen für Ornithologie 16:61-114.

Zink, R.M. and J.V. Remsen. 1986. Evolutionary processes and patterns of geographic variation in birds. *Current Ornithology* 4:1-69.

Ziswiler, V. 1976. *Die Wirbeltiere.* Georg Thieme, Stuttgart.

# 13. Interspecific Hybridization within Cardueline and Estrildid Finches (Passeriformes: Carduelinae and Estrildidae)

Judith Fehrer

**Abstract**

Data on interspecific and intergeneric hybridizations document 145 and 161 such hybrids in the Carduelinae and Estrildidae, respectively. The existence of these hybrids, their partial fertility, the seemingly unlimited potential to hybridize within the respective families, as well as color atavisms of the hybrids, point to a closer genetic relationship among the species than their appearance alone would suggest. The observations can be interpreted within the framework of two gene pools (of the Carduelinae and the Estrildidae), each of which originated from a common ancestor. Multiple parallelisms within the carduelines and estrildids, respectively, can, thus, be explained rather simply. No hybridizations crossing the family borders have been reported. The historic development of the classical taxonomy, too, confirms these two groups to be cohesive taxa with their own unique character complexes. Connections to other avian taxa remain speculative.

## 13.1. Introduction

As far as the classification of organisms is concerned, there can be a considerable variety of opinions because the work of the taxonomist is influenced by a diverse selection and assessment of characteristics considered relevant for taxonomic purposes. Contradictory viewpoints exist especially on the level of genera and families since these taxonomic units are often based on different characters or character complexes. It therefore seems necessary to look for approaches that in some way can holistically comprehend the organisms to be classified.

The technology of DNA-DNA hybridization takes into consideration the total genome of a particular organism. The behavior of single-copy DNA of a particular species when forming heteroduplex-DNA, in comparison with the renaturalization behavior of the DNA of the same species, is regarded as a quantifiable criterion. With the help of this method, numerous groups of birds have been examined and placed into a phylogenetic tree (Sibley *et al.* 1988 and literature quoted therein). This phylogenetic arrangement is partially characterized by blatant contradiction to previous phylogenies. The method is also controversial because of numerous methodical and theoretical problems (Cracraft 1987 and literature quoted therein; Sarich *et al.* 1989). The complexity of eukaryotic genomes, as well as the current ignorance of the connection between similarities (of individual genes, genetic organization etc.) and biological relevancy (for example function and systematic relationships), considerably impedes the interpretation of these results.

Therefore, an alternative method that not only compares complete organisms but also guarantees the biological relevance of similarities is chosen here to unite species into higher taxa. Marsh (1976) introduced the *basic type* as a taxonomic unit. According to his definition, all species which can hybridize directly or indirectly (*i.e.,* via a third species) belong to the same basic type. This approach was used by Scherer and Hilsberg (1982; comp. Scherer 1986), who applied it to the ducks, swans, and geese (Anatidae). The rationale for this approach is that in the case of hybridization, both genomes must be coordinated finely enough for an orderly spatial and temporal expression (morphogenesis) to take place (Scherer 1993, this volume). A successful hybridization, therefore, provides information regarding the genetic relationship of the complete genomes of the two crossbreeding partners without having to assess individual characteristics or complexes of morphological, ethological, or molecular traits as delimiting criteria. This approach also avoids the controversial interpretation of DNA-DNA hybridization data for taxonomic purposes.

The application of the basic type concept to animals living in their natural habitat faces difficulties because biological species are defined by more or less complete reproductive isolation (Mayr 1967). While interspecific hybrids do occasionally occur in nature (Panov 1989), there are numerous reports about hybridizations in captivity. This is especially the case for groups that, for different reasons, are of particular interest to humans: food plants (e.g., crops, fruit trees), domestic animals (e.g., poultry, livestock), as well as ornamental plants and animals (e.g., orchids and exotic birds).

According to Mayr (1967), the biological species is characterized by two main criteria: by an interbreeding community, on the one hand, and by reproductive isolation from other interbreeding communities, on the other hand. Analogously, the basic type can be understood as an expansion of the biological species concept in which the requirements of "natural" conditions and offspring fertility are disregarded.

For the present study, the finch group was chosen as an example. The name "finches" is a collective term for seed-eating songbirds with a conical beak. It includes approximately 950 species belonging to the sparrow order (Passeriformes) and songbird suborder (Oscines). Buntings, weaverbirds, sparrows, true finches, estrildid finches, and goldfinch relatives, among others, belong to this larger group. Over the course of time, this heterogeneous group of finches experienced many contradictory subdivisions. The number of suggested systems of classification was (and is) nearly identical to the number of authors. (For a historical development of the classification of the finches, see Ziswiler 1967.)

As seed-eaters, finches present few problems as far as breeding and care are concerned, so numerous reports exist of

**Figure 1.** Domesticated canaries. Right: a singing cock. Left: the bird displays a partially streaked plumage at its breast, a pattern commonly seen in many cardueline finches, as well as in the wild canary (*Serinus canaria*). Photo by R. Wiskin.

hybridizations in captivity. Beginning with available crossbreeding data (Gray 1958 and others), the applicability of the basic type concept for the finches was examined. In the course of this study it became evident that none of the reports of hybridizations across family boundaries stood the test of a critical examination.[1]

Most of the hybridization data reported for finches involve crossbreedings among goldfinch relatives (subfamily Carduelinae), as well as among the estrildid finches (family Estrildidae). Therefore, in the following section the applicability of the basic type concept for these two groups is investigated.

---

1   A critical evaluation of hybrid reports is particularly important in this taxonomic approach since most of the data are not based on scientific experiments but on observations of breeders. Some criteria for the selection of the listed hybrids should therefore be explained:

Most of the data under discussion (Figures 2 and 6) come from Gray (1958). Claims Gray considered uncertain or questionable were not taken into account. In all other reports, any trace of doubt resulted in the exclusion of that report, for example, if the particular hybrid could not be reproduced. It is, however, still possible that certain hybrid reports included here are based on error. However, such cases hardly have an influence on the conclusions of this paper.

Hybrids of cardueline finches that are displayed in national bird exhibitions can be regarded as rather trustworthy based on strict criteria of judging, especially if these hybrids have been reported more than once. Another dependable source are hybrid specimens in scientific collections. Estrildid finch hybrids are hardly bred anymore due to regulations on the conservation of species, so only a few newer reports exist to confirm or correct the older crossbreeding data.

The sources of all crossbreeding data are quoted in the appendix. Many hybrids were bred in large numbers, so only examples are mentioned from the literature.

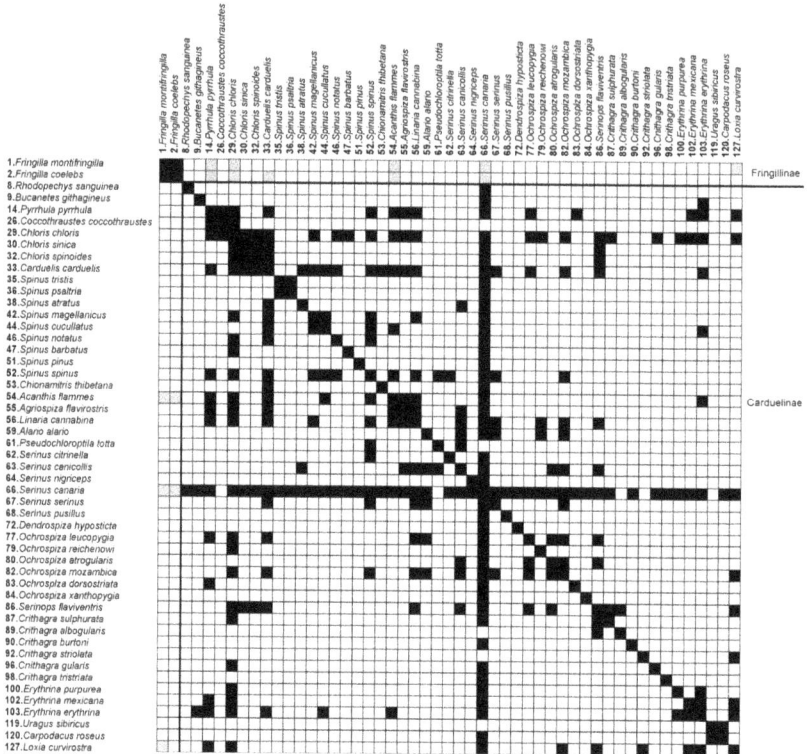

**Figure 2.** Hybridization matrix of the Fringillidae. Interspecific hybrids were summarized from references indicated in the appendix. Species numbers refer to the listing given in Wolters (1975-1982). Hybrids were obtained from two (out of three) species of the Fringillinae and from 49 (out of 125) species of the Carduelinae (except the Hawaiian honeycreepers, see text). The references for each cross are given in the appendix using the same numbering as in this figure. For many crosses, only selected references are given because numerous reports exist. The direction of the crosses is not mentioned since it does not concern the relationship of the species but reflects behavioral constraints. Crosses of the genus *Fringilla* with members of the Carduelinae are shadowed (validity in question, see text and pertinent references). Due to the central role of the domesticated canary in rearing hybrids, canary-hybrids were obtained with nearly all cardueline finches when an attempt was made. Note that there is a distinct correlation between the availability of a species for hybridization and the number of reports instead of a correlation with the expected relationship (*i.e.*, more crosses within genera). This indicates the artificial nature of the genus (compare Scherer 1993 and this volume).

## 13.2. Goldfinch Relatives (Carduelinae)

In the overview of Peters (1931-1970), the goldfinch relatives as the subfamily Carduelinae, together with the true finches (Fringillinae), are placed within the Fringillidae.[2]

---

2      Depending on the author, the "Fringillidae" can also include buntings, cardinals, sparrows, as well as other birds. Wolters (1975-1982), on the other hand, uses this term exclusively for the true finches. In this study, the name is used only for true finches (Fringillinae) and goldfinch relatives (Carduelinae).

Carduelines possess a strongly conical beak, which marks them as seed-eaters. The seeds are cut open with help of the sharp beak edges by moving the beak back and forth—a highly specialized peeling mechanism that enables them to feed on dicotyledonous seeds (true finches use the same mechanism). The juvenile birds are fed with predigested food from the crop. Most species supplement the diets of juveniles with animal protein. Many carduelines live socially in loose flocks. The females build the nests and incubate the eggs alone. The song of the carduelines is a melodic chirping, consisting of many separate elements, which is recited also during flight. The group is represented worldwide with approximately 125 species. Among them are the bullfinches, linnets, goldfinches, hawfinches, siskins, greenfinches, crossbills, and the serins.

The serins include the canary (Figure 1), the domesticated form of the canary serin (*Serinus canaria*). Its potential for breeding makes it a preferred candidate not only for the rearing of breeds but also for generating hybrids.

**13.2.1. Interspecific Hybrids.** The most popular example of hybridization between different species of carduelines (in this case even different genera) is the red canary. It is the result of a successful attempt to introduce the red plumage color of the red siskin (*Spinus cucullatus*) to the canary by crossbreeding them. The majority of the male hybrids are fertile; reproductive females occur first in third generation backcrosses with canaries (Bielfeld 1978).

In order to define the basic type (see above), 145 interspecific hybrids of the Carduelinae were compiled as a hybridization matrix (Figure 2, see Appendix also).

Approximately 40% of the species from 22 of 39 genera of the Carduelinae (as defined by Wolters 1975-1982) were involved in hybridizations. All these species are connected at least indirectly, *i.e.*, through crossing with a third species (in most cases the canary (66)). Hybridizations with canaries (66) are particularly frequent. Additional species that were often crossbred are the greenfinch (29), goldfinch (33), bullfinch (14), linnet (56), common redpoll (54), and the Eurasian siskin (52). There is a clear connection between the hybridization frequency of a particular species and its availability for breeding.[3] This then means that the ability to hybridize does not seem to depend much on genetic or ethological limitations. This seeming arbitrariness of hybridizations

---

3    Many species of this worldwide family are either not being imported at all, or they come from climate zones that limit the possibility of breeding them in moderate climates. This explains why no results are available for about 60% of the species of this family. Apart from the limited availability of these birds (indigenous species, too, are protected; only those specimens which come from breeding facilities can be kept in captivity), the majority of breeders are not interested in systematic aspects but in appearance or singing ability, which in turn influences their choice of hybridization partners. Thus, the hybridization matrix is by no means complete.

is confirmed by Güttinger and van den Elzen (personal communication 1991). (The ability of all Carduelinae to hybridize with each other is regarded as a rule of thumb among breeders [see Massoth 1978]).

Those Carduelinae that were involved in hybridizations belong, by definition, to one and the same basic type. An expansion of the basic type to include those species of the Carduelinae that thus far have not been involved in hybridizations is likely, since the taxonomic coherence of the Carduelinae is well established. However, this is a preliminary hypothesis which should be confirmed by appropriate crossbreeding experiments.

**13.2.2. Interpretation of Atavisms.** The overall similarity of the genetic makeup of the Carduelinae as revealed by the hybridizations is further confirmed by atavisms[4] in interspecific hybrids:

Von Klatt (1901,1902) and Mark (1930) described the following color atavisms in hybrids of finches: a distinctive—usually black streaked—pattern of the dorsal plumage, brown dorsal side and wings, yellow plumage parts, yellow or olive rump and belly, differently colored margins of tail feathers and primaries, special shoulder coloring, and wing bars. Radtke (1981) mentions that, in a hybrid between the twite (*Agriospiza flavirostris*), which does not possess any red, and the red canary, a distribution of the red color occurs corresponding to that of the linnet (*Linaria cannabina*). In this context, the stripe above the eye (supercilium) should be mentioned, which occurs in some hybrids whose parents either do not have such a stripe or it is less distinctive (e.g., greenfinch × red crossbill, chaffinch × greenfinch (see below), greenfinch × bullfinch, linnet × bullfinch; illustrations in Speicher 1970 and Radtke 1981). These characteristics can be found individually or in different combinations in many "pure-bred" cardueline species, especially in females and juvenile birds.

It can be assumed that atavistic characteristics could also include such traits as the song or other behavioral elements of hybrids.

**13.2.3. Systematic Relationships Within the Carduelinae.** According to Mayr et al. (1953), the genus is defined as a group with a common phylogenetic origin that differs from other similar groups by a distinct gap in distinguishing characters. However, as far as the cardueline finches are concerned, Wolters (1967) states that the genera are separated from each other by a mosaic of smaller differences rather

---

4    Atavisms are considered to be potential relapses into previous phylogenetic stages. However, such an interpretation presupposes the knowledge of the ancestor and is relevant only in the case of closely related species (see Steiner 1966). The only genetically-based explanation of atavisms is the existence of genetic information that does not appear in the phenotype but appears due to changed regulatory mechanisms and reorganization of the genetic material as, for example, during the hybridization. The simplest case is the expression of a recessive allele (which, however, is not called an atavism).

than by particularly conspicuous characteristics.

The division of the Carduelinae into different genera is made particularly difficult by frequent parallel developments:

- Güttinger (1978) mentions the occurrence of grayish-white wing tips in different lineages, as well as a complex head pattern that can be found in numerous species that are not closely related to each other. Van den Elzen (1985) interprets this head pattern as a primitive trait that can also reappear in the plumage of juvenile birds of derived species.

- Climbing behavior and narrow, pointed beaks occur as adaptations to feeding on seeds of the Asteraceae in siskins, the African citril, the citril finch, the goldfinch and the common redpoll (Nicolai 1957; Kunkel 1966).

- The serins (mostly comprised formerly in the collective genus *Serinus*), which closely resemble each other in their plumage patterns and beak shapes, were recognized to be an inhomogeneous group and divided into different genera (Nicolai 1960; van den Elzen 1985). In other words, the "serin form" evolved several times independently.

Many authors substantiate the results of their systematic examinations by pointing to hybridizations or the fertility of hybrids (e.g., Mayr *et al.* 1956; Nicolai 1960; Voous 1977; Güttinger 1978; van den Elzen 1985).

However, a look at the hybridization matrix (Figure 2) shows that intrageneric hybridizations do not occur more frequently than intergeneric ones. Also, any other division of genera would not lead to a result in which the affiliation to the same genus would be correlated with a higher frequency of hybridization.

Figure 3. Chaffinch (*Fringilla coelebs*). The conical bill is characteristic of seed-eaters. This fringilline species shows wing bars similar to those of cardueline finches. The relationship of the Fringillinae with Carduelinae still requires elucidation. Photo by R. Junker.

| Number | Species Name | Species Number of Second Parent |
|---|---|---|
| 1 | *Fringilla montifringilla* | 2 |
| 2 | *Fringilla coelebs* | 1 |
| 29 | *Chloris chloris* | 30,**32**,33,66 |
| 30 | *Chloris sinica* | 29,33 |
| 32 | *Chloris spinoides* | **29** |
| 33 | *Carduelis carduelis* | 29,30,54,56,66 |
| 36 | *Spinus psaltria* | 66 |
| 38 | *Spinus atratus* | 66 |
| 42 | *Spinus magellanicus* | **44**,52 |
| 44 | *Spinus cucullatus* | **42**,52,66 |
| 52 | *Spinus spinus* | 42,52,66 |
| 54 | *Acanthis flammea* | 33 |
| 56 | *Linaria cannabina* | 33,**66** |
| 59 | *Alario alario* | 66 |
| 66 | *Serinus canaria* | 29,33,36,38,44,52,56,59,67,80,83,**102** |
| 67 | *Serinus serinus* | 66 |
| 77 | *Ochrospiza leucopygia* | **80** |
| 80 | *Ochrospiza atrogularis* | 66,77 |
| 83 | *Ochrospiza dorsostriata* | 66 |
| 86 | *Serinops flaviventris* | **87,89** |
| 87 | *Crithagra sulphurata* | **86** |
| 89 | *Crithagra albogularis* | **86** |
| 102 | *Erythrina mexicana* | **66** |

Table 1. Fertile hybrids of cardueline and fringilline finches. Generally, fertility is limited to male hybrids. However, in *Alario*-hybrids, only the female hybrids are fertile. Fertility in both sexes is indicated in bold. Refences for the crosses are given in the appendix.

Table 1 contains some examples of fertile hybrids. Usually, the fertility is limited to the homogametic sex, which, in the case of the birds, is the male. Also, the fertility of the $F_1$ hybrids seems to be independent of the generic affiliation of the parents. There are even hybrids from intergeneric hybridizations in which both sexes are reproductive (e.g., 86 × 87, 86 × 89: Nicolai 1960).[5]

It is evident that currently neither the morphology, nor hybridization results, nor fertility of the $F_1$ generation allow for conclusive statements about speciation within the Carduelinae. On the other hand, all three approaches point to a close relationship among members of the group.

**13.2.4. Delimitation of the Carduelinae.** As shown above, the species criterion of "interbreeding community" can in a wider sense (see Introduction) be applied to the Carduelinae. In the following section, the isolation of the group from non-Carduelinae is regarded as a delimiting criterion.

---

5    There are claims regarding the fertility of the hybrid between common rosefinch (103) and canary (66), as well as between the housefinch (102) and the canary (von Boettcher 1944; Bielfeld 1978). But Speicher (1970) is of the opinion that definite proof is still lacking. According to Radtke (1981), an $F_2$ hybrid was supposedly bred somewhere "abroad." Here, one needs to take into account that many crossbreedings, which were once considered either difficult or impossible, are today bred in large number and that with the number of hybrids there is also a growing likelihood that statements can be made regarding their fertility.

**13.2.4.1. Relationship to the Fringillinae.** The Fringillinae consist of only three species of the genus *Fringilla*: chaffinch (Figure 3), brambling, and blue or Teyde finch.

Until recent times there have been reports of occasional hybrids between bramblings or chaffinches with cardueline finches (Figure 2, gray shaded entries). There are indications that at least some of the described hybrids may actually exist:

- Birds that were labeled as, for example, chaffinch × greenfinch were exhibited at national bird shows in Bocholt, Germany (Massoth 1978), and Genoa, Italy (in a 1983 personal communication to S. Scherer, Massoth supported the claim because he observed diagnostic traits of both species expressed in the hybrid in question), as well as the 1985 and 1986 German championships of the German Canary Breeder's Association (DKB 1984-1990). Mayr *et al.* (1956), too, mention the occasional appearance of such hybrids (without reference) as proof of a rather close relationship of the genus *Fringilla* to the Carduelinae.

- In Speicher (1970) can be found an illustration of a hybrid between the chaffinch and the greenfinch; in Radtke (1981) one of goldfinch × chaffinch; both drawn by H. Heinzel after living specimens. The question, whether these pictures actually represent those hybrids, is difficult to answer because no definite predictions can be made regarding the outward appearance of a hybrid. The first case is without doubt a chaffinch hybrid; the second case seems to be a goldfinch hybrid. In each case, the second crossbreeding partner cannot be identified unequivocally on the basis of the drawing.

Despite investigation, no voucher specimen could be found. However, it must be said that the care and breeding of these animals presents considerable difficulties (Radtke 1981), which could explain the rarity of such hybrids. J. Steinbacher (personal communication 1991) is of the opinion that one cannot simply dismiss these reports. However, as long as substantial proof is lacking concerning these debatable crossbreedings, no statement can be made about the relationship between Fringillinae and Carduelinae on that basis.

Comparative systematics does not arrive at an unambiguous answer to this question either: Depending on the evaluation of shared and nonshared traits, the Fringillinae are regarded either as close or more distant relatives of the Carduelinae.

Among the shared traits are the structure of the nest and the courtship behavior (Mayr *et al.* 1956), the pattern of the plumage and the basic structure of the hard palate (Sushkin 1924), seed selection, the highly specialized seed opening mechanism, and some anatomical characters (Ziswiler 1965, 1967). Serological examinations (Mainardi 1957a, 1957b, 1957c) do not suggest any sharp separation between the chaffinch and cardueline finches.[6] The geographical distribution suggests a connection to the Carduelinae rather than to the buntings (to which they are instead assigned). Sibley *et al.* (1988) unite the fringillines and carduelines with the buntings and some other groups into one family (which, for the questions raised in the present study, is not very helpful).

Among the differences between Fringillinae and Carduelinae are: anatomical characters of the skull (Tordoff 1954), no crop feeding in *Fringilla* (Ziswiler 1965), a distinctive *Fringilla* territorial behavior, and relatively stereotyped *Fringilla* song type (in contrast to most carduelines). Based on an electrophoretic separation of hemoglobin, the chaffinch can be distinguished from the carduelines studied (Mainardi 1957c).

A non-contradictory systematic arrangement is therefore not possible at this time.

The status of the genus *Fringilla*, however, is not a matter of debate. Eck (1975) notes that it has remained remarkably stable, as far as its taxonomic delimitation and classification are concerned. The taxonomic results agree with the crossbreeding results: hybrids between chaffinch and brambling are known both from nature and from captivity (Figure 2; for references see Appendix). The island Teyde finch is geographically as well as ecologically isolated from the remaining species of *Fringilla* and has to my knowledge never been bred in captivity, so no crossbreeding data are available. All three species of the genus *Fringilla* presumably belong to the same basic type. Whether they should be united with the Carduelinae, must remain unanswered at this time.

**13.2.4.2. Relationship to Przewalski's Rosefinch and Hawaiian Honeycreepers.** Przewalski's rosefinch (*Urocynchramus pylzowi*) is usually placed among the goldfinch relatives. According to Sushkin (1927), it is a typical finch, considering all characteristics of its external anatomy and its palate, with the exception of the long tenth primary. Wolters (1975-1982) places it in a monotypic family of its own (presumably because of the number of primaries). Also, no hybridization reports exist on this species, so its affiliation with the basic type of the Carduelinae remains uncertain.

Hawaiian honeycreepers (Drepanidinae) live endemically on Hawaii, where they radiated into an enormous diversity of shapes and colors. The debates regarding their systematic position are summarized by

---

6    A comparison with other bird families as a control is lacking. Plus, the different publications cannot be harmonized, which in my opinion calls the investigative method into question.

Sibley and Ahlquist (1982)[7]. Some similarities with the cardueline finches have prompted some researchers to derive the Hawaiian honeycreepers from this group. Wolters (1975-1982) places them without delimitation within the Carduelinae, which raises the question whether they belong to the same basic type. Since, however, the Hawaiian honeycreepers are greatly endangered, it is understandable that no crossbreeding reports about them are available. In such a case it would be an advantage to have other parameters available for the description of basic types than just hybridization. For example, a basic type-specific genetic pattern could be used to make statements about the affiliation of a certain group with other groups without endangering the existence of its representatives. The question whether the Hawaiian honeycreepers belong to the basic type of the Carduelinae must therefore currently remain unanswered.

**13.2.4.3. Relationship to Other Families.** Numerous claims of hybrids between carduelines (especially canaries) and genera from other families can be found in the older literature. Despite all efforts to obtain proof, not one hybrid crossing the boundaries of families could be found. (The hybrid between village indigobird (Viduinae) and canary (illustration in Bielfeld 1978), which thus far has never been reproduced, presumably presents an error, as a comparison with the beaks of the also-portrayed "parents" suggests.) Speicher (1970) and Radtke (1981) usually reject statements like this (in response to taxonomically ignorant breeders) and regard the respective species as inappropriate for breeding with the Carduelinae.

The delimitation of this basic type from other families can therefore be regarded as fairly certain.

### 13.3. Estrildid Finches (Estrildidae)

Estrildid finches are wren- to linnet-sized birds from Australia and the tropics of the Old World. The plumage is usually colorful and, in contrast to all other finches, never shows bunting- or sparrow-like streaks. They build roofed, spherical nests with a lateral entrance. The juvenile birds possess a conspicuous mouth pattern. Estrildid finches are gregarious birds with a distinctive social behavior. Zebra finches (Figure 4) and Bengalese finches (Figure 5) belong to them; both are important breeding birds. The family consists of approximately 130 species.

The Estrildidae today are regarded as a well defined family. Since Delacour (1943) separated the whydahs or widowbirds (Viduinae) and the genera *Anomalospiza* and *Pholidornis* from the estrildid finches, many suggestions have been made regarding their classification, but his work seems to have clarified the question of which birds actually

---

7    Sibley and Ahlquist (1982), on the basis of DNA-DNA hybridizations, place the Hawaiian honeycreepers within the proximity of the Carduelinae. The problems with this method were mentioned in the Introduction.

280                          Basic Types of Life

Figure 4. Zebra finches (*Taeniopygia guttata*). The male zebra finch (right) displays a social behavior typical of estrildid finches. The female (left) is less colored. The plumage of the male bird shows the wave as well as the drop pattern common to many members of the family. Photo by R. Wiskin.

belong to the estrildid finches. Since then, several authors have claimed a distinct family position for the Estrildidae among the songbirds based on miscellaneous comparative studies. (A summary of this development is given by Mayr 1968.) In total contrast to that is the classification of Sibley *et al.* (1988) on the basis of DNA-DNA hybridization (see Introduction). There, the estrildid finches are united with the whydahs, weavers, sparrows, pipits/wagtails and accentors into the sparrow family (Passeridae).

**13.3.1. Interspecific Hybrids Within the Estrildidae.** For the estrildid finches, too, a crossbreeding matrix was constructed (Figure 6). At least 161 hybrids are known within this group. Approximately half of the species from the majority of the genera are involved in these hybridizations and, thus, belong by definition to the same basic type. Based on his experience, Steiner (1952, 1959) thinks that every species of estrildid can be crossbred with every other species. This indicates a seemingly capricious hybridization potential within this family, which in turn demonstrates close relationship.

The species lacking crossbreeding data have thus far not been imported into Europe and, therefore, have never been available for breeding (Bielfeld 1973). Crossbreeding with species of other families have been attempted, but successes have not yet been documented. The distinctiveness of the family therefore agrees with the results of classical taxonomy.

**13.3.2. Atavisms and Phenocopies.** The coherence of the Estrildidae is confirmed by further observations:

In the same way as Mark (1930) did for the Carduelinae, Steiner (1966) united atavisms of interspecific hybrids into a complex of traits that he traces back to the genotype of all estrildid finches: a dark head-breast mask, transverse streaking of the plumage, and red tail coverts. Wolters (1957), too, considers the widespread occurrence of transverse streaking in "pure" estrildid finch species a common trait of estrildid ancestors.

Steiner (1958) further mentions the phenocopy of one estrildid finch species in a hybrid of two other species: the hybrid of the black-headed munia (*Munia atricapilla*) and African silverbill (*Euodice cantans*) resembles the chestnut-breasted munia (*Munia castaneothorax*). Steiner regards this as an example of the reappearance of traits that are common to each of the relevant species. (Also, speciation through hybridization cannot be excluded as a possibility in this case.)

### 13.3.3. Relationships Within the Family.
In contrast to the well-defined delimitation of the estrildid finches with respect to other families, suggestions for classification within this group are divided. (Mayr [1968] and Kakizawa and Watada [1985] discuss the different systems in detail.) The number of genera lies between 15 (Delacour 1943) and 49 (Wolters 1975-1982); the number of species ranges from 108 to 131. Earlier

Figure 5. The Bengalese finch (Estrildidae). The Bengalese finches originated from the white-rumped mannikin (*Lonchura striata*), a species domesticated 300 years ago in Japan. Several differently colored breeds exist. The Bengalese finch is often used for rearing hybrids. Many of them are fertile (see Table 2). Photo by R. Wiskin.

authors mention 50 genera with 137 species (quoted from Delacour 1943). Mayr (1968), since he believes generic divisions should mirror systematic relationships, tried to accommodate to the "real" situation by introducing tribes and subspecies. Since too little is known about the fertility of hybrids, which according to Wolters (1957) would represent the best basis for generic division, other criteria were used (e.g., mouth markings, patterns of the plumage, geographic distribution, behavior, protein-electrophoretic data), but many cases of parallel evolution had to be taken into account (Wolters 1957,1981; Steiner 1960). Among these are, for example, droplet and waved patterns of the plumage, horseshoe and domino patterns of the mouth markings, as well as individual characters of the courtship. Güttinger (1970) mentions a mosaic-like distribution of numerous morphological and ethological traits across all species. The opinion of Immelmann (1962) that individual traits can be highly specialized in primitive forms or can remain primitive in highly specialized species emphasizes the difficulty of interpreting parallel developments in the context of evolutionary history.

What contribution to the classifcation of the family can be made by crossbreeding results? The hybridization matrix of the Estrildidae (Figure 6) shows concentrations of entries in certain areas of the matrix. If one compares different tribal divisions (Delacour, three; Steiner, nine; Kakizawa and Watada, six tribes), one notices that each tribal definition corresponds to a certain concentration of hybrids. Nevertheless, in each tribal classification, hybridization also occurs with species of other tribes. Two explanations (which are not mutually exclusive) are possible:

# Basic Types of Life

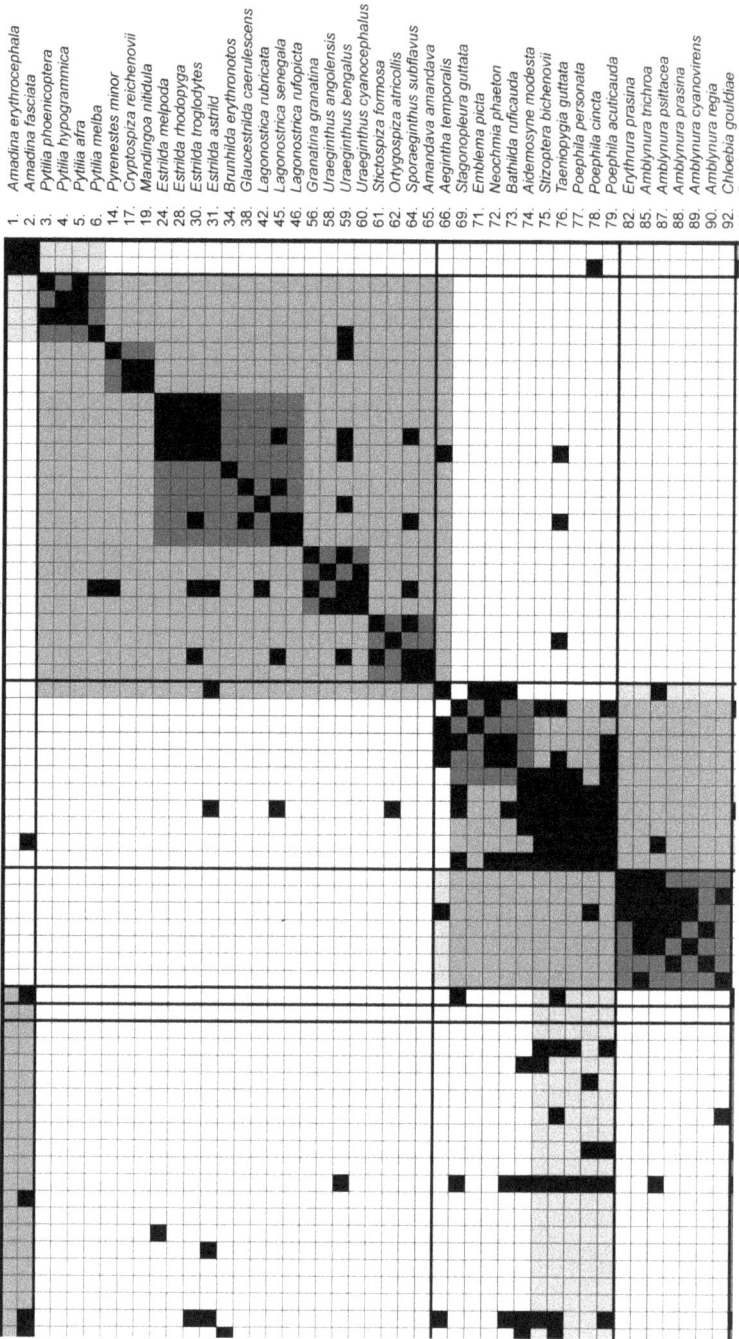

Column headers (left to right):

1. Amadina erythrocephala
2. Amadina fasciata
3. Pytilia phoenicoptera
4. Pytilia hypogrammica
5. Pytilia afra
6. Pytilia melba
14. Pyrenestes minor
17. Cryptospiza reichenovii
19. Mandingoa nitidula
24. Estrilda melpoda
28. Estrilda rhodopyga
30. Estrilda troglodytes
31. Estrilda astrild
34. Brunhilda erythronotos
38. Glaucestrilda caerulescens
42. Lagonostica rubricata
45. Lagonostica senegala
46. Lagonostica rufopicta
56. Granatina granatina
58. Uraeginthus angolensis
59. Uraeginthus bengalus
60. Uraeginthus cyanocephalus
61. Stictospiza formosa
62. Ortygospiza atricollis
64. Sporaeginthus subflavus
65. Amandava amandava
66. Aegintha temporalis
69. Stagonopleura guttata
71. Emblema picta
72. Neochmia phaeton
73. Bathilda ruficauda
74. Aidemosyne modesta
75. Stizoptera bichenovii
76. Taeniopygia guttata
77. Poephila personata
78. Poephila cincta
79. Poephila acuticauda
82. Erythrura prasina
85. Amblynura trichroa
87. Amblynura psittacea
88. Amblynura prasina
89. Amblynura cyanovirens
90. Amblynura regia
92. Chloebia gouldiae

Row labels (top to bottom):

1. Amadina erythrocephala
2. Amadina fasciata
3. Pytilia phoenicoptera
4. Pytilia hypogrammica
5. Pytilia afra
6. Pytilia melba
14. Pyrenestes minor
17. Cryptospiza reichenovii
19. Mandingoa nitidula
24. Estrilda melpoda
28. Estrilda rhodopyga
30. Estrilda troglodytes
31. Estrilda astrild
34. Brunhilda erythronotos
38. Glaucestrilda caerulescens
42. Lagonostica rubricata
45. Lagonostica senegala
46. Lagonostica rufopicta
56. Granatina granatina
58. Uraeginthus angolensis
59. Uraeginthus bengalus
60. Uraeginthus cyanocephalus
61. Stictospiza formosa
62. Ortygospiza atricollis
64. Sporaeginthus subflavus
65. Amandava amandava
66. Aegintha temporalis
69. Stagonopleura guttata
71. Emblema picta
72. Neochmia phaeton
73. Bathilda ruficauda
74. Aidemosyne modesta
75. Stizoptera bichenovii
76. Taeniopygia guttata
77. Poephila personata
78. Poephila cincta
79. Poephila acuticauda
82. Erythrura prasina
85. Amblynura trichroa
87. Amblynura psittacea
88. Amblynura prasina
89. Amblynura cyanovirens
90. Amblynura regia
92. Chloebia gouldiae
93. Padda oryzivora
95. Heteromunia pectoralis
97. Munia teerinki
100. Munia castaneothorax
102. Munia flaviprymna
104. Munia maja
105. Munia ferruginosa
106. Munia malacca
115. Munia qunticolor
118. Lonchura punctulata
121. Lonchura leucogastroides
122. Lonchura striata
123. Lonchura fuscans
124. Lonchura kelaarti
125. Lepidopygia nana
126. Spermestes cucullatus
127. Spermestes bicolor
128. Spermestes fringilloides
129. Odontospiza caniceps
130. Euodice cantans
131. Euodice malabarica

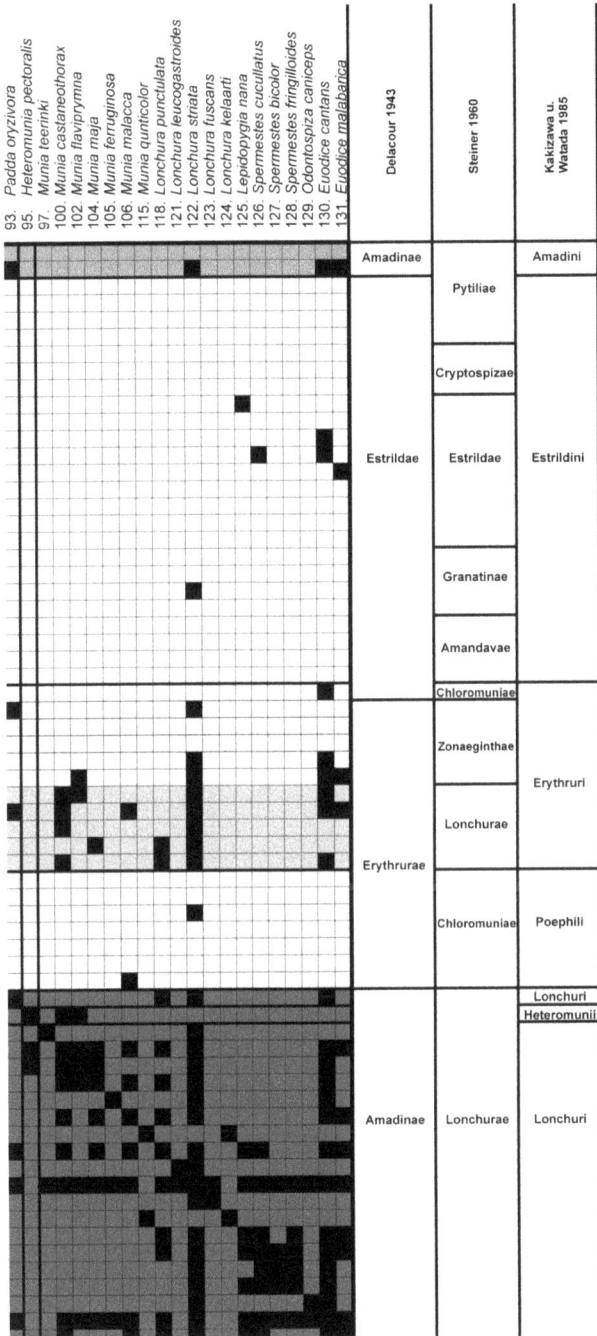

**Figure 6.** Hybridization matrix of the Estrildidae. Hybridization occurred in 34 out of 49 genera (or 65 out of 131 species) of the Estrildidae. References for the crosses are given in the appendix using the same numbering. The list of species is according to Wolters (1975-1982). To the right, different tribes according to Delacour (1943), Steiner (1960) and Kakizawa and Watada (1985) are compared: tribes following Steiner in light shading, those following Delacour in medium shading, and overlap of the two classifications in dark shading; tribes according to Kakizawa and Watada separated by thick lines. Intratribal hybridization occurs most frequently; intertribal crosses are only rarely observed (see discussion in text).

| Number | Species | Species Number of Second Parent |
|---|---|---|
| 1 | Amadina erythrocephala | 2 |
| 2 | Amadina fasciata | 1 |
| 3 | Pytilia phoenicoptera | 5 |
| 4 | Pytilia hypogrammica | 5 |
| 5 | Pytilia afra | 3, 4 |
| 30 | Estrilda troglodytes | 31 |
| 31 | Estrilda astrild | 30, 59 |
| 45 | Lagonostica senegala | 46 |
| 46 | Lagonostica rufopicta | 45 |
| 59 | Uraeginthus bengalus | 31, 60 |
| 60 | Uraeginthus cyanocephalus | 59 |
| 73 | Bathilda ruficauda | 79 |
| 75 | Stizoptera bichenovii | 78 |
| 77 | Poephila personata | 79 |
| 78 | Poephila cincta | 75, 79 |
| 79 | Poephila acuticauda | 73, 77, 78 |
| 82 | Erythrura prasina | 85, 87 |
| 85 | Amblynura trichroa | 82, 87, 88 |
| 87 | Amblynura psittacea | 82, 85 |
| 88 | Amblynura prasina | 85 |
| 93 | Padda oryzivora | 122 |
| 100 | Munia castaneothorax | 104, 106, 122 |
| 102 | Munia flaviprymna | 104, 122 |
| 104 | Munia maja | 100, 102, 122 |
| 105 | Munia ferruginosa | 122 |
| 106 | Munia malacca | 100, 122 |
| 118 | Lonchura punctulata | 122 |
| 121 | Lonchura leucogastroides | 122 |
| 122 | Lonchura striata | 93, 100, 102, 104, 105, 106, 118, 121, 123, 126 |
| 123 | Lonchura fuscans | 122 |
| 126 | Spermestes cucullatus | 122 |
| 127 | Spermestes bicolor | 128 |
| 128 | Spermestes fringilloides | 127 |
| 130 | Euodice cantans | 131 |
| 131 | Euodice malabarica | 130 |

**Table 2.** Fertile hybrids of the Estrildidae. Numbers in bold indicate fertility of hybrids in both sexes. In all other cases, only male hybrids are fertile. References for the crosses are given in the appendix.

1. Within tribes a higher frequency of hybrids is due to graduated relationships within the Estrildidae.

2. Since classifications are ultimately based on similarities, it is obvious that species resembling each other will be more likely subjected to cross-breeding experiments than those that differ from each other (examples are the brown-black-and-white *Lonchura* species, or the magnificently colored green-and-red parrot finches). In this case, species relationships play only a subordinate role in the selection of species for rearing hybrids.

Table 2 lists fertile hybrids of the estrildid finches. Most fertile hybrids come from parents of the same genus. Nevertheless, fertile hybrids (mostly males) occur also across generic boundaries. Intertribal fertility only exists if one uses the tribal divisions of Steiner (1960; see Figure 6. *Estrilda astrild* × *Uraeginthus bengalus, Bathilda ruficauda* × *Poephila acuticauda* or *Poephila cincta,* respectively).

Here, too, neither a comparison of traits and geographical distribution nor interspecific hybridizations and the fertility of the $F_1$-generation result in a conclusive picture of the speciation within the Estrildidae.

## 13.4. Discussion

The number of hybrids documented, including both interspecific and intergeneric, total 145 for cardueline finches and 161 for estrildids. The existence of a largely coherent genetic pattern within each family is obvious, despite distinctive morphological differentiation among the species. An independent orgin of so many similarities as are necessary for hybridization is highly unlikely (see Scherer 1993, this volume). The most obvious explanation is to assume monophyletic descent for both groups.

This assumption is supported by the observation of atavisms in interspecific hybrids: The patterns appearing in such occasions frequently occur also in "pure" species. The existence of a particular genetic potential that might be understood as a kind of gene pool for the basic type is thereby implied. The morphological and ethological characters described as parallel developments find an obvious explanation in this model: according to the composition of the crossbreeding partners, certain phenotypes from among the possibilities that exist in the gene pool are manifested. The frequently observed mosaic-like distribution of traits (Güttinger 1970) would, according to this model, virtually have to be expected. Depending on the degree of specialization (which normally is accompanied by a decrease of genetic potential) of the species involved in the hybridization, lost or hidden potential can be restored in the phenotype through recombination during the hybridization.

Therefore it is natural to use atavisms to speculate about what the ancestral forms may have looked like:

If one summarizes the atavistic characteristics of the Carduelinae (see above), a bird with a serin-like plumage pattern emerges. The observation that many hybrids among the carduelines have a serin-like appearance (Radtke 1981), as well as the above-mentioned parallel developments of the serin pattern in different lineages of the Carduelinae, supports the ancestral form having such an appearance.

Analogously, the ancestor of the Estrildidae could have had a dark head and breast mask, transversely streaked plumage, and red tail-coverts, based on atavistic traits and characters that are widespread within this family. The different patterns of the mouth markings could either be part of the given range of variability of the estrildid gene pool, or could be derived from one another. The same is applicable to the streaked and droplet patterns of the plumage. Since, however, thus far nothing is known about the connection between the genetic constitution and the resulting phenotype, we cannot speculate any further.

Cardueline and estrildid finches can therefore be understood as two well-defined groups ("basic types"), each with a monophyletic descent. Each displays a unique combination of traits, as well as a particular range of variability, within boundaries of which the diversification into the present-day species took place through microevolutionary processes. The families differ from each other not only in certain characters but also in their range of variability. Some examples are:

- *Plumage patterns:*[8] Both groups possess a wide range of variability. In the estrildid finches, colors and patterns are more diverse.
- *Tail feathers:* In cardueline finches they are about equally long; the tail is usually forked or notched, varying little. In the estrildid finches, the tail is graduated and varies strongly in form and length.
- A *wing bar* exists in most carduelines; it varies in shape and color. It is missing in estrildid finches.
- The *beak* is generally known as a very plastic trait, being closely associated with the feeding strategy. It varies in both groups, in the Carduelinae more in its shape, in the Estrildidae more in its color.
- *Mouth markings* are absent or inconspicuous in the Carduelinae. In the Estrildidae considerable differences exist among the individual species.
- *Courtship rituals* of the Estrildidae consist of many different elements in several variations and combinations. They are more uniform and consist of fewer elements in the Carduelinae.
- The *song*[8] differs from species to species. In cardueline finches, it is always structurally varied, richly modulated, and consists of many separate elements. In estrildid finches, it is always relatively monotonous and consists of fewer elements.
- As far as the *size* is concerned, there is a wide range of variability among the cardueline finches—their body weight varies as much as an order of magnitude[9]—whereas there are hardly any differences in estrildid finches. (A connection to the geographic distribution is likely: Carduelinae inhabit all climate zones world-wide; estrildid finches are found exclusively in the tropics and subtropics of Africa, Southeast Asia, and Australia.)

It is surprising that, despite strong species-level differentiation, the existing isolating mechanisms collapse rather quickly in certain surroundings (e.g., in captivity), so that an almost unlimited hybridization within the groups can take place. This brings up the question of the length of time that has passed since the splitting into different species. Since the

---

8   Due to the strongly developed optical and acoustic abilities, colors and sounds play a crucial role in the speciation of all songbirds. Therefore, a high variability of these abilities is to be expected in all groups. Nonetheless, the range of variability seems to be fixed in each group.

9   Among the extremes of the carduelines are the Lawrence's goldfinch (*Carduelis lawrencei*, 8-11 g) and the black-and-yellow grosbeak (*Mycerobas icterioides*, up to 100 g).

fossil data do not provide any clues regarding the speciation within the Carduelinae and Estrildidae, a comparison between molecular-biological and historical data might contribute to the clarification of this question.

Microevolutionary speciation should ideally be reflected in systematic categories between the levels of species and family (normally in genera). In this context, hybridizations and the fertility of the $F_1$ generation are frequently used as data (e.g., Wolters 1957). There are, however, reasons to suspect that these criteria do not provide any help in solving lower-level systematic problems:

In interspecific hybridizations, the development of the offspring can terminate in various ontogenetic stages (fertilized eggs to almost hatched chicks) and result in variously successful hybrids (restricted to fully viabile; sterile hybrids to partially fertile hybrids). In certain cases, the ontogenetic stage during which the disruption occurs seems to be identical for a given type of hybridization—which is not a rare phenomenon in the animal kingdom.[10] Examples of varying degrees of viability and fertility in hybrids of goldfinch relatives and estrildid finches are:

- Hawfinch (*Coccothraustes coccothraustes*) × bullfinch (*Pyrrhula pyrrhula*), where in several broods well-developed chicks died in the same stage a few days before hatching (pers. comm. A. Hübschen 1988);
- Hybrids involving the bullfinch (*P. pyrrhula*) or zebra finch (*Taeniopygia guttata*), where in spite of a high frequency of hybridizations, not a single fertile hybrid is known;
- Black-headed canary (*Alario alario*) × canary (*Serinus canaria*), where the female hybrids are always fertile, but the males never appear to be fertile;
- In pairings of munias (*Munia* spp.) with the Bengalese finch (*Lonchura striata* var. *domestica*), almost all males are fertile, and almost all females sterile (Steinbacher and Wolters 1965);
- Red siskin (*Spinus cucullatus*) × canary (*Serinus canaria*), where approximately 65% of the male hybrids are fertile (Nicolai 1960).

However, caution is suggested in utilizing the developmental stage that was reached in a particular hybridization as a criterion of degree of systematic relationship. All these ontogenic disruptions also occur in pure breeds. It seems that relatively insignificant causes can lead to drastic restrictions in viability or fertility. In human beings, for example, only approximately 40% of all fertilized eggs survive the first

---

10   Similar observations were reported in hybrids of ducks and pheasants (Poll 1910, quoted in Nicolai 1960), as well as in breeds of the leopard frog (*Rana pipiens*) (after Hadorn and Wehner 1974, p. 525).

three weeks of pregnancy, which is traced back to the high complexity of early embryonic development (Zankl 1980, p. 97-98). In order to be able to make solid statements about degrees of relationship in any given case, the nature of the particular disruption that leads to the lower viability or fertility would have to be clarified. Apart from the lack of quantitative data, the limits of current biological knowledge seem to have been reached when it comes to many of these cases. Hybrid sterility and similar disruptions may, therefore, provide little information on relationships. On the other hand, the informative value of successful hybridizations and fertile hybrids is high. (For this reason, the basic type concept does not require hybrid fertility as a criterion. It is an exclusively positive definition, postulating the affiliation of relevant species to the same basic type on the basis of documented hybridization.)

There are further considerations that jeopardize lower-level systematics of this nature: Steiner (1952) points out that, on one hand, fertile offspring occur across generic or higher boundaries, whereas on the other hand, hybrid sterility and crossing barriers (which are part of the definition of a "good species") are in rare cases possible even between different breeds of the same species. Renaming species on the basis of the $F_1$ hybrid sterility, which seems to be the solution of this problem, is, however, not justifiable, given the entire biological context. Clear character differences certainly justify splitting them into different genera in spite of successful hybridization between certain species.

This general problem of the generic divisions is made especially clear in the following two quotes: Nicolai (1960) advocates a strong "splitting" of the genera of the Carduelinae in order to avoid uniting almost all goldfinch relatives on the basis of hybridization in a single genus (!); van den Elzen (1985) claims that the uniting of several species groups ("lumping") into a single genus is preferable nowadays for practical reasons (!).

"... It is virtually impossible to design a linear order of genera and species that even tentatively reflect the evolutionary history of the family's diversification" (Voous 1977). These considerations again demonstrate that phylogenetically the genus is an artificial category, whereas it is completely meaningful in a descriptive taxonomic sense. We, therefore, advocate a practical arrangement of genera that primarily serves clarity without claiming to reflect the phylogeny of the Carduelinae or Estrildidae. In contrast, the basic type concept is plausible as a second natural category above the level of the species.

If one combines the hybridization results with the results of the comparative taxonomy, using morphological and ethological data, the phylogenetic tree of Sibley et al. (1988), which is based on DNA-DNA hybridizations, is all the more surprising. Unfortunately, the authors do not discuss any reasons for the discrepancies. Without wanting to deny the value of molecular approaches to taxonomic questions, a comparison with results gained on the organismal level, as well as a more cautious attitude towards the value and limits of one's own methods would be very desirable. The newest system is not necessarily the best.

## Acknowledgements

I want to express my gratitude to all who contributed to the completion of this paper: Professor S. Scherer for the initiation of this study; Professor H. R. Güttinger, Professor S. Scherer, and S. Hartwig-Scherer for evaluating the manuscript; N. Marzlin (a referee in the DKB) for his untiring assistance in questions regarding the breeding of hybrids; R. Wiskin for photographs, D. Fehrer for patient training at the computer, J. Steinbacher and R. van den Elzen for their friendly support.

## Appendix: documentation of hybrids

No importance was placed on the question of which was the male and which the female crossbreeding partner. Even though the bypassing of behavioral barriers often influences the direction of the hybridization, this does not have any impact on the systematic relationship. Among breeders, it is customary to name the male partner first.

The crossbreeding partners are mentioned with the same numbers as in Figure 2 and 6. Numbers in [ ] correspond to the number of the quotation in the list of references. Bold numbers refer to quotations in which statements are made regarding the fertility of the hybrids. Crossbreeding reports of the genus *Fringilla* (1, 2) with Carduelinae are to be regarded with reservation (see Figure 2, discussion see text). Each hybridization is mentioned with both crossbreeding partners; however, only the first mentioning includes the literature reference.

Documented Hybrids of Fringillinae and Carduelinae (Figure 2)

1    2 [10, **14**, 19, Brit. Museum], 14 [14], 29 [14, 47], 54 [14], 66 [14], 127 [47]

2    1, 14 [14], 29 [9 (1984, 1985), 14, 60], 33 [14, 47, 62], 54 [14], 66 [14, 72]

8    66 [9 (1984)]

9    66 [ (9 (1984-1990), 14], 103 [9 (1987)]

14   1, 2, 26 [14, pers. communication: A. Hübschen 1988], 29 [2, 14, 19] 33 [2, 9 (1984-1987, 1989, 1990), 14, 19], 52 [14], 54 [2, 9 (1985), [14], 55 [34], 56 [2, 9 (1984, 1985, 1987, 1990), 14] 66 [9 (1984-1986, 1988-1990, 14, 19], 77 [9 (1986), 60], 83 [9 (1984-1986], 102 [3, 9 (1984), 14], 103 [2, 9 (1986, 1987), 127 [9 (1985)]

26   14, 29 [14]

29    1, 2, 14, 26, 30 [9 (1984, 1985, 1987), **14**], 32 [9 (1984), **14**], 33 [9 (1984-
      1987), **14**], 42 [9 (1984, 1989)], 46 [9 (1986)], 47 [9 (1990)], 52 [9 (1984,
      1990), 14], 54 [14], 55 [14], 56 [9 (1989), 14], 66 [9 (1984-1990), **14**, 31,
      **47, 60**], 77 [3], 79 [9 (1986)], 82 [9 (1985, 1986), 14], 86 [9 (1986, 1988,
      1989)], 87 [9 (1985)], 96 [14], 100 [14], 102 [3, 9 (1986), 14], 103 [2, 9
      (1984-1986), 14], 127 [9 (1985-1990)]

30    29, 32 [14], 33 [9 (1984, 1986), **14**], 66 [9 (1985, 1990)], 86 [9 (1986)]

31    33 [14], 66 [9(1987)]

32    29, 30, 33 [14], 66 [9 (1984, 1986, 1988, 1990), 14], 86 [9 (1989)]

33    2, 14, 29, 30, 31, 32, 38 [9 (1985)], 42 [9 (1986, 1987), 14, 61], 44 [3, 9
      (1990), 14, 47, 61], 46 [61], 52 [9 (1989), 14], 54 [9 (1985, 1986, 1990),
      **14**], 55 [14], 56 [9 (1985, 1986), **14**], 66 [2, 9 (1984-1990), **14**, 31], 67 [14],
      77 [3], 82 [14, 60], 86 [9 (1984), 14] 103 [9 (1985)]

35    36 [14], 66 [3, 9 (1984), 14]

36    35, 66 [9 (1985), **43**]

38    33, 63 [pers. communication: R. v. d. Elzen 1991], 66 [9 (1984-1990), **14**,
      **47**]

42    29, 33, 44 [Senckenberg Museum, **41**], 52 [**13**, 14], 56 [9 (1986)], 66 [9
      (1984-1990), 14]

44    33, 42, 52 [9 (1984) **47**, 61], 54 [9 (1984-1990)], 66 [9 (1984-1990, **1986**),
      **14, 43**], 103 [9 (1987)]

46    29, 33, 52 [14], 66 [9 (1984-1966, 1988), 14]

47    29, 66 [9(1986),14]

51    66 [9 (1989, 1990)]

52    14, 29, 33, 42, 44, 46, 54 [9 (1987), 14, 48], 56 [14, Senekenberg Museum],
      61 [14], 62 [14], 66 [9 (1984-1990), **14, 47, 60**], 67 [9 (1990)], 82 [14]

54    1, 2, 14, 29, 33, 44, 52, 55 [14], 56 [14], 66 [9 (1984-1987, 1989), 14], 103
      [9 (1985)]

55    14, 29, 33. 54, 56 [14], 63 [14], 66 [9 (1990), 14, 48, 62]

56    14, 29, 33, 42, 52, 54, 55, 63 [14], 66 [2, 9 (1984-1990), **14**, 31, **43, 47**], 67
      [14], 77 [14, 27], 82 [14], 86 [14]

59    63 [14], 66 [9 (1984-1990), **12, 14, 43**], 67 [48], 77 [14], 82 [14]

61    52, 63 [14]

62    52, 66 [14, 43]

63    38, 55, 56, 59, 61, 66 [9 (1985), 14, 43], 80 [14], 82 [14], 86 [14]

64    66 [9 (1986)]

66   1, 2, 8, 9, 14, 29, 30, 31, 32, 33, 35, 36, 38, 42, 44, 46, 47, 51, 52, 54, 55, 56, 59, 62, 63, 64, 67 [3, 9 (1984, **1985**, 1986-1988, 1990), **14**, 31, **43**, **47**], 68 [9 (1990)], 72 [9 (1989)], 77 [3, 9 (1985, 1988), 14, 43], 79 [9 (1986)], 80 [**14**, 43], 82 [9 (1984-1986, 1988, 1990), 14, 43], 83 [9 (1984, 1990), **60**, 61], 84 [61], 86 [3, 9 (1984-1990), 14, 43], 87 [9 (1985, 1988, 1989), 14, 60], 90 [9 (1988)], 96 [60], 98 [18], 100 [14], 102 (3, **6**, 9 (1984, 1986, 1987, 1989), 14], 103 [9 (1984-1990), 14], 120 [9 (1990)], 127 [9 (1984-1986, 1989)]

67   33, 52, 56, 59, 66, 82 [14]

68   66

72   66

77   14, 29, 33, 56, 59, 66, 80 [3, 14, **43**], 82 [3, 14, 43], 86 [14]

79   29, 66

80   63, 66, 77, 82 [9 (1984), 11], 86 [14]

82   29, 33, 52, 56, 59, 63, 66, 67, 77, 80, 127 [pers. communication: A. Hübschen 1988]

83   14, 66

84   66

86   29, 30, 32, 33, 56, 63, 66, 77, 80, 87 [12, **43**], 89 [12, **43**], 127 [9 (1988-1990)]

87   29, 66, 86

89   86

90   66

92   127 [9 (1989)]

96   29, 66

98   66

100   29, 66, 103 [14]

102   14, 29, 66, 103 [2], 127 [9 (1988, 1989)]

103   9, 14, 29, 33, 44, 54, 66, 100, 102, 127 [9 (1989)]

119   120 [9 (1989)]

120   66, 119

127   1, 14, 29, 66, 82, 86, 92, 102, 103

## Documented Hybrids of Estrildidae (Figure 6)

1   2 [**14**, 50, **63**]

2   1, 78 [63], 93 [63], 122 [63], 130 [63], 131 [63]

3   5 [14, **51**, **63**]

4   5 [**49**]

5   3, 4

| | |
|---|---|
| 6 | 59 [14] |
| 14 | 59 [63] |
| 17 | 19 [63] |
| 19 | 17 |
| 24 | 28 [14, 63], 30 [14, 63], 31 [14, Mus. A.Koenig, Bonn], 125 [63] |
| 28 | 24, 30 [14, 63], 31 [14] |
| 30 | 24, 28, 31 [**14**], 45 [14], 59 [14], 64 [14], 130 [14, 63] |
| 31 | 24, 28, 30, 59 [14, 51], 66 [14, 63], 76 [14], 126 [14, 63], 130 [14, 63] |
| 34 | 131 [61] |
| 38 | 45 [14] |
| 42 | 59 [50, 56, 63] |
| 45 | 30, 38, 46 [14, **63**], 64 [14], 76 [14, 73, Mus, A. Koenig, Bonn] |
| 46 | 45 |
| 56 | 59 [14, 63] |
| 58 | 60 [63] |
| 59 | 6, 14, 30, 31, 42, 56, 60 [**14**], 64 [14], 122 [14, 63] |
| 60 | 58, 59 |
| 61 | 64 [14] |
| 62 | 76 [14] |
| 64 | 30, 45, 59, 61, 65 [14] |
| 65 | 64 |
| 66 | 31, 71 [63], 72 [14, 63], 73 [63], 87 [14, 63], 130 [63] |
| 69 | 72 [14, 63], 75 [44], 76 [14, 44, 51, 63], 79 [61, 63], 93 [14], 122 [14, 63] |
| 71 | 66 |
| 72 | 66, 69, 73 [14, 63], 79 [14, 63] |
| 73 | 66, 72, 76 [14, 63], 79 [61, **63**], 122 [63], 130 [1] |
| 74 | 75 [14, 63], 76 [14, 63, 68], 77 [63], 79 [14, 63], 102 [14], 122 [61, 63], 130 [63], 131 [14, 63, 68] |
| 75 | 69, 74, 76 [14, 44, 50, 63], 77 [14, 44, 63], 78 [14, **44**, 63], 79 [14, 44, 63], 100 [44], 102 [14, 63], 122 [63], 130 [35] |
| 76 | 31, 45, 62, 69, 73, 74, 75, 77 [14, 63], 78 [14, 63], 79 [14, 63], 93 [63], 100 [14, 63], 106 [63], 122 [14, 50, 63], 130 [14, 63], 131 [63] |
| 77 | 74, 75, 76, 78 [63], 79 [14, **63**], 100 [14, 63], 122 [14] |
| 78 | 2, 75, 76, 77, 79 [**14, 50, 63**], 87 [14, 63], 104 [14, 63], 118 [63], 122 [14, 61, 63] |
| 79 | 69, 72, 73, 74, 75, 76, 77, 78, 100 [23, 63], 118 [14, 63], 122 [14, 63], 130 [35, 63] |

| | |
|---|---|
| 82 | 85 [**14**, 63], 87 [14, **63**] |
| 85 | 82, 87 [**14**, 50, **63**], 88 [**14**], 89 [63], 92 [63] |
| 87 | 66, 78, 82, 85, 88 [14], 89 [14], 122 [14] |
| 88 | 85, 87, 90 [63] |
| 89 | 85, 87 |
| 90 | 88 |
| 92 | 85, 106 [63] |
| 93 | 2, 69, 76, 118 [63], 122 [**14**, 50, **63**], 130 [14, 63] |
| 95 | 100 [14], 102 [63] |
| 97 | 122 [61] |
| 100 | 75, 76, 77, 79, 95, 102 [63], 104 [**14**, **50**, **63**], 106 [14, **22**, 63], 118 [14, 63), 122 [14, **22**, 50, **63**], 130 [14, 63], 131 [14] |
| 102 | 74, 75, 95, 100, 104 [**63**], 122 [61, **63**], 130 [63] |
| 104 | 78, 100, 102, 106 [14, 63], 118 [14], 122 [**14**, 22, 63], 130 [14] |
| 105 | 122 [**22**, 63], 130 [63] |
| 106 | 76, 92, 100, 104, 118 [14, 63], 122 (14, 15, **22**, **59**, 61, **63**), 130 [14, 63, 68] 131 [14] |
| 115 | 124 [36] |
| 118 | 78, 79, 93, 100, 104, 106, 122 [**14**, 50, 59, 61, 63, 64], 125 [63, 64,] 126 [14, 63, 64], 130 [14, 51, 63, 68], 131 [14, 63] |
| 121 | 122 [**14**, **63**] |
| 122 | 2, 59, 69, 73, 74, 75, 76, 77, 78, 79, 87, 93, 97, 100, 102, 104 105, 106, 118, 121, 123 [**14**, **63**], 125 [14, 63, 64], 126 [**14**, 59, 63, 64], 127 [14, 15, 63, 64], 128 [63], 129 [63], 130 [14, 59, 63, 68], 131 [14, 63] |
| 123 | 122 |
| 124 | 115 |
| 125 | 24, 118, 122, 126 [14, 63, 64], 128 [63], 130 [14], 131 [14, 63, 64] |
| 126 | 31, 118, 122, 125, 127 [14, 63, 64], 128 [14, 51, 63, 64], 130 [14, 63, 64], 131 [14, 63, 64] |
| 127 | 122, 126, 128 [**14**, 63, **64**], 130 [14, 63, 64] |
| 128 | 122, 125, 126, 127, 130 [14, 63, 64] |
| 129 | 122, 130 [63] |
| 130 | 2, 30, 31, 66, 73, 74. 75, 76, 79, 93, 100, 102, 104, 105, 106, 118, 122, 125, 126, 127, 128, 129, 131 [**14**, **63**] |
| 131 | 2, 34, 74, 76, 100, 106, 118, 122, 125, 126, 130 |

# References

[1]   André, G. 1984. Binsenastrild × Silberschnäbelchen ungewollt gezüchtet. *Gefiederte Welt* 108:24.
[2]   Armani, G.C. 1983. *Guide des Passereaux Granivores*. Delachaux and Niestlé, Neuchâtel, Switzerland.
[3]   Aschenborn, C. 1966. *Fremdländische Stubenvögel*. Albrecht Philler, Minden, Germany.
[4]   Bielfeld, H. 1973. *Prachtfinken*. Ulmer, Stuttgart.
[5]   Bielfeld, H. 1978. *Kanarien*. Ulmer, Stuttgart.
[6]   Boetticher, H. von. 1944. Mischlingszucht mexikanischer Karmingimpel-Männchen × Kanarienvogel-Weibchen. *Ornithologische Monatsberichte* 52:157.
[7]   Cracraft, J. 1987. DNA hybridization and avian phylogenetics. *Evolutionary Biology* 21:47-96.
[8]   Delacour, J. 1943. A revision of the subfamily Estrildinae of the family Ploceidae. *Zoologica* 28:69-86.
[9]   DKB. 1984-1990. *Catalogues of the German Championship 37-43*. Deutscher Kanarienzüchterbund (German Canary Breeder's Association), Freudenberg, Germany.
[10]  Eck, S. 1975. Evolutive Radiation in der Gattung *Fringilla*, L. Eine vergleichend-morphologische Untersuchung (Aves: Fringillidae). *Zoologische Abhandlungen Staatliche Naturhistorische Sammlungen Dresden Museum für Tierkunde* 33:278-302.
[11]  Elzen, R. van den. 1981. Zwei Bastarde Mosambikgirlitz (*Ochrospiza mozambica caniceps*) × Angolagirlitz (*Ochrospiza a. atrogularis*); Aves, Carduelidae. *Bonner Zoologische Beiträge* 32:127-135.
[12]  Elzen, R. van den. 1985. Systematics and evolution of African canaries and seedeaters (Aves: Carduelidae). In: *Proceedings of the International Symposium on African Vertebrates*. Alexander Koenig Museum, Bonn, pp. 435-451.
[13]  Friedrichs, E. and W. Friedrichs. 1981. Zuchterfolg Magellan- × Erlenzeisig (*Spinus magellanicus* × *Spinus spinus*). *Gefiederte Welt* 105:49-50.
[14]  Gray, A.P. 1958. *Bird Hybrids: A Check-List with Bibliography*. Commonwealth Agricultural Bureaux, Farnham Royal, England.
[15]  Güttinger, H.R. 1970. Zur Evolution von Verhaltensweisen und Lautäußerungen bei Prachtfinken (Estrildidae). *Zeitschrift für Tierpsychologie* 27:1011-1075.
[16]  Güttinger, H.R. 1978. Verwandtschaftsbeziehungen und Gesangsaufbau bei Stieglitz (*Carduelis carduelis*) und Grünlingsverwandten (*Chloris* spec.). *Journal für Ornithologie* 119:172-190.
[17]  Hadorn, E. and R. Wehner. 1974. *Allgemeine Zoologie*. Thieme, Stuttgart.
[18]  Hemmer, W. 1976. Mischlingszucht Rüppellsgirlitz × Kanarienvogel (*Crithagra tristriata* × *Serinus canaria dom.* var. *isabell*). *Gefiederte Welt* 100:220-222.

[19] Hinde, R.A. 1956. A comparative study of the courtship of certain finches (Fringillidae). *Ibis* 98:1-23.

[20] Immelmann, K. 1962. Beiträge zu einer vergleichenden Biologie australischer Prachtfinken (Spermestidae). *Zoologische Jahrbücher Systematik* 90:1-196.

[21] Kakizawa, R. and M. Watada. 1985. The evolutionary genetics of the Estrildidae. *Journal of the Yamashina Institute for Ornithology* 17:143-158.

[22] Kirschke, S. 1963. Probleme der Mischlingszucht zwischen Mövchen, Nonnen und Schilffinken. *Gefiederte Welt* 87:141-144.

[23] Kirschke, S. 1980. Nochmals über die Zucht von Mövchen-Zebrafinken-Mischlingen - ein verständlicher Irrtum. *Gefiederte Welt* 104:124-125.

[24] Klatt, G.T. von. 1901. Über den Bastard von Stieglitz und Kanarienvogel. *Archiv für Entwicklungsmechanik* 12:414-453, 471-528.

[25] Klatt, G.T. von. 1902. Über den Bastard von Stieglitz und Kanarienvogel. *Zoologischer Garten* 43:285-293.

[26] Kunkel, P. 1966. Beiträge zur Biologie und Ethologie einiger zentralafrikanischer Girlitze I. *"Serinus" citrinelloides* Rüppell. *Journal für Ornithologie* 107:257-277.

[27] Märzhäuser, H. 1979. Mischlingszucht Grauedelsänger × Bluthänfling. *Gefiederte Welt* 103:91-92.

[28] Mainardi, D. 1957a. Sulle possibilità di ricavare una serie filetica da dati sulla affinità sierologica. Ricerche sui Fringillidi. *Istituto Lombardo* 91:565-569.

[29] Mainardi, D. 1957b. Affinità sierologiche e filogenesi nei Fringillidi. Rapporti sierologici tra il Verdone (*Chloris chloris*), il Fringuello (*Fringilla coelebs*) e il Cardellino (*Carduelis carduelis*). *Archivo Zoologico Italiano* 42:151-159.

[30] Mainardi, D. 1957c. L'evoluzione nei Fringillidi. Concordanza tra una "mappa sierologica" e i dati dell'analisi elettroforetica delle emoglobine. *Istituto Lombardo* 92:180-186.

[31] Mark, R. 1930. Untersuchungen an Bastarden zwischen Kanarien und Wildfinken. *Zeitschrift für Wissenschaftliche Zoologie* 137:476-549.

[32] Marsh, F.L. 1976. *Variation and Fixity in Nature*. Pacific Press, Mountain View, CA.

[33] Massoth, K.H. 1978. Cardueline Finken. *Kanarienfreund*. Hanke, Pforzheim, Germany, p. 234.

[34] Mau, K.G. 1977. Berghänfling × Gimpel. *Gefiederte Welt* 101:100.

[35] Mayer, H. 1981. Neue Prachtfinkenmischlinge: Ringelastrild × Silberschnäbelchen und Spitzschwanzamadine × Silberschnäbelchen. *Gefiederte Welt* 105:84-85.

[36] Mayer, H. 1986. Von meinen Bergbronzemännchen (*Lonchura kelaarti* [Jerdon 1863]) und Mischlingszucht mit der Fünffarbennonne (*Lonchura quinticolor* [Vieillot 1807]). *Gefiederte Welt* 110:97-98.

[37] Mayr, E. 1967. *Artbegriff und Evolution*. Parey, Hamburg.

[38] Mayr, E. 1968. The sequence of genera in the Estrildidae (Aves). *Breviora* 287:1-14.

[39] Mayr, E., R.J. Andrew, and R.A. Hinde. 1956. Die systematische Stellung der Gattung *Fringilla*. *Journal für Ornithologie* 97:258-273.

[40] Mayr, E., *E.g.*, Linsley, and R.L. Usinger. 1953. *Methods and Principles of Systematic Zoology*. McGraw-Hill, New York.

[41] Mowe, H. 1979. Erfahrungen mit amerikanischen Zeisigen. *Gefiederte Welt* 103:63-64.

[42] Nicolai, J. 1957. Die systematische Stellung des Zitronenzeisigs ("*Carduelis*" *citrinella* L.). *Journal für Ornithologie* 98:363-371.

[43] Nicolai, J. 1960. Verhaltensstudien an einigen afrikanischen und paläarktischen Girlitzen. *Zoologische Jahrbücher* 87:317-362.

[44] Oppenborn, G. 1971. Mischlinge unter australischen Prachtfinken. *Gefiederte Welt* 95:145-147.

[45] Panov, E.N. 1989. *Natural Hybridization and Ethological Isolation in Birds*. Nauka, Moscow.

[46] Peters, J.L. 1931-1970. *Checklist of Birds of the World, Volumes XIII-XV*. Harvard University Press, Cambridge, MA.

[47] Radtke, G. 1981. *Positurkanarien und Mischlinge*. Albrecht Philler, Minden, Germany.

[48] Radtke, G. 1983. Seltene Mischlinge — unbeabsichtigt. *Gefiederte Welt* 107:282-284.

[49] Röder, J. 1984. Der rotflügelige Rotmaskenastrild—ein Mischling? *Gefiederte Welt* 108:253-255.

[50] Robiller, F. 1978. *Prachtfinken*. Neumann-Neudamm, Berlin.

[51] Rokitansky, G. and H. Schifter. 1968. Über einige Hybriden der Vogelsammlung des Wiener Naturhistorischen Museums. *Annalen des Naturhistorischen Museums in Wien* 72:213-230.

[52] Sarich, V.M., C.W. Schmid, and J. Marks. 1989. DNA hybridization as a guide to phylogeny: A critical analysis. *Cladistics* 5:3-32.

[53] Scherer, S. 1986. On the limits of variability: Evidence and speculation from morphology, genetics and molecular biology. In: Andrews E.H., W. Gitt and W.J. Ouweneel, eds. *Concepts in Creationism*. Evangelical Press, Welwyn, England, pp. 219-240.

[54] Scherer, S. 1993. Basic Types of Life. In: Scherer, S. ed. *Typen des Lebens*, Pascal, Berlin, pp. 11-30.

[55] Scherer, S. and T. Hilsberg. 1982. Hybridisierung und Verwandtschaftsgrade innerhalb der Anatidae—eine systematische und evolutionstheoretische Betrachtung. *Journal für Ornithologie* 123:357-380.

[56] Schiffer, H. 1968. Weitere Angaben zur Kreuzung Dunkelroter Amarant × Schmetterlingsfink. *Gefiederte Welt* 92:199.

[57] Sibley, C.G. and J.E. Ahlquist. 1982. The relationships of the Hawaiian honeycreepers (Drepanidinii) as indicated by DNA-DNA hybridization. *Auk* 99:131-140.

[58] Sibley, C.G, J.E. Ahlquist, and B.L. Monroe Jr. 1988. A classification of the living birds of the world based on DNA-DNA hybridization studies. *Auk* 105:409-423.

[59] Siroki, Z. 1971. Mischlingszucht mit Prachtfinken. *Gefiederte Welt*

95:115-116.

[60] Speicher, K. 1970. *Finkenmischlinge*. Kosmos, Stuttgart.

[61] Speicher, K. 1971. Bericht von der COM-Schau Basel 1971. *Gefiederte Welt* 95:67-71.

[62] Steinbacher, J. 1979. Mischlingszuchten europäischer Vögel 1977. *Gefiederte Welt* 103:58.

[63] Steinbacher, J. and H.E. Wolters. 1965. *Vögel in Käfig und Voliere. Prachtfinken*. Second Edition. Hans Limberg, Aachen, Germany.

[64] Steiner, H. 1952. Vererbungsstudien an Vogelbastarden. *Archiv der Julius Klaus-Stiftung für Vererbungsforschung, Sozialanthropologie und Rassenhygiene* 27:121-137.

[65] Steiner, H. 1958. Artspezifische Merkmalsphänokopien bei australischen Prachtfinken, Spermestidae, insbesondere beim Zebrafinken, *Taeniopygia castanotis* Gould. *Archiv der Julius Klaus-Stiftung für Vererbungsforschung* 33:62-70.

[66] Steiner, H. 1959. Kreuzungsversuche zur Vererbung artspezifischer Merkmale: Die Rachenzeichnung der Nestlinge der Prachtfinken, Spermestidae. *Archiv der Julius Klaus-Stiftung für Vererbungsforschung* 34:220-228.

[67] Steiner, H. 1960. Die Klassifikation der Prachtfinken, Spermestidae, auf Grund der Rachenzeichnung ihrer Nestlinge. *Journal für Ornithologie* 101:92-112.

[68] Steiner, H. 1966. Atavismen bei Artbastarden und ihre Bedeutung zur Feststellung von Verwandtschaftsbeziehungen: Kreuzungsergebnisse innerhalb der Singvogelfamilie der Spermestidae. *Revue Suisse de Zoologie* 73(17):321-337.

[69] Sushkin, P.P. 1924. [On the Fringillidae and allied groups]. *Bulletin of the British Ornithologists' Club* 45:36-39.

[70] Sushkin, P.P. 1927. On the anatomy and classification of the weaver-birds. *Bulletin of the American Museum of Natural History* 57:1-32.

[71] Tordoff, H.B. 1954. Relationships in the New World Nine-Primaried Oscines. *Auk* 71:273-284.

[72] Voous, K.G. 1977. List of recent holarctic bird species: Passerines. *Ibis* 117:376-406.

[73] Wolter, J. 1977. Ein neuer Prachtfinken-Mischling. *Gefiederte Welt* 101:162-167.

[74] Wolters, H.E. 1957. Die Klassifikation der Webefinken (Estrildidae). *Bonner Zoologische Beiträge* 8:90-129.

[75] Wolters, H.E. 1967. Über einige asiatische Carduelidae. *Bonner Zoologische Beiträge* 18:169-172.

[76] Wolters, H.E. 1981. Die systematische Stellung des Dornastrilds, *Aegintha temporalis* (Latham) (Aves, Estrildidae). *Bonner Zoologische Beiträge* 32:137-144.

[77] Wolters, H.E. 1975-1982. *Die Vogelarten der Erde*. Parey, Hamburg.

[78] Zankl, H. 1980. *Humanbiologie*. Fischer, Stuttgart.

[79] Ziswiler, V. 1965. Zur Kenntnis des Samenöffnens und der Struktur des hörnernen Gaumens bei körnerfressenden Oscines. *Journal für*

*Ornithologie* 106:1-48.

[80] Ziswiler, V. 1967. Vergleichend morphologische Untersuchungen am Verdauungstrakt körnerfressender Singvögel zur Abklärung ihrer systematischen Stellung. *Zoologische Jahrbücher Systematik* 94:427-520.

# 14. Further Candidates for Basic Types of Birds - A Resumé

Sheena Tyler

**Abstract**

This chapter reports 12 groups of birds which are further candidates for basic type status, resulting primarily from analysis of hybridization data and morphological characters. The prospective basic types are also evident from their earliest fossil representatives, which bear clear morphological affinity with their modern day relatives, The 12 candidate types include the flamingos, cranes, pelicans, loons and divers, herons and egrets, parrots, spoonbills and ibises, and birds of paradise. A further 35 groups are included for further consideration. These taxa, ranging from the genus to family level, include the gulls, stilts and avocets, toucans, storks, hummingbirds, manakins and pigeons. In all of these groups, no hybridization is found between the groups delineated and their nearest proposed outgroups, indicating clear discontinuities, and providing further empirical evidence for the validity of the basic type concept.

## 14.1. Analysis Of Hybridization Data

The following is a summary of an analysis of avian hybridization and other data. Published avian hybridization records were examined from citations chiefly by Annie Gray (1958) and Eugene McCarthy (2006). From analysis of this data, firstly, if the majority of the reports are accepted, up to 2081 bird species may hybridize out of a world total of 10,021, *i.e.* 1 in 5 species of birds. However, the documentation ranges from weaker evidence from personal communications, secondary sources and brief reports to much stronger evidence from publications by ornithologists or zoologists providing detailed descriptions of hybrids. Nonetheless, even if only well-documented reports are accepted, it confirms the Grant and Grant (1992) prediction that 1 in 10 world species can hybridize, and may be even more common than 1 in 10[1].

Secondly, the initial analysis revealed 90 families in which at least 20 % of the members hybridize between species. In certain taxa, predominantly families, relatively more of the species hybridize with one another. These are reviewed in Section 2. A further 20 groups and several monotypic taxa also worthy of further consideration are examined in Section 3.

---

1    If formerly good species merged simply due to hybridization are included, the world total (*i.e.*, number of species hybridizing) rises to 2616.

The minimum value of 20 % provides an initial screening tool from which prospective types may be recognized, giving a preliminary indication of the degree of hybridization throughout the taxon. The degree of interconnectivity was assessed between the member species and, if applicable, higher taxonomic categories, and where appropriate by the use of Venn diagrams. As a consequence some of these 90 groups were discounted due to low interspecific hybridization connectivity (for instance, in the Dendrocolaptidae). In small taxa, notably in Rheidae comprising just two species, connectivity criteria alone might appear to exaggerate the prospective basic type status, although, equally, the basic type may comprise these species alone. Dubious or challenged reports were discounted, but not necessarily data from antiquity, for instance with the birds of paradise, in which interspecific and intergeneric hybrids are well-authenticated. In certain cases, the need for further data to substantiate the validity of a cross is stated. Species status and nomenclature was referenced chiefly in relation to the Howard and Moore complete checklist of the birds of the world (4th edition) (Volume 1: non-passerines [Dickinson and Remsden 2013; Volume 2: passerines [Dickinson and Christidis 2013]) and the Clements checklist (Clements et al. 2023).

## 14.2. Candidates For Basic Type Status

In certain taxa (primarily families) many or even all of the species hybridize with one another. In some of these groups, all of the genera are also united by hybridization. The degree of connectivity between the member species is generally high, often extending to intergeneric hybrids or even hybrids between families. The number of species hybridizing in some families is high, for instance up to 165 species (46 %) throughout the parrot-like birds, with 41 genera linked by hybrids. These criteria provide strong indicators for prospective basic type status.

Taken together, these taxa form 12 groups (Basic Types). The data for each of these groups is now reviewed, including key distinguishing criteria.

**14.2.1 Phoenicopteridae (Flamingos).** The flamingo form is unmistakable (del Hoyo 1992). They have a characteristic bill, with a central downward bend, and trough-like lower mandible covered with rows of lamellae for sieving fine food particles (reviewed by del Hoyo 1992). They have a large, oval-shaped body, with pink plumage and black flight feathers, and long, sinuous necks. Flamingos are also amongst the most ancient of birds, with fossil forms being found from mid-Eocene periods. Current taxonomies (Dickinson and Remsen 2013; Clements et al. 2023) recognise 3 genera within the flamingo family, although Delacour considers that all flamingos are sufficiently alike to be placed within a single genus, Phoenicopterus (Johnson and Cezilly 2009). Five of the six species (80%) of flamingos hybridize with one another but with no other group. Intergeneric hybrids link all three genera together (Figure 1). Within the genus *Phoenicopterus*, a hybrid between *Phoenicopterus ruber* (Caribbean flamingo) and *Phoenicopterus roseus* (Greater flamingo) was observed and ringed by the Flamingo Specialist Group at Zwillbrocker Venn, Germany, in 2005 (Flamingo Specialist Group, 2005). At the same location, 6 sightings of *Phoenicopterus chilensis x Phoenicopterus roseus* (Chilean x Greater flamingo) hybrids were logged by 5 observers from 2021 – 2022 (https://observation.org). Intergeneric hybrids from the cross *Phoenicopterus chilensis x Phoeniconaias minor* have been observed, again at Zwillbrocker Venn (Blair 2000). There is a zoo report of hybridization between *Phoenicoparrus andinus* and *Phoenicopterus ruber* (Andean x American flamingo) (International Zoo Yearbook [IZY] 1974). The intergeneric crosses require further confirmation, although if Delacour is followed, the flamingos are united within a single genus, negating the need for inter-generic hybrids to indicate connectivity. These various records are brief, but the morphology of a cross *Phoenicopterus chilensis x Phoenicopterus ruber*, observed in the Camargue of France was described in more detail (Cezilly and Johnson 1992).

Some authors place flamingos within the Ciconiiformes (according

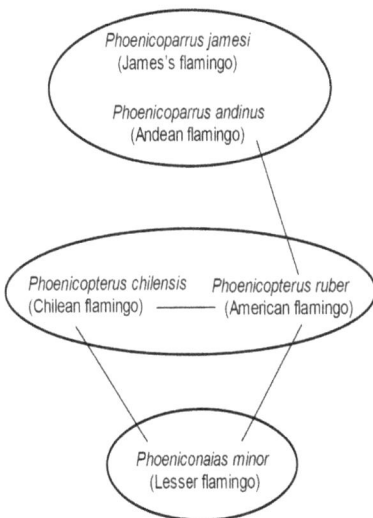

Figure 1. The correlation between the hybridization data and morphology is often strikingly high in prospective basic types, for instance, in the flamingo. A. Five of the six species of flamingos hybridize (83 % of all flamingos). Intergeneric hybrids (indicated by straight lines) link all three genera (circled) together. B. The flamingo form is unmistakable, and clearly distinguished from proposed relatives such as duck-like birds or grebes.

to DNA-DNA hybridization), whilst others ally them variously with Anseriformes (based on the structure of biliary acids), or with grebes, or as the sole members of their own order, Phoenicopteriformes. No hybrids are known between flamingos and species within any of these other orders. The distinctive morphology distinguish the flamingos as a basic type candidate, supported by a small body of hybridization data, including several reports from flamingo specialists.

**14.2.2 Gruidae (Cranes).** There are records of hybridization between 14 of the 15 species of cranes (93% of all cranes). Some inter-generic crosses are documented in detail. In an artificial insemination (AI) study at a crane breeding centre in Russia, a hybrid between a Siberian crane and a White-naped crane was produced, as an experimental model to assess cryoconservation of semen from the endangered Siberian crane (Maksudov and Panchenko 2002). In another AI study, a hybrid was produced from crossing a Eurasian and Siberian crane (Kashentseva and Postelnykh 2013). The hybrid showed both Siberian and Eurasian crane features, but also its vocalization differed from both parents. The Whooping crane is also an endangered species. As an attempt to repopulate the species in North America, a Whooping crane chick was fostered with a Sandhill crane. However, the chick imprinted on the Sandhill, and subsequently mated with a Sandhill, which resulted in a hybrid (https://savingcranes.org/whoopsie-the-whooping-sandhill-crane-chick). Other hybrids are also known from this cross (Olsen and Derrickson 1980; https://www.audubon.org).

The crosses just described link 4 of the 6 genera recognised in the Clements checklist. These genera may also be linked by hybrids to the remaining 2 genera, according to brief reports (Gray 1958, Johnsgard 1983), which require further confirmation. There is no hybridization with the rails, their nearest relatives.

Cranes possess quite long and straight bills, elongated and slender necks and legs, and long inner secondary feathers overhanging the tail. The large height (from 90 to 150 cm) and distinctive, graceful form of cranes is easily recognized (Archibald and Meine 1996). Cranes first appeared in the Eocene, with 11 species found in the fossil record from that period. Thus there is a correlation between morphology—in extant and fossil forms—and hybridization data within the Gruidae, which together distinguish the cranes as a basic type candidate. Added to this is the behavioural feature of dancing. Although some other bird groups exhibit dancing too, the extensive dancing of cranes is highly characteristic of the taxon, with graceful jumps and miniature soars with outstretched wings, bowing and bending their legs.

**14.2.3 Pelicanidae (Pelicans).** There are reports of hybridization between up to six out of the eight species of pelicans[2] (75 % of all pelicans) (Gray 1958; McCarthy 2006) (Figure 2). The records are often brief and lacking morphological detail. Examples include IZY reports of hybrids

2    Formerly 7 species were recognised (Dickinson 2003), in which Pelecanus thagus was merged with P. occidentalis.

in zoo captivity: *Pelicanus crispus* x *P. rufescens* (Dalmatian x Pink-backed pelican) (Klos 1969); *P. onocrotalus* x *P. occidentalis* (Great white x Brown pelican) (IZY 1998). Fossil pelicans date back to the Oligocene, mostly being placed within the same genus (*Pelicanus*) and evidently pelican structure has changed little since these

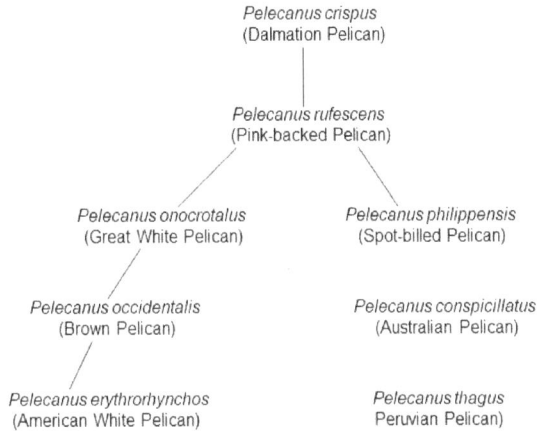

Pelecanus crispus
(Dalmation Pelican)

Pelecanus rufescens
(Pink-backed Pelican)

Pelecanus onocrotalus
(Great White Pelican)

Pelecanus philippensis
(Spot-billed Pelican)

Pelecanus occidentalis
(Brown Pelican)

Pelecanus conspicillatus
(Australian Pelican)

Pelecanus erythrorhynchos
(American White Pelican)

Pelecanus thagus
Peruvian Pelican)

Figure 2. Hybridizing species within the Pelicanidae (pelicans). Six of the eight species of pelicans hybridize (75 % of all pelicans) in this single-genus taxon. All six species are interlinked via hybridization (straight lines).

fossil forms. Proposed relatives (none of which are known to hybridize with pelicans) include the gannets and boobies (Sulidae), cormorants (Phalacrocoracidae) and darters (Anhingidae). However, the pelican form is clearly recognisable, with its huge, distensible skin pouch and long bill (Elliott 1992a). A loose articulation of the lower mandible facilitates the distension of the pouch for large numbers of fish capture. Thus there is a correlation between morphological form and hybridization data within Pelicanidae, which together distinguish the pelicans as a basic type candidate. However, more detailed evidence is needed to verify the hybridization data.

**14.2.4 Gaviidae (Loons And Divers).** Records of hybridization (summarized by McCarthy 2006) link together up to four out of the five species (80%) of all loons and divers, but with no other group. The sole family within the Gaviiformes, the systematic relationship of divers with other birds is far from clear (Carboneras 1992). Speculated nearest relatives include the grebes (Podicipedidae), gulls (Laridae), auks (Alcidae), or, using DNA studies, frigate birds and penguins. The Gaviidae are amongst the most ancient of birds, with fossils represented as far back as the Paleocene. Diversification is thought to have stemmed from the Black-throated diver (*Gavia arctica*) which colonised North America, which in turn was cut off, leading to speciation there. It is then thought that the Red-throated and Black-throated divers then underwent further diversifications throughout the Nearctic. This can be explained in terms of an ancestral basic type within the Gaviidae, which underwent a radiation as it colonized these extensive geographical regions.

The Gaviidae have a longish, thick-set neck, elongated head and strong, pointed beak. Carboneras (1992) describes them as exquisitely designed for swimming: with well set-back legs to aid streamlined propulsion; laterally compressed tarsi to reduce water resistance; webbed toes; narrow nostrils; and the innate ability to dive, expressed in newly hatched chicks.

Therefore hybridization data can be correlated with morphological form, and distinguish the Gaviidae as a basic type candidate.

**14.2.5 Ardeinae (Herons And Egrets).** The members of this group are also ancient, with representatives dating back to the lower Eocene. The family Ardeidae has undergone many taxonomic revisions over the years, with placement within the order Ciconiiformes or a separate order, Ardeiformes.

Within Ardeinae 23 species hybridize with one another (52% of all herons and egrets). Hybridization also links 6 of the 10 genera: *Ardea* (e.g. Grey heron, Great egret), *Nycticorax* (night herons), *Egretta* (e.g. Little blue heron, Little egret), *Bubulcus* (Cattle egret), *Nyctanassa* (Yellow-crowned night-heron) and *Ardeola* (Indian pond-heron). There is no hybridization with any of the three other subfamilies (the Botaurinae (bitterns), Tigrosomatinae (Forest bittern and tiger herons), and *Cochlearius* (Boat-billed heron). No hybrids link them to neighbouring taxonomic outgroups.

The Ardeinae are long-legged wading birds with long, broad, wings, short tail and a long, slender neck with a highly distinctive S-shaped kink resulting from a peculiar articulation of the sixth cervical vertebra (Martinez-Vilalta and Motis 1992). This feature acts as a spring to provide added thrust when the bird lunges with its spear-like bill to harpoon prey. In contrast, within neighbouring subfamily Botaurinae, bitterns have a more squat body and behavioural peculiarities including their exceptionally low-pitched call, and, when threatened, frozen stance with skyward pose of bill to facilitate its cryptic features. Also in contrast, the boat-billed heron possesses a wide bill and associated skull modifications for support (Cameron and Harrison 1978) leading some authors to place it in a separate monotypic family. Hybridization data are lacking between Ardeinae and Tigrosomatinae, and therefore do not inform on their relationship. Dickinson (2003) recognized the subfamilies Tigrosomatinae, Cochleariinae, Botaurinae and Ardeinae, but Dickinson and Remsen (2013) merge all these subfamilies together. Payne and Risley (1976), by applying numerical taxonomy based on 33 skeletal characters, proposed four equidistant subfamilies: Ardeinae (day-herons); Nycticoracinae (night-herons and Boat-billed heron); Tigrosomatinae (tiger-herons) and Botaurinae (bitterns). However, night-herons are able to hybridize with day-herons, suggesting their close affinity, and that the night-herons should remain within the Ardeinae. Thus there is a clear correlation between hybridization data and morphology within the Ardeinae, together giving credence to its prospective basic type status.

**14.2.6 Psittacidae And Cacatuidae (Parrot-Like Birds).** Up to 166 out of the 364 species (46 %) of all members of the parrot-like birds that exist throughout the order Psittaciformes are linked by interspecific hybridization (summarized in McCarthy 2006). This is a large number of reports, although these await systematic further appraisal for validity. No hybrids are known to link them to any proposed nearest relatives. Ninety species hybridize between genera, with intergeneric hybridization linking up to 42 of the 92 genera[3], depending upon the degree of acceptance of the hybridization records. Parrots and allies have been treated as one family, Psittacidae (Dickinson 2003), yet historically up to eight families have been perceived (reviewed by Collar 1997). Dickinson and Remsen (2013) and Clements et al. (2023) recognize four families: Cacatuidae, Psittacidae, Psittaculidae and Strigopidae. However, hybridization links Cacatuidae and Psittaculidae together. Moreover, all three subfamilies within Cacatuidae are united by hybridization. This includes a reported hybrid between the Galah and Cockatiel (Marshall 2015), which unites the subfamilies Cacatuinae and Nymphicinae:

*Eolophus roseicappilus* (Galah) [Cacatuinae] x *Nymphicus hollandicus* (Cockatiel) [Nymphicinae].

Another intergeneric hybrid links Psittaculidae with Cacatuidae:

*Neophema chrysostoma* (Blue-winged parrot) [Psittaculidae] x *Nymphicus hollandicus* (Cockatiel) [Cacatuidae]

Yet another intergeneric hybrid (although from old data) links these families:

*Nymphicus hollandicus* (Cockatiel) [Cacatuidae] x *Psephotus haematonotus* (Red-rumped parrot) [Psittaculidae].

On the other hand Collar (1997), on the basis of considerable evidence recognized just two families, Psittacidae (in which Psittaculids are subsumed within Psittacidae) and Cacatuidae. However, the above crosses unite these two families. Further hybridization data are needed to confirm these links.

Figure 3. One hundred and sixty-five species of parrot-like birds hybridize, with intergeneric hybridization linking up to 42 genera together. Hybrids with non-parrot-like birds are unknown. The data correlates with numerous musculo-skeletal features unique to these birds.

---

3        The Howard and Moore checklist 4th edition (Dickinson and Remsen, 2013) recognises 92 genera. Cornell Lab of Ornithology recognises 99 genera (https://birdsoftheworld.org/, accessed 15.11.23).

Parrots and cockatoos are the sole members of the order Psittaciformes. The form is homogeneous within the order, but there appear to be no close relatives, and according to Collar (1997), parrots have exhibited their distinctive parrot form for a long history, with the earliest uncontested fossils to date being found in the lower Eocene. The parrot-like birds all have a unique cranial morphology comprising strong, decurved bill with a fleshy upper mandible; deep, scoop-shaped lower mandible; large blade-like vertical palatine bones, and suborbital process projecting posteriorly below the eye; and unique jaw muscle, the ethmomandibularis, for strong jaw action (Zusi 1993). They have a short neck, rounded wings and body form (Figure 3).

These morphological peculiarities correlate with the hybridization data, which together is suggestive of a basic type status, with one to several basic types, depending upon the taxonomic treatment.[4] Moreover, their shrill, sharp squalks and whistles, often announced in a cacophony, in group foraging journeys to the skies, sets them apart from vocalizations of proposed relatives within the orders comprising either the pigeons (Columbiformes) or cuckoos (Cuculiformes). A more detailed analysis of the parrot-like birds in relation to basic type status can be found in Landgren et al. 2011.

**14.2.7 Rheidae (Rheas).** The family Rheidae (rheas) comprise just two species, both of which hybridize with one another (Folch 1992b; Delsuc et al. 2007), but with no other bird groups. In the Delsuc study, molecular genotyping confirmed the hybrid origin of the chicks. The powerful flight muscles of birds require a strong keel-like attachment to the sternum in contrast to the flightless birds, which are often termed ratites (from the Latin ratis, [raft]) to denote the flat, raft-like keel). Some authors place rheas in their own order (Rheiformes), whilst others merge all ratites (including the ostrich, emu, cassowaries and kiwi) into the order Struthioniformes (Folch 1992b). Rheas have been placed variously between family and subfamily status. Hybrids are unknown between the ostrich (a taxon comprising a single species) and rheas, although it is not surprising that there is a lack of natural hybrids, since the rheas are exclusively Neotropical endemics and the ostrich is endemic to Africa. There are no known hybrids between any other members of the Struthioniformes. Fossil rheas date back to the Eocene.

In contrast to Struthio (ostrich), which has two hind toes, rheas have three, and the rhea head and neck are feathered, unlike the ostrich (Folch 1992a). In the rhea the postorbital process develops from the pleurosphenoid, in contrast to the ostrich, in which it develops from the frontal bone (Zusi 1993). Thus there is a correlation between the hybridization data and morphology of rheas. Morphological features

---

4        For instance, with the Collar (1997) treatment of an enlarged Psittacidae containing the Psittaculids, since hybridization unites the latter with the Cacatuidae, the parrot-like birds could be a single basic type.

shared with other members of the order include long, strong neck and legs, although in the kiwis the neck is shorter and the legs short and stout.

Thus the data to date are suggestive that the rheas comprise a basic type, but future data may shed light on whether or not this is a more inclusive group encompassing the ostrich and other members of the order.

**14.2.8 Podicipedidae (Grebes).** Eleven of the 22 species of grebes hybridize (50% of all grebes), with intergeneric hybrids linking two of the six genera (*Podiceps* and *Tachybaptus*). No hybrids link them to any proposed outgroups.

The sole members of the order Podicipediformes, grebes are taxonomically closely related to one another but not to any other group (Cameron and Harrison

Figure 4. Bird behaviour can also be characteristic of the basic type, as in the grebes, with their unique, synchronised courtship dance over the water.

1978; Llimona and del Hoyo 1992). They represent one of the oldest of all birds, with fossils found as far back as the Upper Cretaceous, commonly understood to be dated over 80 million years ago. Their subsequent adaptive radiation is considered to have centred on the Neotropical regions, where half of the 22 species are found today. Their diagnostic morphology includes a pointed bill, lobed toes, reduced tail and unique feather hooks of only one in every two to three barbules on the inner barb. Moreover their three front toes are unusual in being independently lobed, and with a membranous interconnection. They are highly specialized for diving, with the front of the body laterally compressed and pointed, to aid water penetration, and legs set far back for streamlining and propulsion.

Thus there is a good correlation between the hybridization data and the morphology of grebes, suggestive of a preliminary designation as a basic type. This is further substantiated by behaviour, unique to grebes, of courtship rituals, amongst the most spectacular of all birds (Llimona and del Hoyo 1992). These include the rushing ceremony, in which the pair rush powerfully over the water, with necks bent stiffly, followed by a synchronous dive. Another is the weed ceremony, in which the pair swim away from each other in a highly ceremonious manner, then submerge, and reappear, synchronously, carrying a bill-full of weeds, and rise towards each other, treading water and entering into a beautiful dance (Figure 4).

**14.2.9 Paradisaeidae (Birds Of Paradise).** No other family of birds, apart from the Anatidae, has exhibited such prolific interspecific and intergeneric hybridization in the wild, according to Frith and Beehler (1998). In 1901 the ornithologist Reichenow first proposed that a specimen he had previously described as a new species might in fact be a hybrid, but his ideas were strongly rejected by other ornithologists. It was not until Streseman in the 1930s confirmed Reichenow, making the then revolutionary deduction that 17 species were of hybrid origin. Although now universally accepted, the hybrid hypothesis continued to meet with incredulity, Iredale dismissing it as a "fantasy" until the 1950s.

Streseman's student, Mayr, was, however, convinced of their hybrid origin, reinforced by the presence of museum specimens.

Nineteen of the 41 birds of paradise hybridize (46 %) between species, which also link 7 of the 13 genera. Frith and Beehler (1998), in an extensive review of these hybrids, comment firstly that the "remarkable diversity of intergeneric hybrid birds of paradise emphasises genetic compatibility and thus the presumed close relationships within the subfamily Paradisaeinae." Secondly, they add that some of the hybrids exhibit characters of three genera, concluding that hybrids may be exhibiting ancestral

Figure 5. Bird of paradise wild hybrid (Long-tailed paradigalla x Superb bird of paradise) showing characters of three genera. Such hybrids may be exhibiting ancestral (atavistic) characters not apparent in the two parent species. This hybrid is known as Rothschild's lobe-billed bird of paradise.

(atavistic) characters not apparent in the two parent species. For instance, Rothschild's lobe-billed bird of paradise (Figure 5) is recognized as the wild hybrid between the parents *Astrapia carunculata* (Long-tailed paradigalla) x *Lophorina superba* (Superb bird of paradise). It exhibits a shape of crown, plus the iridescent green throat, face and mantle (upper back), which are all suggestive of the superb bird. The tail and gape are suggestive of the paradigallas. Other structures such as the bill are almost perfectly intermediate between the two putative parents. However, the hybrid's purple crown and breast-shield, with iridescent green beneath, purple-edged, fan-like outer pectoral feathers, deep iridescent purple wing feathers and deep plum coloured underparts are reminiscent of the Black sicklebill and Magnificent riflebird! (see http://www.biodiversitylibrary. org/item/53699#page/89/mode/1up for colour image).

Hybridization data also correlates with morphology within Paradisaeidae. Birds of paradise are superficially crow-like in form, with strong bills (varying in length, width and curvature) and highly ornate and strongly iridescent plumes on the flank and tail and when also on the head, intricate feathers are inserted into appropriate skull modifications. The plumage is often highly modified for nuptial display, especially in the central retrices. In the Ribbon-tailed astrapia, these can be over a metre in length. Certain plumes may be wire like, as in the erective occipital plumes of the Six-wired bird of paradise (*Parotia carolae*), and in the flank plumes of the Twelve-wired bird of paradise (*Seleucidis melanoleuca*).

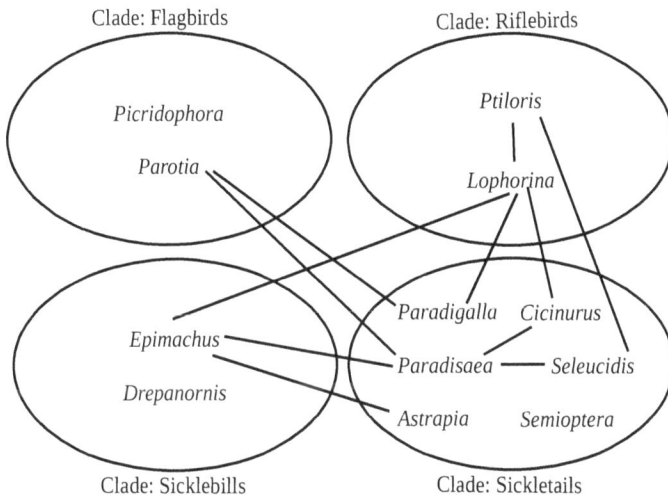

Clade: Flagbirds    Clade: Riflebirds

*Picridophora*    *Ptiloris*

*Parotia*    *Lophorina*

*Paradigalla*    *Cicinurus*

*Epimachus*    *Paradisaea* — *Seleucidis*

*Drepanornis*

*Astrapia*    *Semioptera*

Clade: Sicklebills    Clade: Sickletails

Figure 6. Genera within the four clades of the plumed birds of paradise. Typical plumed birds of paradise are included with Sickletails. The four clades, each with widely divergent morphology, are all linked by hybridization (straight lines).

Although some authors place them together with the satin-birds on the basis of certain shared osteological characters, the satin birds also possess traits such as weak feet and bill, and very wide gape, not found in the true birds of paradise, so other authors (such as Dickinson and Christidis 2013) place satin birds in a separate family (Cnemophilidae), which is also confirmed by biomolecular studies which ally them closer to the bowerbirds, which in turn are no longer thought to be allied to the birds of paradise. The birds of paradise can be sub-divided into two lineages: manucodes and the paradise-crow (six species in all); and the remaining 35 species, the plumed paradisaeine birds. The true, plumed paradisaeine birds form four morphologically distinct clades: the flagbirds, the riflebirds, the sicklebills and the sickletails/typically plumed birds, all of which seem to have resulted from a single radiation (Frith and Frith 2009a). All of these clades are linked by hybridization (Figure 6). As to whether the manucodes and paradise-crow should be included is so far not informed by hybridization data. No hybrids are known with nearest proposed outgroups, within which hybridization too is common, such as within the Corvidae (crows, ravens, magpies and jays), in which 28 % of the species hybridize with one another, and in the Ptilorhynchidae (bowerbirds) (42 %). Thus the extensive hybridization (at least within the plumed birds of paradise), along with the clear lack of hybridization with any proposed relatives, and their distinctive morphology together indicate their prospective basic type status.

**14.2.10 Haematopodidae (Oystercatchers).** Six out of the 12 species (50 %) hybridize, with four of the species interconnected by hybridization in this single-genus family. McCarthy (2006) indicates that most members of this family are known to hybridize where their ranges overlap.

Medium-sized shorebirds (40–53 cm long), oystercatchers have either black or pied plumage, pinkish legs and a longish, laterally compressed dagger-like bill which is orange-red during the breeding season (Hockey 1996). They have a worldwide, mainly coastal, distribution, but with a greater number of species in the Southern Hemisphere, which has led to the suggestion of their radiation from that region. However, as to whether the black form or the pied form represents the prototypical oyster is a matter of debate. The oldest known fossil, from the lower Pliocene, has affinities with the Magellanic oystercatcher, which is pied. However, both pied and black forms tend to co-exist sympatrically throughout numerous regions worldwide. Some individuals of the variable oystercatcher are intermediate in form ("smudgies") between these two extremes, which may indicate a polygenic basis for these characters.

The Ibisbill (*Ibidorhyncha struthersii*) has also been included in this family according to Dickinson and Remsen (2103). However, Dickinson (2003) and Clements et al. (2023) place the Ibisbill in Ibidorhynchidae, as did Hockey (1996), on the basis of distinctive morphology and unique breeding habits. The Ibisbill is also not known to hybridise with oystercatchers to date. Since the position of the ibisbill is a matter of debate, this chapter follows its separation from Haematopodidae.

Oystercatchers are morphologically clearly distinguishable from other families within the order Charadriiformes (Hockey 1996), such as the gulls (Laridae), and between which there are no known hybrids.

This provides initial evidence to ascribe putative basic type status to the Haematopodidae.

**14.2.11 Spheniscidae (Penguins).** Eleven species of penguins hybridize out of a total of 16 to 18 species[5], although hybrids do not link the five genera. There are reports of hybridization between several species in wild populations. For instance, positive evidence of hybridization between *Spheniscus magellanicus* (Magellanic penguin) and *Spheniscus humboldti* (Humboldt penguin) in the wild was provided by genetic markers, in which Humboldt mitochondrial DNA and Magellanic species-specific alleles were detected in hybrids (Hibbets *et al.* 2020). Hybrids of the cross *Eudyptes chrysocome chrysocome x E. chrysolophus* (Rockhopper x Macaroni penguins) were reported on the Falkland Islands (White and Clausen 2002). The pink bill margins were indicative of Rockhopper parentage, but the golden-yellow crown feathers are features of Macaroni penguins. The Rockhopper is also reported to hybridize with the Royal penguin (Simpson 1985) and the Erect-crested penguin (Napier 1968). There are also reports of other penguin species hybridizing readily in captivity (citations in McCarthy 2006).

The sole taxon within the order Sphenisciformes, penguins hybridize with members of no other taxa, including proposed nearest allies the divers (Gaviidae), petrels (Procellaridae) or frigate birds (Fregatidae), although no fossil data informs such proposals. It is the only order in which member species are both flightless and aquatic. The earliest fossils date from the late Eocene. Fossils already bear the short and thick fused tarsus and metatarsus peculiar to penguins.

The hybridization data distinctive to the penguins also correlates with their distinctive morphology: thick-set bodies with highly streamlined shape for swift aquatic locomotion, and wings modified to form flat scimitar-shaped paddles for high acceleration propulsion and manoeuvres (Cameron and Harrison 1978; Shen *et al.* 2020), in contrast to other aquatic birds, in which aquatic locomotion stems from webbed feet. Their feathers are highly specialized for efficient heat insulation, water-proofing and wind-proofing, particularly for environments of extreme cold (Martinez 1992). Most other birds have alternating feathered and bare tracts, but in penguins there are uniform, stiff, small coverts, with a hyporachis to increase feather length to compensate for the absence of down.

The penguins thus provide an example of a taxon which is a candidate for basic type status with evidence from highly characteristic morphology, with additional evidence from interspecific hybridization (also see Figure 12 in Section 4 Discussion).

---

5       Dickinson and Remsen (2013) recognise 16 species. Clements *et al.* (2023) recognise 18 species.

Figure 7. Ibis (left) and spoonbill (right). Two species of ibis can hybridize with two species of spoonbill (see text). This is in spite of their morphological differences, indicating their morphogenetic closeness.

**14.2.12 Threskionithidae (Spoonbills And Ibises).** Based on easily discernable external features, the Threskiornithidae is divided into two subfamilies (Matheu and del Hoyo 1992), Threskiornithinae (ibises) and Plataleinae (spoonbills). Twelve out of 34 species[6] (35 % of all spoonbills and ibises) hybridize, and the hybrids can be fertile. Four of the 13 genera are linked by intergeneric hybridization, although these are either brief or based on several old reports[7]:

1. *Threskiornis aethiopicus* (Sacred ibis) x *Platalea alba* (African spoonbill)

2. *Threskiornis melanocephala* (Black-headed ibis) x *Platalea minor* (Black-faced spoonbill)

These link together the spoonbills and ibises, providing modest evidence that spoonbills and ibises share a common putative basic type (Figure 7).

Fossil representatives first appeared in the Eocene, and radiation appeared early, with fossil deposits in the Pleistocene virtually indistinguishable from numerous modern-day counterparts in several genera. Ibises are distinguished by a long, slender and decurved bill, whilst spoonbills have a long, straight, flattened bill with a broader distal end. Both groups are medium-sized to large birds (50–110 cm in length), with an elongated body, and quite long neck and legs. They are related to the storks (Ciconiidae), but only superficially (Matheu and del Hoyo 1992), and indeed the storks constitute one of the most distinctive of families (Elliott 1992c). Neither do members of Threskionithidae <u>hybridize with</u> members of Ciconiidae, although hybridization within

6        Clements *et al.* (2023) recognise 36 species.
7        *Threskiornis aethiopicus* x *Platalea alba*: IZY 1991; *Threskiornis melanocephala* x *Platalea minor*: 6 old reports (cited in Gray 1958).

the respective families is common. Thus there is a correlation between interspecific hybridization data and morphology, providing evidence that spoonbills and ibises are members of a common putative basic type.

## 14.3. Other Potential Candidates For Basic Type Status

The following taxa (again primarily families) contain core groups (united by hybridization and/or from other data) which also merit further consideration for possible basic type status.

### 14.3.1. Trochilinae (Typical Hummingbirds).

Hummingbirds are a morphologically homogeneous group, with their adaptations for nectar-drinking: a specialized wing shape for hovering over the flower, and typically a long, fine bill, often with tube-like tongue to suck the nectar. The hovering flight habit is enabled by numerous musculo-skeletal adaptations, including a deep, elongated keel; supernumerary ribs for flight stability; a strengthened coracoid-sternum joint; modified wing skeleton to enable optimal rotation; a highly developed *Musculus pectoralis major* and *Musculus supracoracoideus* flight muscles (reviewed by Schuchmann 1999). The cardiovascular and respiratory system are adapted to the exceptionally high oxygen demands of flight, with a heart rate ranging from 500 to 1000 beats per minute, and a breathing rate from 300 to 500 times per minute. They are small to tiny, sometimes with tail-streamers, iridescent plumage, and a mainly neotropical distribution. Two subfamilies are recognized: the majority of species (90 %) are within Trochilinae (typical hummingbirds: 301 species) and the remainder in Phaethornithinae (hermits). Morphology, behaviour and nest shape criteria enable distinct trochiline clades to be distinguished, such as the toothbills, coquettes, woodnymphs and sapphires, true emeralds, amaziline emeralds, mountain gems, brilliants, starthroats and woodstars, and an Andean clade (e.g. sunbeams, sylphs, trainbearers and velvetbreasts) (reviewed by Schuchmann 1999).

Of the 304 species recognized within Trochilinae, 100 of them hybridize (33 % of all typical hummingbirds). Interspecific hybridization is found in 38 genera linking 31 of the 105 genera together. Interspecific hybrids are represented in both subfamilies, although the majority (including the intergeneric hybrids) are confined to the Trochilinae, and to date the two subfamilies are not linked via hybridization. Within the Trochilinae, most of the 11 clades recognized by Schuchmann are linked via hybridization into two groups (Figure 8). Intergeneric hybridization also links six of the eight principal trochiline clades recognized by McGuire et al. (2009), namely topazes, brilliants, coquettes, mountain gems, bees, and emeralds.

Thus interconnectedness of species and genera is good within the typical hummingbirds. Moreover, the various intergeneric hybrids, particularly those between the clades, link birds with widely divergent

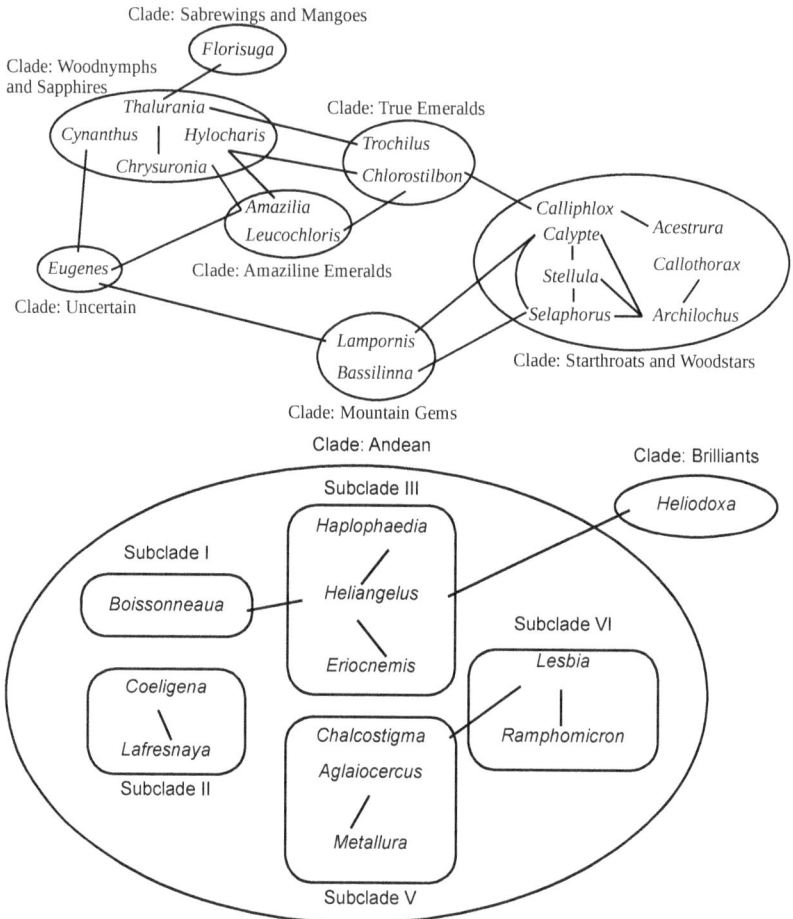

Figure 8. Hybridizing genera within and between Trochilinae (hummingbird) clades. Most of the eleven clades are linked via hybridization. The intergeneric hybrids link birds with widely divergent morphology. Only hybridizing genera are indicated.

morphology within the basic hummingbird form. Schuchmann (1999) also noted evidence for atavisms in certain enigmatic hummingbirds. Thus the typical hummingbirds can be ascribed prospective BT status. This would not include the hermits, given the lack of hybridization with the trochiline hummingbirds (whilst within the latter it is common), and modest differences in morphology between these two sub-families. However, both groups still share a recognizably "hummingbird" form, including the distinctive bill adapted to nectivory, suggestive of a BT inclusive of hermits. Thus further data are needed to inform more definitively upon the BT placement of the hermits.

There is no record of hybridization between members of Trochilinae and other proposed relatives, such as Apodidae and Hemiprocnidae.

**14.3.2. Bucerotidae (Hornbills).** The hornbills are "one of the most recognizable groups of all birds", according to Kemp (2001). Traditionally placed within the Coraciiformes, the hornbills are increasingly viewed as so divergent from other families that some authors ascribe them their own order, Bucerotiformes. However, wherever they are placed, there is no doubt as to what constitutes a hornbill, with their deep, curved great bill surmounted by a unique hollow casque, typically reinforced internally with bony ridges and rods, variously employed as a resonating chamber, chiselling tool, and head-butting weapon in contests between rival males. This large bill and associated musculature has led to fusion of the first two neck vertebrae and articulation with a supraoccipital and basioccipital condyle, unique amongst birds. They are medium to large birds with rounded wings and a long tail.

Eight species hybridize with one another (15 % of all hornbills), but at least 16 (26 %) if species merged primarily due to hybridization are re-split. The hybrids are found in three of the 15 genera. For instance, hybridization in the wild between *Buceros bicornis* (Great hornbill) and *Buceros rhinoceros* (Rhinoceros hornbill) was confirmed by genetic analysis (Chamutpong *et al.* 2013). Phenotypic characteristics of the chicks mostly resembled the Great hornbill with a few features shared between both hornbills. There are no intergeneric hybrids to date. Connectedness by hybridization among the hornbills is modest, although in captivity hybridization between species is known to be quite common (McCarthy 2006, p. 115).

**14.3.3. Cracticinae (Butcherbirds).** The butcherbirds are so-named by their habit of storing pieces of prey on a spike. They share skull features with the woodswallows (Artaminae), including the presence of two zygomatic processes instead of the one found in other corvoid lineages (Russell and Rowley 2009). They have a stout bill which is hooked at the tip. They are ground and arboreal foragers, with longish wings and a long tail. They produce amongst the most melodious songs to be heard in the Australasian bush and forests, including the flute-like yodelling called Carolling, practised singly, in duets or even Carolling groups.

From three to four genera are recognized in this family/subfamily (depending on the perception of the author) of medium to large-sized passerines. Three out of 11 species hybridize (27 % of all butcherbirds). If species merged primarily due to hybridization are re-split, at least nine species hybridize (53 % of all butcherbirds). Two of the four genera (*Cracticus* and *Gymnorhina*) are linked by hybrids, although this is based on only a single record (Donato and Potts 2004). However, the record describes the hybrid morphology in detail. From three to five species are linked together by hybridization, depending on the treatment of species. Initially they were allied with known Old-world families, namely the shrikes (Lanidae) and crows (Corvidae). Molecular studies have subsequently led authors to ally them variously with bush-shrikes, helmet-shrikes, ioras and vangas, but most closely as a subfamily in the Artamidae (*i.e.* with the woodswallows (reviewed by Russell and Rowley 2009), from which they can be clearly distinguished morphologically (see below).

**14.3.4. Artaminae (Woodswallows).** Woodswallows are fast-flying, aerial hunters of insects, rivalling the true swallows in their aerobatic gliding prowess (Rowley and Russell 2009). They are small, (12–21 cm), with long, pointed wings and short tail. Their brush-like tongue also enables nectar-feeding. Dickinson (2003) placed the Woodswallows as a separate family, as do Rowley and Russell (2009). The butcherbirds were placed separately, in Cracticidae, based on their distinctly different morphology. Dickinson and Christidis (2013) and Clements *et al.* (2023) merge woodswallows and the butcherbirds into a wider Artamidae.

Four of the 11 species hybridize (36 % of all woodswallows) in this single genus taxon. Interconnectedness of species via hybridization is low, but the hybridization is between species distinct from one another, as opposed to subspecies of each other. There are no hybrids with their closest proposed allies, the butcherbirds.

**14.3.5. Ptilonorhynchidae (Bowerbirds).** In the bowerbirds, their unique behaviour particularly defines them. They build a court or "bower" used exclusively for courtship and mating, with constructions by some species almost beyond belief (reviewed by Frith and Frith 2009b). The maypole bower is a tower of sticks, sometimes also roofed or hut-like. The avenue builders construct walls of sticks or stems, sometimes with side-avenues at right angles to the main avenue. Bowers are decorated with an array of objects. Yet other species also "paint" their walls with a charcoal and vegetable matter. Over 5000 such objects have been found within a bower, with some females showing a preference for items rare in nature.

Figure 9. The Black-lored tit (*Parus xanthogenys*), defies classification, possessing features which together are atypical of the various tit genera. For instance, a belly stripe and nuchal spot are typical of the genus Parus, but adult yellow cheeks are atypical of Parus but typical of Periparus. However, this is well explained in terms of a polyvalent ancestral gene pool within a basic type, which has then diversified, producing a mosaic distribution of characters in the descendents.

They are medium-sized (22–37 cm) birds, mostly with stout and powerful bills and the males with diverse and striking plumage between the various species, with a greater number of secondary feathers (11 to 14) compared with most other passerines (which have only nine or ten).

Originally they were considered to be most closely related to the birds of paradise, and even placed with them, but several molecular studies confirm their distinct status as a family, and distant from the birds of paradise. Eight out of the 19 species hybridize in the wild (42 % of all bowerbirds), emphasizing the genetic closeness between the species, according to Frith and Frith (2009a). Two of the eight genera (*Sericulus* and *Ptilonorhynchus*) are linked by hybridization, via the cross Regent x Satin bowerbirds (Blunt and Frith 2005; Frith 2016), described in detail. In general the connectivity via interspecific hybridization is modest.

**14.3.6. Ciconiidae (Storks).** Of the three stork tribes recognised (Clancy 2009), Storks are notable in that hybridization unites two of these tribes, via linking members of genera (*Leptoptilos* and *Mycteria*) respectively from each of these tribes (Forest 2006: 3 reports). This is of interest because each of these tribes has morphological peculiarities. Members of *Mycteria* have specialized bills which are long, tapered, slightly decurved and with sensitive regions near the tip to enable fishing in difficult conditions. Members of *Leptoptilos* have heavy, conical bills and an enormous wingspan (up to 152 cm). Such intertribal distinguishing features are clearly not a barrier to hybridization. Overall seven out of the 19 species of storks hybridize (37 % of all storks) between the species. However, there are no known hybrids with any neighbouring taxonomic groups.

Storks are also one of the most distinctive of families (Elliot 1992c), with their medium to very large size, long legs, necks and often bills. They were already so-distinguished from the early Tertiary, with the earliest storks emerging from the Upper Eocene, and mainly radiating during the Oligocene.

**14.3.7. Laniidae (Shrikes).** Shrikes are slender-bodied predatory birds with a slightly hooked bill and tomial tooth (similar to that of falcons), rounded wings and longish tail (Josef 2008). Thirty to 32[8] out of the 34 species of shrikes are found within the genus *Lanius*. Within this genus, 14 species hybridize with each other (43 %), and hybridization interconnects 11 of these species. Hybridization with neighbouring taxa is unknown.

**14.3.8. Paridae (Tits And Chickadees).** Twenty-two out of 59 species (37 % of all tits) hybridize. If merging due to hybridization is taken into account, *i.e.* if the salient species are split, 29 out of 63 species (44 %) hybridize. Seven of the 14 genera are linked together via hybridization.

---

8        Dickinson and Remsen (2013) recognise 4 genera. Clements *et al.*, 2023 place 2 of these (Corvinella and Urolestes) within Lanius. In the latter case, Lanius thus comprises 32 of the 34 species within Laniidae.

Separate family status has been recognized to distinguish the true titmice (e.g. *Parus*) [Paridae], the long-tailed tits [Aegithalidae], and the penduline tits [Remizidae]. The DNA-DNA hybridization studies summarized by Sibley and Ahlquist (1990) concluded that Paridae and Aegithalidae could be clearly distinguished. In a review of subsequent genetic studies (Jønsson and Fjeldså 2006), the Aegithalidae were placed along with the Paridae within the super-family Syvioidea, although each family was ascribed to separate clades not deemed to be closely related.

Thus an account of the history of the systematics reflects equivocal treatment of Aegithalidae, particularly in relation to Paridae Two old records report a cross between *Parus* and *Aegithalos* (cited by McCarthy 2006), which would unite Aegithaidae and Paridae, but no more recent reports have verified this remote cross. Hybrids with other taxa are unknown.

The tits are small to medium-sized (9–21 cm) passerines with shortish, typically black bill, strong legs and short to medium-length tail, and with great variety of plumage colour (Gosler and Clement 2007). The intrafamilial classification has been controversial and confusing for over a hundred years. This is exemplified (Figure 9[9]) by an oriental species, *Parus xanthogenys* (the Black-lored tit), which possesses a belly stripe and nuchal spot typical of the genus *Parus*, but retains yellow cheeks into adulthood (atypical of *Parus* but typical of the genus *Periparus*). Moreover, it exhibits an eyestripe typical of *Cyanistes*, a black cap typical of *Poecile*, and a crest typical of *Lophophanes*. The subsequent consensus was to lump most the species into a greatly enlarged *Parus*. In 2005 mitochondrial DNA studies led to the division of Paridae again into 9 genera, with sister groups including the Remizidae (penduline tits) (which had often been submerged into an expanded Paridae) and the long-tailed tits (Aegithalidae). Dickinson (2003) recognized just three genera. However, Dickinson and Christidis (2013) have again contracted *Parus*, re-assigning various members of *Parus* to other genera, and expanding the number of genera to 14.

As Gosler and Clement (2007, p.673) comment, "the problem of traditional parid systematics, then, is that plumage characters that seem almost to define intrafamilial taxa do not quite do so. The wingbars, eyestripes, nuchal spots and white outer tail feathers crop up in odd species here and there, irrespective of generic affiliation. This undoubtedly reflects the great age of the family and, therefore, the fact that many of these species represent relicts from older taxa." Indeed, this can easily be understood in terms of a polyvalent ancestral gene pool within a basic type, which has undergone diversification, leading to a mosaic distribution of characters throughout the descendents.

**14.3.9. Viduidae (Indigobirds/Whydahs).** These are small Afrotropical passerines with a short, stubby bill. They are brood parasites,

9          © Shailendra Patil, reproduced with permission..

laying their eggs in nests of other species. The tenth primary feather is not apparent, being very small. Breeding males have extensive black plumage. Some have highly elongated central tail retrices. The family contains two genera, the single species *Anomalospiza*, (Cuckoo finch), and *Vidua* (whydahs and indigobirds), with a number of unique skeletal features distinguishing them from other old world finches (reviewed by Payne 2010).

Within *Vidua* 12 out of 20 species hybridize (60 % of all indigobirds and whydahs), producing 19 various interspecific hybrids. These species are also all linked together via hybridization. The Viduidae are considered to have diverged from the Estrildidae, with evidence of their sister clade relationship (Olsson and Alström 2020), suggesting some affinity with the wider finch-like assemblage.

**14.3.10. Larinae (Gulls).** Hybridization between the herring gull and relatives has been a major challenge to the biological species concept (Burger and Gochfeld 1996). For instance, certain species do not hybridize within France (although co-existing there), but do so elsewhere in Europe. Gulls are uniformly shaped, small to large (25–79 cm) heavily built seabirds with stout bills, long wings, webbed feet and rounded tail (Burger and Gochfeld 1996). The gulls, terns and allies are often considered to form distinctly separate families: Laridae (gulls), Sternidae (terns), Stercorariidae (skuas) and Rhynchopidae (skimmers). Alternatively, other authors combine some or all of them as subfamilies within an expanded Laridae. Dickinson and Remsen (2013) for instance treat them as subfamilies within Laridae, yet recognize separate family status for the skuas. Whatever the treatment, the various authors are agreed that there are four distinct groups, with the differing treatments reflecting the perceptions of the author. Moreover, the morphological differences are, according to Burger and Gochfeld (1996) sufficient to warrant their separate family status.

Within Larinae (with composition as recognized by Dickinson and Remsen [2013]), 27 out of 52 species (52 %) hybridize, interconnecting one another and 4 out of 11 genera. For instance, hybridization between *Ichthyaetus melanocephalus* (Mediterranean gull) and *Croicocephalus ridibundus* (Black-headed gull) links the 2 genera. The hybrids show features such as wing plumage patterns intermediate between the parent species, as indicated in detailed descriptions and photographic records of birds in the hand (Zieliński *et al.* 2019). Multiple records are documented for many of the crosses. For instance, there are 39 published reports of hybridization between *Larus argentatus* (Herring gull) and *Larus hyperboreus* (Glaucous gull) (McCarthy 2006). At a Belgian gull colony, Lesser black-backed x Herring gull and Yellow-legged gull x Herring gull hybrids were described, recognised by careful assessment of mantle colour and width of wing white tertial and trailing edges (Adriaens *et al.* 2012). From analysis of 33 colonies along the western seaboard of the USA, a population of *Larus glaucescens* (Glaucous-winged gull) x *L. occidentalis* (Western gull) hybrids were identified by mantle and primary tip plumage melanism, colour of bill, eye-ring and iris pigmentation (Bell 1996). These hybrids formed an expanding population zone, which could provide a source of rapid speciation (Abbott *et al.* 2013). There are no known hybrids between gulls and any other proposed relatives.

**14.3.11. Sterninae (Terns).** Terns are small to medium (20 – 56 cm) sized birds with homogeneous features including slender, elongated bodies, and typically a deeply forked tail; although allied to the gulls, they have a longer bill, plus more long, narrow wings suited to their agile plunge-diving foraging (Gochfeld and Burger 1996). Most terns are migratory, with the arctic tern having the longest migration of any bird, ranging from the arctic to Antarctic regions, and other species moving around the globe in relation to the trade winds. Seventeen species (45 %) of terns hybridize, with 12 of these species inter-connected by the hybridization. For instance, *Sterna paradisaea* (Arctic tern) x *S. hirundo* (Common tern) hybrids studied at Massachusetts, USA demonstrated intermediate characteristics as well as features similar to the parents. Hybrid status was confirmed by molecular analyses. The hybrids produced fertile offspring, with reproductive performance comparable to that of common terns and higher than that of Arctic terns at the site (Mostello *et al.* 2016).

**14.3.12. Recurvirostridae (Avocets And Stilts).** Avocets and stilts are slim waders with long legs, long and upcurved or medium-length and straight bills, and generally pied plumage (Pierce 1996).

From three to six of the 9 species of avocets and stilts hybridize, depending upon the degree of merging accepted. Hybrids with other taxa are unknown. Dickinson and Remsen (2013) recognize two subfamilies (formerly one). These two subfamilies (*Recurvirostrinae* (avocets) and *Himantopodinae* (stilts) and two of the three genera within the family are united by hybridization:

For instance, *Himantopus mexicanus* (Black-necked stilt) x *Recurvirostra americana* (American avocet) hybrids demonstrated a combination of intermediate characters, such as a relatively straight and shorter bill than the distinctively upturned bill of the avocet (Principe 1977; Morlan *et al.* 2004).

**14.3.13. Parulidae (New World Warblers).** These birds are generally small to very small New World nine-primaried oscines; usually with a quite slender bill suited to foraging for insects; and short to quite long tail, which is sometimes cocked upwards (reviewed by Curson 2010). The systematics has undergone much revision (reviewed by Curson 2010). They were long-treated as a separate family together with the Olive warbler (as, for instance, in the Peters checklist, according to Lowery and Monroe 1968). Ridgely and Tudor (1989) removed the conebills and the Bananaquit (*Coereba*) from the New World warblers, placing them respectively with the tanagers and in their own family (Coerebidae). Also in the 1980s the parulids were subsumed within the Emberizidae along with the tanagers (Thraupidae) and New World blackbirds (Icteridae). The Sibley-Ahlquist DNA–DNA hybridization studies led to the merging of this expanded Emberizidae with the Fringillidae. However, this too has been superseded: the parulids have again been elevated to separate family status as the Parulidae; *Coereba* has been aligned with the tanagers; and the olive warbler too is separated into its own family, the

Figure 10. A barbet (left) cannot be mistaken for a toucan (right), although some authors place them together in one family according to molecular data. Hybridization occurs among toucans, and among barbets, but is unknown between barbets and toucans, which, along with their distinguishable forms, are suggestive of separate basic type status.

Peucedramidae. Yet another study has led to further revisions, including *Parula* (the type genus *i.e.* the typical representative genus of the family), being dismantled and all its members placed within *Setophaga* (Lovette et al. (2010).

However, some species and genera have continued to be difficult to place. For instance, although parulids in general are morphologically distinct from the tanagers, this is less clear in the tropics (Curson 2010). Skeletal studies suggest *Xenoligea montana* (White-winged warbler) and *Microligea palustris* (Green-tailed warbler) (parulids placed in *Genera Incertae Sedis* in Dickinson 2003) are closer to the tanagers than to the New World warblers: these have now been relocated into Phaenicophilidae (Warbler tanagers) according to Dickinson and Christidis (2013) and Clements et al. (2023). Other species (treated as parulids in Dickinson 2003) include *Rhodinocichla rosea* (Rosy thrush-tanager), now placed in Rhodinocichlidae, and *Icteria virens* (Yellow-breasted chat) now in Icteridae (Clements et al. (2023). None of these variable treatments affect the hybridization data below. However, in view of their being formerly placed with the emberizid finches, tanagers and icterids, at least on morphological grounds, there remains the possibility that they are part of the finch assemblage.

Forty-five species hybridize (42 % of all New World warblers). If species merged due to hybridization are re-split, up to 55 species (47 %) hybridize, producing 60 various hybrids. At least 39 species and nine of the 18 genera are linked together by hybridization. For instance, inter-generic hybrids of the cross between *Mniotilta varia* (Black-and-white warbler) and *Setophaga coronata* (Yellow-rumped warbler) showed a number of characters and song intermediate between the parental species, and the hybrid status was confirmed by DNA analysis (Vallender *et al.* 2009). Another study described a three-species hybridization event, in which genetic, morphometric and bioacoustic data confirmed the hybrid status of a cross between Setophaga pensylvanica (Chestnut-sided warbler) and a Vermivora species, which itself was a hybrid between *Vermivora chrysoptera* (Golden-winged warbler) and *Vermivora cyanoptera* (Blue-winged warbler (Toews *et al.* 2018).The taxonomy of Golden-winged and Blue-winged warblers has been debated for over a century (Confer *et al.* 2020). Initially the cross was considered a new species, Brewster's warbler (*Vermivora leucobronchialis*), but was then found to be a hybrid between the Blue-winged and Golden-winged warblers (https://avianhybrids.wordpress.com/2021/02/24/genomic-study-unveils-the-true-identity-of-brewsters-and-lawrences-warbler).

**14.3.14. Ramphastidae (Toucans).** The toucans are a morphologically uniform Neotropical family, all sharing a very large, laterally compressed bill; numerous unique skull characteristics (Höfling 1998); distinctive short, rounded wings; and fused rear three vertebrae attached via a ball-and-socket joint (reviewed by Short and Horne 2002b). Toucans have been placed together in one family along with either all barbets or the Neotropical barbets, particularly on the basis of DNA-DNA hybridization studies. However, their current treatment as a separate family is recognized on morphological grounds: according to Short and Horne (2002a), a barbet could not be mistaken for a toucan by anyone. In spite of this, Dickinson and Christidis (2013) merge the toucans and barbets, having formerly treated them separately (Dickinson 2003). However, Clements *et al.* (2023) also treat them as separate families.

Twelve of the 34 species hybridize (35 % of all aracaris, toucans and toucanets). If species merged primarily due to hybridization are re-split, at least 26 species hybridize (51 %). Within the genera there is moderate interconnectivity via interspecific hybridization but more so if species merged due to hybridization are re-split (7 to 9 species within *Pteroglossus*; 3 to 6 in *Ramphastos*), but only within these two genera. No hybrids are recorded between Toucans and other members of the order piciformes.

**14.3.15. Picidae (Woodpeckers And Allies).** The family comprises three sub-families: Jynginae (wrynecks), Picumninae (piculets) and

Picinae (true woodpeckers). Woodpeckers are small to largish birds with a straight bill, reinforced skull, long, barbed tongue, feet modified for climbing and a stiffened tail (Winkler and Christie 2002). All of these features enable the habit of excavating wood, boring and probing crevices, and climbing vertical surfaces. Added to this, specialized neck muscles and insertions into modified vertebrae enable the hammering action.

Hybridization is reported in 74 out of 233 species (32 %) within the family, with up to 88 species (38 %) hybridizing if merged ones are re-split (McCarthy 2006). Interspecific hybrids are represented within 15 of the 33 genera. For instance, a hybrid specimen from the cross *Campephilus leucopogon x C. melanoleucus* (Cream-backed woodpecker x Crimson-crested woodpecker) were analysed at Pilar University, Paraguay. The authors reported a surprising mixture of characteristics clearly attributable to both parent species rather than a series of largely intermediate characteristics. The wing upper surface and body were reminiscent of *C. leucopogon* whilst the wing undersurface recalled *C. melanoleucus*. The authors found it remarkable that the specimen displayed a capucine buff barring body, only normally only found in a different species, *C. pollens*, providing another example of a prospective ancestral (atavistic) character hidden in both parents becoming expressed in the hybrid (Chialchia and Smith 2014). Connectivity between species via hybridization within the genera is moderate. Up to eight species are linked in and between *Colaptes* and *Melanerpes*; six in *Dendrocopus*; five in *Picoides*; seven in *Picumnus* and four in *Sphyrapicus*. Two of the three subfamilies (Picumninae and Picinae) are represented by interspecific hybrids, but to date there is no hybridization between members of the respective subfamilies. Hybrids between woodpeckers and other proposed outgroups are unknown.

**14.3.16. Pipridae (Manakins).** Fourteen out of 53 species are known to hybridize (26 % of all manakins). If species merged due to hybridization are split, there are at least 17 species (32 %). From one to two subfamilies are recognized. Six intergeneric hybrids link six of the 13 genera. For instance, the hybrid status of offspring from the cross *Ilicura militaris* (Pin-tailed manakin) x *Chiroxiphia caudata* (Swallow- tailed manakin) was confirmed by multiple character systems including DNA sequence data (Marini and Hackett 2002. A core group of 11 species are all linked together via interspecific hybridization across 3 genera, which also unite the tribes Piprini (via *Heterocercus* and *Manacus*) and Ilicurini within the sub-family Piprinae. These small neo-tropical forest-dwelling brightly-coloured birds have short, broad-gaped and slightly hooked bills. The tail is short but a few species are wire-tailed. In some species the feathers are modified to produce loud mechanical snapping sounds involved in courtship. Manakins are highly agile. When perching they do not hop but make distinctive rapid sideways slides ("moonwalking"), and aerobatic courtship flights. In *Chiroxiphia caudata* (Blue manakin) three or more males may participate in an aerial "cartwheel" dance, one of the most amazing spectacles in the natural world (Snow 2004). Hybrids with any other proposed allies are unknown.

**14.3.17. Hirundinidae (Swallows & Martins).** Twenty-two species of the family (26 % of all swallows and martins) hybridize, producing 18 various interspecific hybrids. Ten species link six of the 20 genera via hybridization. Hybridization between *Hirundo rustica* (Barn swallow) and *Delichon urbicum* (House martin) are well known, with dozens of reports. For instance, in a swallow ringing project in Finland in 1999-2000, one in 489 barn swallows was found to be a hybrid with the house martin (Saurola 2001). A captured hybrid of a Barn swallow (*Hirundo rustica*) x Sand martin (*Riparia riparia*) cross was described in detail (Dunning *et al.* 2014).This distinctive family are morphologically similar, and superbly adapted to the aerial foraging of insects (Turner 2004). They have long, pointed wings, streamlined body, and often a forked tail, all of which aid aerial manoeuvres and gliding at low flight cost relative to other comparable passerines. No hybrids link them with any other taxa.

**14.3.18. Columbidae (Pigeons And Doves).** Pigeons and doves are cosmopolitan birds ranging from 15 to 75 cm in size, with a small head, short bill and legs, compact, stocky body, a large amount of flight muscle (up to 44 % of total body mass) compared with other birds, and the young are fed with a nutritious crop-milk. They are easily distinguished from other avian taxa, with no agreement as to nearest relatives, and DNA hybridization data suggesting they have no close relatives (reviewed by Baptista *et al.* 1997). Five extant subfamilies are widely recognized, the majority being within Columbinae (generally seed-eating: 181 species), and Treroninae (fruit-doves: 123 species).

Sixty-six species out of a total of 348 are known to hybridize (19 % of all pigeons and doves). If species merged primarily due to hybridization are again split, 75 species hybridize (21 %). Interestingly, Dickinson (2003) recognized 4 subfamilies (Columbinae, Gourinae, Otidiphibinae and Treroninae). Intergeneric hybridization was restricted to within the Columbinae. However, Dickinson and Remsen (2013) recognizes a different arrangement of subfamilies: Columbinae (118 species); Raphinae (175 species); and Peristernae (17 species). Several of the genera formerly in Columbinae are now placed in Raphinae. As a result, Columbinae and Raphinae (the two largest subfamilies) are now linked together by several records of hybridization between member genera. Within the remaining subfamily (Peristernae), *Columbina* and *Metropelia* are linked by hybridization. Thus a core of nine genera across two of the three subfamilies are linked by hybridization, which raises the possibility that the Columbidae are a prospective basic type.

Hybrids linking them to other taxa are unknown.

**14.3.19. Sturnidae (Starlings).** Starlings are small to medium-sized birds, often with black plumage, and frequently with striking iridescence (Craig and Feare 2009). Most species associate in flocks, with roosts known of up to two million common starlings, famed for their co-ordinated aerial mass manoeuvres. Their vocal mimicry is highly developed, and some starlings and mynas can mimic human speech and artificial sounds such as bells, machines and telephones.

There are records of inter-specific hybridization for 22 species (or up to 25 if species merged due to hybridization are re-split) (between 18 to 20 %) of all starlings and mynas. Although the three subfamilies are not linked by hybrids, 6 of the 35 genera are linked by intergeneric hybrids. This includes hybridization between mynas and starlings. For instance, there are brief zoo reports of hybridization between *Acridotheres* and *Sturnia*, e.g.:

Figure 11. Some birds are the sole species within their genus and family, such as the Hamerkop, shown here. Their form is unique and distinctive, and they hybridize with no other birds, suggesting that they too are prospective basic type candidates.

*A c r i d o t h e r e s fuscus* (Jungle myna) x *Sturnia erythropygius andamanensis* (Andaman white-headed starling) (IZY 1972).

There are no known hybrids linking them to nearest proposed allies.

**14.3.20. Regulidae (Crests And Kinglets).** These tiny conifer-dwellers possess a needle-like bill, incised tail tip, brilliant yellow to red crown stripe, and, in some species, an additional lateral crown stripe (hence the name Regulidae: "little king"). They possess deeply furrowed toe pads, enabling them to perch on thin conifer twigs and, in some regulids, even individual needles. Fossil representatives first appeared during the Pliocene (Boev 1999). The family comprises a single genus, *Regulus*. Four out of the six species hybridize (67 % of all goldcrests and kinglets), as follows:

*Regulus calendula* (Ruby-crowned kinglet) x *Regulus satrapa* (Golden-crowned kinglet)

*Regulus ignicapilla* (Firecrest) x *Regulus regulus* (Goldcrest)

The species have a similar morphology, with the exception of the ruby-crowned kinglet, leading to its placement in a separate genus, *Corthylio*, by some authors[10]. In view of this, it is interesting to note that the ruby-crowned kinglet hybridizes with another regulid species, as indicated above, indicating genetic closeness.

However, the hybrids do not interconnect all four species together, perhaps due to antagonistic and aggressive courtship behaviour between sympatric species leading to pre-mating isolation barriers between them (Thaler 1986). Thus from the hybridization data alone it is not certain if the kinglets may be a separate BT, but they do share numerous morphological features with the other regulid species, suggestive of their close affinity. For instance, although the golden-crowned kinglet and firecrest do not hybridize, they both possess a black eyestripe. All regulids also have common and characteristic elements in their song pattern. Hybrids with any other outgroups are unknown.

## 14.4. Monotypic Taxa

In addition, certain taxa are monotypic (*i.e.* in these cases they are the sole species and sole genus within its family). These include *Upupa epops* (Hoopoe) [Upupidae]; *Balaeniceps rex* (Shoebill) [Balaenicipitidae]; *Opisthocomus hoazin* (Hoatzin) [Opisthocomidae]; *Pedionomus torquatus* (Plains-wanderer) [Pedionomidae]; and *Scopus umbretta* (Hamerkop) [Scopidae].

The Hoopoe is unlikely to be mistaken for any other bird (Krištin 2001), with its rufous plumage marked with broad black and white bands, striking erectile fan-like crest and long, slightly decurved bill. Equally distinctive is its butterfly-like flapping flight pattern and the male call, a resonant "hoop-hoop-hoop" from which the bird takes its name.

The Hoopoe has no close relatives, its nearest allies being thought to be the woodhoopoes, although DNA-DNA hybridization studies suggest they diverged from the hornbills. The most common treatment comprises a single species or superspecies with nine subspecies, but Clements *et al.* (2023) follows a trend towards splitting them into three distinct species.

McCarthy (2006) recognizes *Upupa africana* (African hoopoe) and *Upupa epops* (Eurasian hoopoe) as distinct species, which have a putative hybrid zone between them. There is no known hybridization with proposed relatives such as the woodhoopoes, scimitarbills or hornbills, all of which possess unique, distinctive aspects of morphology (see Section 3.2: Bucerotidae).

The Hoatzin does not fit satisfactorily within any other group of birds, with its stiffly shafted rufous crest, bright blue orbital area, and other

---

10          Thus the Clements 2023 checklist recognises the ruby-crowned kinglet as *Corthylio calendula*

features (Thomas 1996), which contradict proposed affinities with the cuckoos and allies. The Plains-wanderer, found only in Eastern Australia, is superficially quail-like, but with finer bill, longer neck and legs, and with a reversal of the usual sexual dimorphism, in which here it is the female plumage that is more striking, with a black and white collar and rufous upper breast (Baker-Gabb 1996). Its nearest relative may be the seedsnipes (Thinocoridae) of South America, but earliest fossils consist of specimens of both the genera that are still extant, shedding no light on a hypothetical common origin. The Shoebill (Elliott 1992d) is superficially stork- or heron- like, but with a long, broad, flattened bill with a massive terminal hook. The Hamerkop (Figure 11) (Elliott 1992b) is a medium-sized brownish bird with a "hammer-like" long, horizontally orientated feather crest on the back of its head. There is no known hybridization between these monotypic taxa and any proposed outgroups.

## 14.5. Supplementary Criteria: Bioacoustics

Future directions of prospective value in revealing BTs can explore additional criteria, such as bird sound analysis (bioacoustics). Bioacoustics data is being increasingly employed in bird taxonomy. For instance, the hybrid status of a cross between two highly phenotypically divergent species within the Cardinalidae, *Pheucticus ludovicianus* (Rose-breasted grosbeak) and *Piranga olivacea* (Scarlet tanager), was confirmed by bioacoustics analysis (Toews *et al.* 2022). Bioacoustics analysis was employed to identify a new taxon of *Myzomela* honeyeater in Indonesia (Prawiradilaga *et al.* 2018). Vocal characteristics can be recognized on sonograms, a pictorial frequency-time waveform graph, complemented by digital software dedicated to sound analysis. In musical notation, notes are located on a staff, while the sonogram indicates the frequency in kilohertz (kHz), but both read from left to right in measurement of time, and place high-pitched notes at the top and low notes at the bottom. Bird species have unique combinations of seven tone qualities. For instance, the burry sound results from rapid rises and falls in pitch (https://earbirding.com/blog/specs).

The typical calls and territorial songs of birds are often clearly recognizable and approximate to distinct taxonomic groups. For example, members of the family Paridae (chickadees and tits) demonstrate calls composed of multiple acoustically distinct notes (Suzuki *et al.* 2019), with syntactical rules (Hailman and Ficken 1986). Distinctive sounds are also shown in owls, flycatchers and sparrows (McCallum 2011). For example, flycatchers within the genus *Empidonax* demonstrate characteristic song-types and calls shared between the species, as well as ones unique to a particular species (Figure 12)[11]. For instance, in "Hammond's" flycatcher (*Empidonax hammondi*), the call system (E,F in the sonogram) resembles that of the Dusky flycatcher (E.F).

11        © Arch McCallum. Used with permission.

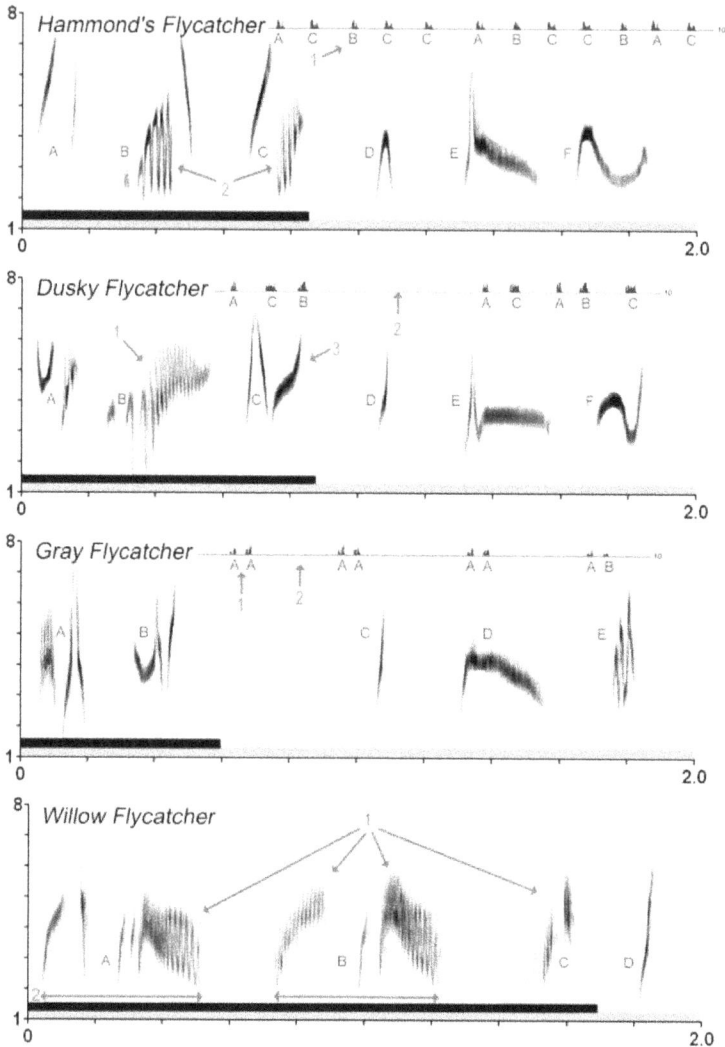

Figure 12. Sonograms to compare sound repertoire of flycatchers within the genus *Empidonax* ("empids") showing features in common and ones distinct to a species. Shown are major sounds (letters) with song-types to left, followed by call note and other sounds. The y-axis represents sound pitch (kilohertz), and x-axis time (seconds). Black bar at bottom of sonogram indicates sound in pre-dawn singing; grey after dawn. Across top is 10 second segment of dawn song to show pacing and syntax.

"Hammond's" Flycatcher *Empidonax hammondi* Two of 3 song-types have burry note (ie rapidly changing in pitch) (2) vs. 1 in Dusky. Contact note D only shared with Alder flycatcher. Call system (E,F) resembles Dusky (E,F). C sounds similar to A of Gray flycatcher.

"Dusky" Flycatcher *Empidonax oberholseri* 3 song types (A,B and C) with only 1 burry. Pauses (2) every 2 – 6 songs (contra Hammond's). Upslur (3) diagnostic. Contact note D shared with Gray and Willow. Post-dawn song (E,F) resembling Hammond's (E,F).

"Gray" Flycatcher *Empidonax wrightii* Two song-types (A,B) often given in fast doublets (1), mostly AA, separated by silence (2). Contact note C shared with Dusky and Willow. D may be unique. Seldom-heard E similar to Hammond's C.

"Willow" Flycatcher *Empidonax traillii* Three song types (1) all burry and longer than most other empids.

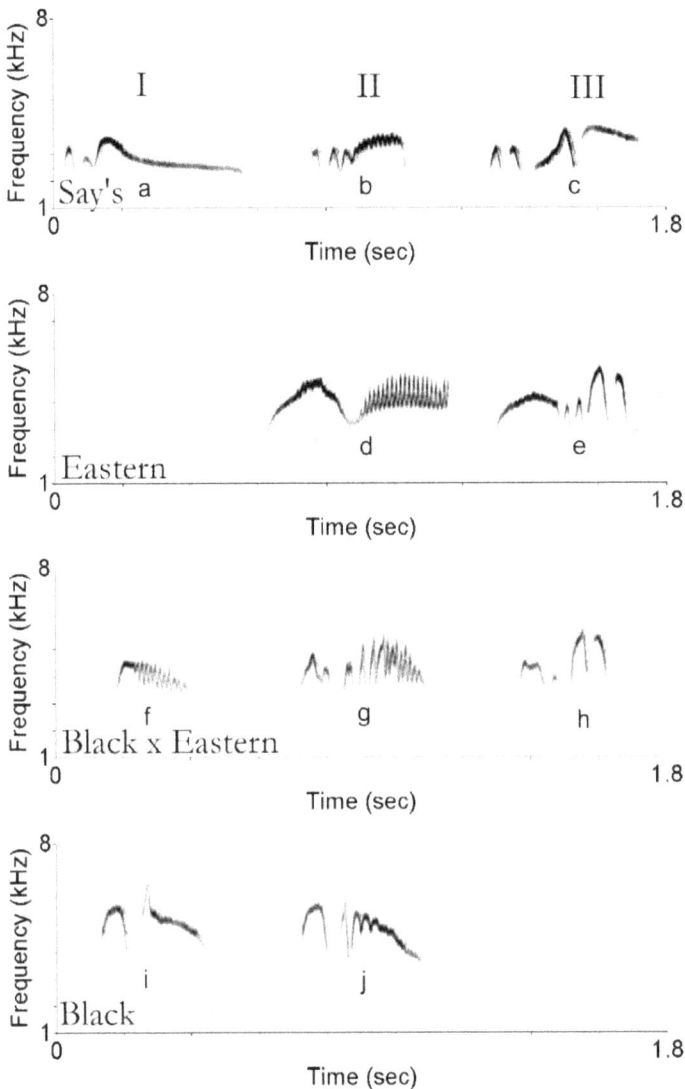

Figure 13. Sonograms of song types and syntax in phoebes, demonstrating phylogenetic signals and evidence of atavism. Highly stereotyped songs are arranged vertically to indicate similarity of song type between species (Roman numerals I to III). The sonograms reveal both similarities and species-specific differences. For instance, each of the seven song types (a, b, c, d, e, i, j) of the genus begins with a species-specific prefix note, known as the "pip." In all three species the frequency trend of the pip is a similar overslur (a rising and then falling trace on the sonogram). In the Black phoebe the pip is clearly separate from the terminal portion of the song (the song phrase). In the Eastern, the pip of song type II is continuous with the song phrase (d). The clearest homology within the phrase types is between the burry phrases of the Say's (b) and of the Eastern (d). Both feature a frequency that initially rises and then levels off and is frequency-modulated at a constant rate. The hybrid Eastern × Black sang with three song types rather than two, and in accordance with the syntax pattern of Say's phoebe, in the order [[I] II [I] III], rather than that of the parents, suggesting an atavistic trait inherited from the common ancestor of all three phoebes.

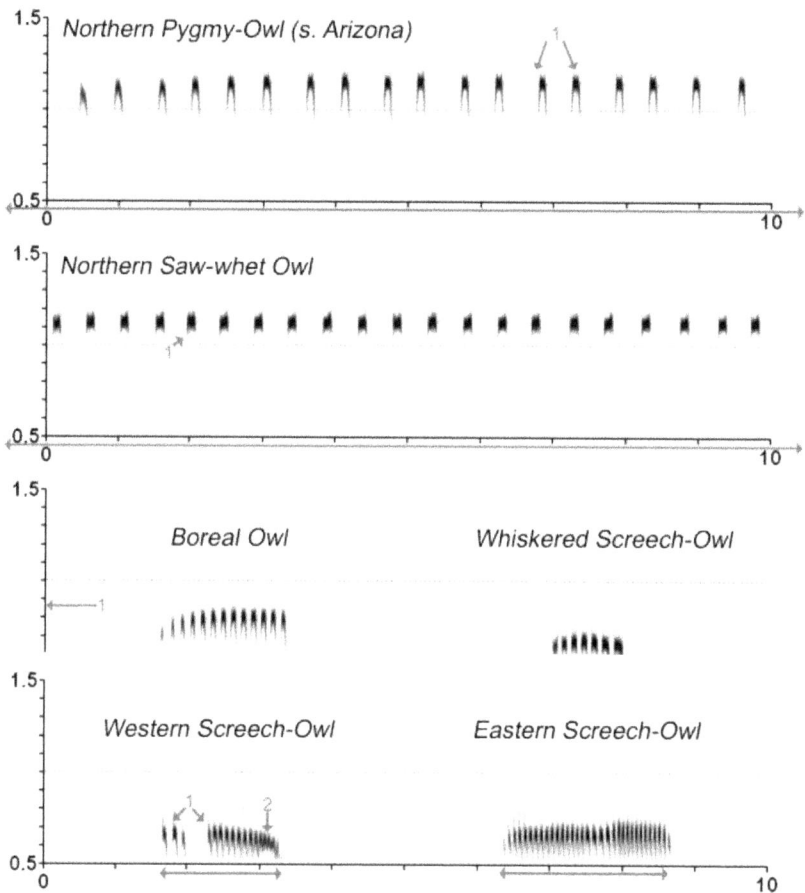

Figure 14. Sonograms of territorial song hoots and trills in small owl species. Sonogram trace shows that hoots/toots (single or two-note clusters) and trills (clusters of notes) are very similar, with characteristic "two-legged" notes, differing in "leg" length, pitch and tempo between the species. Original sounds courtesy of Macaulay Library (ML) at Cornell Lab of Ornithology, from *Voices of North American Owls* [VNAO] (2006). Original sound recordists, VNAO and ML catalogue numbers indicated below.

Northern pygmy owl. *Glaucidium gnoma.* Double-headed arrow alongside x-axis [time] shows indefinite duration of hoot pattern. A pair of double-legged notes are indicated (1). VNAO 2-25, Irby Davis, ML 9418.

Northern saw-whet owl. *Aegolius acadicus.* Again indefinite duration of pattern but basic note has shorter "legs" (1), compared with Northern pigmy owl. VNAO 2-61, Geoffrey Keller, ML42199.

Boreal owl. *Aegolius funereus.* Faster and lower pitch (1) than sympatric saw-whet owl. VNAO 2-53, Leonard Peyton, ML 49540.

Whiskered screech owl. *Megascops trichopsis.* Trill shorter and slower than sympatric western screech. VNAO 1-35, Geoffrey Keller, ML 40588.

Western screech owl. *Megascops kennikottii.* Trill usually comes in two parts (1) and accelerates towards end (2). VNAO 1-20, David Herr, ML 47692.

Eastern screech owl. *Megascops asio.* Continuous trill longer than Western's and more constant tempo. VNAO 1-26, Wilbur Hershberger, ML 107446.

In another flycatcher study, there was not only evidence of homologous song and phrase types shared between species and species-specific ones, but also atavism. Phoebes (from the genus *Sayornis*), were found to sing with two or three stereotyped song types, designated I, II and III, all three of which are used in Say's phoebe (*Sayornis saya*) (McCallum and Pieplow 2010). Only two of these (types I and II) are used in the Black phoebe (*S. nigricans*), and only types II and III are used in the Eastern phoebe (*S. phoebe*). A hybrid Black x Eastern phoebe used all three phrase types and sang like another species, Say's phoebe (Figure 13)[12]. The authors expected the hybrid bird to have inherited the syntax (the non-random order in which phrases are sung) of one parent and/or the other (*i.e.* predominantly [I–II] pattern of the Black, and [II–III] of the Eastern). However, to the authors' surprise, the hybrid instead sang syntactically in the stereotyped order [[I] II [I] III] of Say's Phoebe. The authors considered that the genes responsible for the neurological circuitry directing the characteristic song performances have been inherited from the common ancestor of all three phoebes, Around 20 % of the 400+ species of Flycatchers hybridise throughout the family Tyrannidae, but connectivity between genera is limited, so it is difficult to assess its possible BT status. It is thus salient that the introductory pip note of the phoebes is seen not only in species within *Sayornis* but also in the genera *Contopus* (peewees) and *Empidonax* (McCallum and Pieplow 2010), hinting at a phylogenetic affinity between the genera.

Other bird groups, such as owls and sparrows, are also characterised by distinctive bioacoustic signatures. For instance, the toots and trill sonograms of numerous small owl species are very similar (Figure 14)[13], even between pygmy owl and screech owl genera, but differing in pitch and tempo between the species (McCallum 2011). To the human ear they all sound easily recognisable as owl-like, as is evident from listening to the linked sound recordings.

The examples from these different bird groups suggest that bioacoustic characteristics can provide empirical supplementary data of taxonomic value. Future studies could analyse bioacoustics information across various families, to assess whether they may have a role in helping to discern further prospective BTs.

## 14.6. Discussion

In summary, 12 bird groups can be distinguished primarily from hybridization data. These are contenders for BT status: many of the species hybridize with one another, and inter-connectivity between these species is high. In several taxa, several or all of the genera, and even subfamilies are also united by hybridization. A further 20 groups provide more modest evidence for basic type status, sufficient to warrant

12      © Arch McCallum. Used with permission.
13      Sonograms © Arch McCallum. Used with permission.

Basic Types of Life

further study. For the majority of these BTs, morphological criteria are not controversial, and the correlation between the hybridization data and morphology is often strikingly high. For instance, the penguins (Figure 15) have a highly streamlined shape and wings modified to form flat scimitar-shaped paddles for propulsion. Sixty-nine percent of the penguin species hybridize with one another, yet with no other taxon, including proposed nearest allies such as the divers or petrels, from which their form is easily distinguished. Similarly, the flamingo form is unmistakable, with its downward bending bill and trough-like lower mandible. All three flamingo genera are linked by hybridization. Its form is clearly distinguished from proposed relatives such as duck-like birds or grebes. Pelicans, of which 75% of the species hybridize together, cannot be mistaken for nearest proposed relatives such as gannets or cormorants.

Figure 15. Penguins are another example of a basic type showing a strong correlation between distinctive form (e.g. streamlined form and paddle-like wings) and hybridization (69 % of all penguin species hybridize). These correlations suggest not just genetic but morphogenetic closeness, i.e., a developmental machinery in common between the hybridizing parents, from which the distinctive form of the basic type is generated. At least 32 basic types of birds can be distinguished in this way.

Together these 32 prospective basic types of birds show clear discontinuities between each basic type and proposed nearest relatives, both in hybridization and morphology. This suggests that the members of each type possess a developmental program sufficiently in common to allow hybridization to occur and produce viable offspring, and to generate their common form. In addition, each of at least five monotypic taxa may be prospective BTs. Thus 37 groups of birds may merit BT status. According to Mayr (2005, 2009) the earliest records of European fossil birds closely resemble their modern counterparts. This is reiterated for the 12 contenders for BT status in Section 2: fossil data shows that their ancestral forms from their earliest emergence are identifiable with their respective basic type of form, indicating that the basic forms are ancient, and typically underwent adaptive radiations, often rapidly.

In some families formerly good species have been subject to merging due to hybridization. This is sometimes a significant issue in assessing BT status, potentially leading to the number of BTs being under-reported (Appendix 2). For instance, there are 61 races of *Pachycephala pectoralis* (Golden whistler), treated variously as races or separate species, many of which hybridize, and which Dickinson and Christidis (2013) recognize have been over-merged. McCarthy documents numerous examples of this problem. For instance, flame-rumped and yellow-rumped tanagers "although long treated as separate species, due to hybridization, [both] birds are now often lumped" [McCarthy 2006, p.329].

In at least ten further families, several species are linked together into a core group, but, in relation to the family as a whole the data appears more moderate (Appendix 2). How many species should be linked by hybridization to constitute a basic type? This is difficult to answer. When just two species of rhea hybridize, this appears to be a strong contender for BT status, since the family comprises only two species. In contrast, although appearing modest hybridization in relation to the larger group size, hybridization nonetheless links together a core group of the following: nine species of corvids (crows, ravens, magpies and jays); four species within Coracidae (rollers); from six to 13 species of motacillids (pipits and wagtails); 11 species of turacos (Musophagidae); 11 species of bulbuls (Pycnonotidae); five species of owls within Strigidae; four species of white-eyes (Zosteropidae); four species of nuthatch (Sittidae); and within Turdidae all three species of bluebirds (*Sialia* spp.) and 17 species of thrushes within *Turdus*. Although not included in Sections 2 or 3, such groups could be described as basic type cores, in anticipation of the groups becoming enlarged or more established with the advent of further data. If these groups are added, the list of prospective BTs is now 49. This provides a shortlist of candidate groups of promise for future further studies. An additional 23 families demonstrate more modest interspecific hybridization data (Appendix 2), but with supplementary criteria these may reveal yet further future BT candidates.

Employment of degrees of intergeneric hybridization to reveal prospective BTs is not without problems, because species allocation within genera is subjective, as is evident within Paridae (Section 3 8: Paridae) in which the number of genera linked has ranged from 0 to 7, depending on the treatment of the author. In contrast, the interspecific hybridization rests upon an empirical foundation, revealing real morphogenetic affinities between the species. Therefore it is important that interspecific hybridization data are examined with sufficient scrutiny so as to not under-report the existence of core groups that may exist exclusively within one genus, but be overlooked because the researcher is focussed on intergeneric crosses. As an example, within Laniidae (shrikes) there are no intergeneric hybrids between the four genera of the family. However, 30 of the 34 species are within the genus *Lanius*, so greater weight needs to be given to assessing the degree of hybridization within *Lanius*.

The prospective BTs outlined above are commonly identifiable with their respective family taxon. In general these avian forms are well established; easily recognizable even to the layman. However, the traditional taxonomies are being subject to phylogeny-sensitive revisions. In these, morphological or molecular data are linked to homology hypotheses, which involve circular reasoning: the homology relations are reconstructed from distribution patterns of similar morphological traits, which in turn presupposes homology of traits (Vogt 2008). For example, Livezey and Zusi (2007) made an avian phylogenetic reconstruction based on analysis of 2954 morphological characters. This was critiqued by Mayr (2008), who concluded that the study had unacceptable generalizations on character distribution based on unverified assumptions. Nowadays DNA-DNA hybridization methods have become replaced with large-scale and ambitious DNA sequence analysis projects (Fjeldså 2013). However, these still yield conflicting and unlikely phylogenetic relationships. In contrast, morphologies grounded in observation and experimentation represent empirical data, and so, according to Vogt (2008), morphological descriptions should be defined without reference to homology relations, and be purely descriptive and free of evolutionary or other explanatory connotations.

Another trend relates to the number of avian species recognized. R.B. Sharpe's catalogue (1899–1909 [cited in Fjeldså 2013]) recognized 18,939 species, based on the British Museum Bird data. In the following decades, many species and subspecies were merged, so that the Peter's checklist (1951) which still remains a reference template for avian taxonomic arrangements, recognized just over 9,000 species. However, this trend has been reversing, and the list may well return to 19,000 species recognized by Sharpe or at least 12,000 (Fjeldså 2013)[14]. Although BT methodologies aim to be taxon-independent, to reiterate, the above trends are relevant to searches for further prospective BTs, and indicate that species merged due to hybridization should not necessarily be discounted from BT analyses.

## 14.7. Conclusions

In birds, a rich source of hybridization and other data enable the basic type concept to be tested and validated. Wherever the hybridization data are sufficiently abundant, basic types regularly emerge, connected by hybridization, common morphology and other attributes such as behaviour; and separated from neighbouring groups by discontinuities in these attributes. This points to there being natural groupings, with real distinctions separating them. This is just an initial study, in which the list of prospective BTs is not likely to be exhaustive. Birds thus provide a rich treasure trove of data, with much remaining to be tapped. Although there is an ever-growing trend towards avian taxonomies linked to phylogenetic tree determinations, these are increasingly dependent upon molecular characters which have no proven morphogenetic significance (and are thus potentially irrelevant). However, BT analyses

14    The Clements checklist (Clements et al. 2023) recognises 11,017 species.

can and do expose where phylogenetically-linked taxonomies are suspect. If members of a group can hybridize, they should be placed together, even if certain molecular characters have indicated placement separately. If taxonomists employing molecular data want to overcome phylogenies which are endlessly in conflict between primitive and advance characters, they should look to basic types of birds. Most excitingly, such empirically determined basic types must each share the elusive molecular determinants of their respective morphologies. Basic types thus promise a way to discovering the nature and mode of action of these determinants. Bird basic types are sufficiently common to be emerging as a flagship group to study such problems.

### 4.7. Acknowledgements

I would like to thank David Tyler for helpful comments on the manuscript; Luke and Jason Tyler for assistance with data analysis; Lorenz Landgren and Herfried Kutzelnigg for providing data; to Roger Sanders for constructive editorial support; and to Jersey Zoo (Durrell Wildlife Park) for assistance with bird images.

## References

Abbott, R., Albach, D., Ansell, S., Arntzen, J.W., Baird, S.J., Bierne, N., Boughman, J., Brelsford, A., Buerkle, C.A., Buggs, R. and Butlin, R.K., 2013. Hybridization and speciation. *Journal of evolutionary biology*, 26(2), pp.229-246.

Adriaens, P., Vercruijsse, H.J. and Stienen, E.W., 2012. Hybrid gulls in Belgium—an update. *British Birds*, 105(9), p.530.

Archibald, G.W. and C.D. Meine., 1996. Gruidae (Cranes). *Handbook of the Birds of the World* 3:60-89.

Baker-Gabb, D. J., 1996. Pedionomidae (Plains-wanderer). *Handbook of the Birds of the World* 3:534-537.

Baptista, L.F., P.W. Trail and Horblitt, H.M., 1997. Columbidae (Pigeons and Doves). *Handbook of the Birds of the World* 4:60-245.

Bell, D.A., 1996. Genetic differentiation, geographic variation and hybridization in gulls of the Larus glaucescens-occidentalis complex. *The Condor*, 98(3), pp.527-546.

Blair, M. J. McKay, H., Musgrove, A. J., Rehfisch, M. M., 2000. Review of the status of introduced non-native waterbird species in the agreement area of the African-Eurasian waterbird agreement research contract CR0219. *British Trust for Ornithology, Research Report* no. 229.

Blunt, D. and Frith, C.R., 2005. A living 'Rawnsley's Bowerbird' -an adult male resulting from a Hybridisation in the wild between a Regent 'Sericulus chrysocephalus' and a Satin 'Ptilonorhynchus violaceus' Bowerbird. *Australian Field Ornithology*, 22(2), pp.53-57.

Boev, Z., 1999. *Regulus bulgaricus* sp. n. the first fossil Kinglet (Aves:

Sylviidae) from the late Pliocene of Varshets, Bulgaria. *Historia naturalis bulgarica* 10:109-115.

Burger, J. and Gochfeld, M. 1996. Laridae (Gulls). *Handbook of the Birds of the World* 3:572-623.

Cameron, A.D. and Harrison, C.J.O., 1978. *Bird Families of the World*. Elsevier-Phaidon, Oxford.

Carboneras, J., 1992. Family Gaviidae (Divers). *Handbook of the Birds of the World* 1:162-172.

Cezilly, F. and Johnson, A.R., 1992. Exotic Flamingos in the Western Mediterranean Region: A Case for Concern? *Colonial Waterbirds*, pp.261-263.

Chialchia, A.O.C. and Smith, P., 2014. A notable hybrid woodpecker (*Campephilus leucopogon x C. melanoleucus*) (Aves: Picidae) from Paraguay. *Ornitologia Neotropical*, 25, pp.459-464.

Chamutpong, S., Ponglikitmongkol, M., Charoennitikul, W., Mudsri, S. and Poonswad, P., 2013. Hybridisation in the wild between the great hornbill (*Buceros bicornis*) and the rhinoceros hornbill (*Buceros rhinoceros*) in Thailand and its genetic assessment. *Raffles Bulletin of Zoology*, 61(1).

Clancy, G.P., 2009. Species Review: The Black-necked Stork' *Ephippiorhynchus asiaticus*'-an Overview. *Australian Field Ornithology*, 26(4), pp.110-115.

Clements, J. F., P. C. Rasmussen, T. S. Schulenberg, M. J. Iliff, T. A. Fredericks, J. A. Gerbracht, D. Lepage, A. Spencer, S. M. Billerman, B. L. Sullivan, and Wood, C.L., 2023. The eBird/Clements checklist of Birds of the World: v2023. Downloaded from https://www.birds.cornell.edu/clementschecklist/download/

Collar, N.J., 1997. Psittacidae (Parrots). *Handbook of the Birds of the World* 4:280-477.

Confer, J.L., Porter, C., Aldinger, K.R., Canterbury, R.A., Larkin, J.L. and Mcneil Jr, D.J., 2020. Implications for evolutionary trends from the pairing frequencies among golden-winged and blue-winged warblers and their hybrids. *Ecology and Evolution*, 10(19), pp.10633-10644.

Craig, A. J. F. K. and Feare, C.J., 2009. Sturnidae (Starlings). *Handbook of the Birds of the World* 14:654-758.

Curson, J.M., 2010. Parulidae (New World Warblers). *Handbook of the Birds of the World* 15:666-800.

del Hoyo, J., 1992. Phoenicopteridae (Flamingos). *Handbook of the Birds of the World* 1:508–526.

Delsuc, F., Superina, M., Ferraris, G., Tilak, M.K. and Douzery, E.J., 2007. Molecular evidence for hybridisation between the two living species of South American ratites: potential conservation implications. *Conservation Genetics*, 8(2), pp.503-507.

Donato, D.B. and Potts, R.T., 2004. An Australian Magpie '*Gymnorhina tibicen*' x Pied Butcherbird '*Cracticus nigrogularis*' Hybrid in the Tanami Desert, Northern Territory. *Australian Field Ornithology*, 21(2), pp.79-80.

Dickinson, E.C., ed. 2003. *Howard and Moore Complete Checklist of the Birds of the World*, 3rd edition. Christopher Helm, London.

Dickinson , E.C. and L. Christidis, eds. 2013. *Howard and Moore Complete Checklist of the Birds of the World*, 4th edition, Vol. 2. Aves Press, Eastbourne, UK.

Dickinson , E.C. and J.V. Remsen, eds. 2013. *Howard and Moore Complete Checklist of the Birds of the World*, 4th edition, Vol. 1. Aves Press, Eastbourne, UK.

Dunning, J., Hanmer, H. and Christmas, S.E., 2014. Hybridisation between House Martin *Delichon urbicum* and Sand Martin *Riparia riparia*: a new observation and review of past occurrences as a case study into hybrid reporting rates. *Ringing & Migration*, 29(2), pp.86-89.

Elliott, A., 1992a. Pelicanidae (Pelicans). *Handbook of the Birds of the World* 1:290–311.

Elliott, A., 1992b. Scopidae (Hamerkop). *Handbook of the Birds of the World* 1:430-435.

Elliott, A., 1992c. Family Ciconiidae (Storks). *Handbook of the Birds of the World* 1:436–465.

Elliott, A., 1992d. Balaenicipitidae (Shoebill). *Handbook of the Birds of the World* 1:466-471.

Fjeldså, J., 2013. Avian classification in flux. *Handbook of the Birds of the World* Special Volume: 77-146.

Flamingo Specialist Group, 2005. *Flamingo* 13: 11.

Folch, A., 1992a. Family Struthionidae (Ostrich). *Handbook of the Birds of the World* 1:76-83

Folch, A., 1992b. Family Rheidae (Rheas). *Handbook of the Birds of the World* 1:84-89.

Frith, C.B., 2016. A second living'Rawnsley's Bowerbird'-a wild adult male hybrid from a Regent Bowerbird 'Sericulus chrysocephalus' x Satin Bowerbird 'Ptilonorhynchus violaceus' cross. *Australian Field Ornithology*, 33, pp.14-15.

Forest, M.M., 2006. *Status Overview and Recommendations for the Conservation of Milky Stork Mycteria cinerea in Malaysia*. Wetlands International and the Department of Wildlife and National Parks, Peninsular Malaysia.

Frith, C.B., and Beehler B.M., 1998. *The Birds of Paradise (Paradisaeidae)*. Oxford University Press.

Frith, C.B. and Frith, D.W., 2009a. Paradisaeidae (Birds of Paradise). *Handbook of the Birds of the World* 14:404–493.

Frith, C.B. and Frith, D.W., 2009b. Ptilonorhynchidae (Bower Birds). *Handbook of the Birds of the World* 14:350–403.

Gochfeld, M. and Burger, J., 1996. Sternidae (Terns). *Handbook of the Birds of the World* 3:624-667.

Gosler, A.G. and Clement, P., 2007. Paridae (Tits and Chickadees). *Handbook of the Birds of the World* 12:662–750.

Grant, P.R. and Grant, B.R., 1992. Hybridization of Bird Species. *Science* 256:193-197.

Gray, A., 1958. *Bird Hybrids*. Commonwealth Agricultural Bureaux, England.

Hailman, J.P. and Ficken, M.S., 1986. Combinatorial animal communication with computable syntax: Chick-a-dee calling qualifies as" language" by structural linguistics. *Animal Behaviour*. 34(6): 1899–1901.

Hibbets, E.M., Schumacher, K.I., Scheppler, H.B., Boersma, P.D. and Bouzat, J.L., 2020. Genetic evidence of hybridization between Magellanic (*Sphensicus magellanicus*) and Humboldt (*Spheniscus humboldti*) penguins in the wild. *Genetica*, 148(5), pp.215-228.

Hockey, P.A.R., 1996. Haematopodidae (Oystercatchers). *Handbook of the Birds of the World* 3:308-325.

Höfling, E., 1998. Comparative cranial anatomy of Ramphastidae and Capitonidae. *Ostrich* 69(3-4):389-390.

*International Zoo Yearbook*, 1972. Zoological Society of London, London, UK, Vol. 12 (1). p. 371.

*International Zoo Yearbook*, 1974. Zoological Society of London, London, UK, Vol. 14 (1). p. 337.

*International Zoo Yearbook*, 1991. Zoological Society of London, London, UK, Vol. 30 (1). p. 350.

*International Zoo Yearbook*, 1998. Zoological Society of London, London, UK, Vol. 36 (1), p. 403.

Johnsgard, P.A., 1983. Aviculture and Hybridization. In: *Cranes of the world*. Indiana University Press, 1983; electronic edition: University of Nebraska-Lincoln, 2008. https://digitalcommons.unl.edu/bioscicranes/3/

Johnson, A. and Cézilly, F., 2009. *The greater flamingo*. A&C Black, London.

Jønsson, K. A., and Fjeldså, J., 2006. A phylogenetic supertree of oscine passerine birds (Aves: Passeri). *Zoologica Scripta* 35(2):149-186.

Josef, R., 2008. Laniidae (Shrikes). *Handbook of the Birds of the World* 13:732 - 796.

Kashentseva, Г. and Postelnykh, K., 2013. The morphology of hybrid of Eurasian and Siberian cranes. In *Proc. VIIth European Crane Conference "Breeding, Resting, Migration, and Biology,"* Stralsund, Germany, October 14–17 (pp. 109-113). Crane Conserv. Germ.

Kemp, A.C., 2001. Bucerotidae (Hornbills). *Handbook of the Birds of the World* 6:436–523.

Klos, U., 1969. A brief note on the incubation and rearing of a pelican hybrid at West Berlin Zoo. *International Zoo Yearbook* 9: 121–122.

Krištin, A., 2001. Upupidae (Hoopoes). *Handbook of the Birds of the World* 6:396–411.

Landgren L, Gustafsson L and Kutzelnigg, H., 2011. Grundtypstudien an Papageien. *Stud. Integrale J.* 18 (1), 4-16.

Livezey, B. C. and Zusi, R.L., 2007. Higher order phylogeny of modern birds (Theropoda, Aves: Neornithes) based on comparative anatomy. II. Analysis and discussion. *Zoological Journal of the Linnean Society* 1 49(1):1-95.

Llimona, F. and del Hoyo, J., 1992. Podicipedidae (Grebes). *Handbook of* the Birds of the World 1:174 –196.

Lovette, I.J., Pérez-Emán, J.L., Sullivan, J.P., Banks, R.C., Fiorentino, I., Córdoba-Córdoba, S., Echeverry-Galvis, M., Barker, F.K., Burns, K.J., Klicka, J. and Lanyon, S.M., 2010. A comprehensive multilocus phylogeny for the wood-warblers and a revised classification of the Parulidae (Aves). *Molecular phylogenetics and evolution*, 57(2), pp.753-770.

Lowery, G.H., Jr. and Monroe, B.L.Jr., 1968. Family Parulidae. In: Paynter, R.A., Jr. ed. *Check-list of Birds of the World*, Vol. 14. Museum of Comparative Zoology, Cambridge, Massachusetts, pp. 3–93.

Maksudov, G.Y. and Panchenko, V.G., 2002. Obtaining an interspecific hybrid of cranes by artificial insemination with frozen–thawed semen. *Biology Bulletin of the Russian Academy of Sciences*, 29(3), pp.311-314.

Marini, M.Â. and Hackett, S.J., 2002. A multifaceted approach to the characterization of an intergeneric hybrid manakin (Pipridae) from Brazil. *The Auk*, 119(4), pp.1114-1120.

Matheu, E. and Del Hoyo, J., 1992. Threskiornithidae (Ibises and Spoonbills). *Handbook of the Birds of the World* 1:472– 506.

Marshall, L. 2015. World first—Galah breeds with Cockatiel. T*alking Birds: Australia's Avian Newsmagazine*. Accessed 19 August 2014 at http://talkingbirds.com.au/world-firsts/galatiel-php/ and http://talkingbirds.com.au/world-firsts/galatiel-php/world-first-galah-breeds-with-cockatiel/.

Martinez, I., 1992. Spheniscidae (Penguins). *Handbook of the Birds of the World* 1:140–160.

Martinez-Vilalta, A. and A. Motis. 1992. Ardeidae (Herons). *Handbook of the Birds of the World* 1:376-429.

Mayr, G., 2005. The Paleogene fossil record of birds in Europe. *Biological Reviews* 80(4):515-542.

Mayr, G., 2008. Avian higher-level phylogeny: well-supported clades and what we can learn from a phylogenetic analysis of 2954 morphological characters. *Journal of Zoological Systematics and Evolutionary Research* 46(1):63–72.

Mayr, G., 2009. *Paleogene Fossil Birds*. Springer Berlin Heidelberg, pp. 169-204.

McCallum, D.A. and Pieplow, N.D., 2010. A reassessment of homologies in the vocal repertoires of phoebes. *Western Birds*, 41(1), pp.26-43.

McCarthy, E., 2006. *Handbook of avian hybrids of the world*. Oxford University Press, England.

McGuire, J.A., Witt, C.C., Remsen, J.V., Dudley, R. and Altshuler, D.L., 2009. A higher-level taxonomy for hummingbirds. *Journal of Ornithology*, 150, pp.155-165.

Morlan,J., Dakin, R. E. and Rosso, J., 2004. Apparent hybrids between the American Avocet and Black-necked Stilt in California. *Western Birds* 35:57-59.

Mostello, C.S., LaFlamme, D. and Szczys, P., 2016. Common Tern *Sterna hirundo* and Arctic Tern *S. paradisaea* hybridization produces fertile offspring. *Seabird*, 29, pp.39-65.

Napier, R., 1968. Erect-crested and Rockhopper Penguins interbreeding in the Falkland Islands. *British Antarctic Survey Bulletin* 16: 71-72.

Olsen, D.L. and Derrickson, S.R., 1980. *Whooping Crane Recovery Plan, January 1980*. US Fish and Wildlife Service.

Olsson, U. and Alström, P., 2020. A comprehensive phylogeny and taxonomic evaluation of the waxbills (Aves: Estrildidae). *Molecular Phylogenetics and Evolution*, 146, p.106757.

Payne, R. B., 2010. Viduidae (Whydahs and Indigobirds). *Handbook of the Birds of the World* 15:198-232.

Payne, R.B. and Risley, C.J., 1976. Systematics and Evolutionary relationships among the Herons (Ardeidae). *Miscellaneous Publications Museum of Zoology University of Michigan* 150:1-150.

Peters, J. L., 1951. *Check-list of Birds of the World*, Volume 7. Museum of Comparative Zoology, Cambridge, Massachusetts.

Pierce, R.J., 1996. Recurvirostridae (Stilts and Avocets). *Handbook of the Birds of the World* 3:332- 347.

Prawiradilaga, D.M., Baveja, P., Suparno, S., Ashari, H., Ng, N.S.R., Gwee, C.Y., Verbelen, P. and Rheindt, F.E., 2018. A colourful new species of *Myzomela* honeyeater from Rote Island in Eastern Indonesia. *Treubia*, 44, pp.77-100.

Principe, W. L. J., 1977. A hybrid American Avocet × Black-necked Stilt. *The Condor* 79: 128–129.

Ridgely, R.S. and Tudor, G. 1989. *The birds of South America*, Vol.1. University of Texas Press, Austin.

Rowley, I.C.R. and Russell, E.M., 2009. Artamide (Woodswallows). *Handbook of the Birds of the World* 14:286–307.

Russell, E.M. and Rowley, I.C.R., 2009. Cracticidae (Butcherbirds). *Handbook of the Birds of the World* 14:308–343.

Saurola, P., 2001. The EURING swallow project in Finland years 1999–2000. Euring Newsl, 3:35. https://euring.org/files/documents/newsletters/euring3complete.pdf

Schuchmann, K.L., 1999. Trochilidae (Hummingbirds). *Handbook of the Birds of the World* 5:468-535.

Shen, Y., Harada, N., Katagiri, S. and Tanaka, H., 2020. Biomimetic realization of a robotic penguin wing: Design and thrust characteristics. *IEEE/ASME Transactions on Mechatronics*, 26(5), pp.2350-2361.

Short, L.L. and Horne, J.F.M., 2002a. Capitonidae (Barbets). *Handbook of the Birds of the World* 7:140-219.

Short, L.L. and Horne, J.F.M., 2002b. Family Ramphastidae (Toucans). *Handbook of the Birds of the World* 7:220-272.

Sibley, C.G., and Ahlquist, J.E., 1990. *Phylogeny and classification of birds: a study in molecular evolution*. Yale University Press.

Simpson, K.N.G., 1985. A rockhopper x royal penguin hybrid from Macquarie Island. *Australian Bird Watcher*, 11(2): pp.35-45.

Snow, D.W., 2004. Pipridae (Manakins). *Handbook of the Birds of the*

*World* 9:110–169.

Suzuki, T.N., Griesser, M. and Wheatcroft, D., 2019. Syntactic rules in avian vocal sequences as a window into the evolution of compositionality. *Animal Behaviour*, 151, pp.267-274.

Thaler, E., 1986. Zum Verhalten von Winter- und Sommergoldhähnchen (Regulus regulus, Regulus ignicapillus) — etho-ökologische Differenzierung und Anpassung an den Lebensraum. *Der Ornithologische Beobachter* 84:281–289.

Thomas, B.T., 1996. Opisthocomidae (Hoatzin). *Handbook of the Birds of the World* 3:24-32.

Toews, D.P., Streby, H.M., Burket, L. and Taylor, S.A., 2018. A wood-warbler produced through both interspecific and intergeneric hybridization. *Biology Letters*, 14(11), p.20180557.

Toews, D.P., Rhinehart, T.A., Mulvihill, R., Galen, S., Gosser, S.M., Johnson, T., Williamson, J.L., Wood, A.W. and Latta, S.C., 2022. Genetic confirmation of a hybrid between two highly divergent cardinalid species: A rose-breasted grosbeak (*Pheucticus ludovicianus*) and a scarlet tanager (*Piranga olivacea*). *Ecology and Evolution*, 12(8), p.e9152.

Turner, A.K., 2004. Hirundinidae (Swallows and Martins). *Handbook of the Birds of the World* 9:602-685.

Vallender, R., Gagnon, J.P. and Lovette, I., 2009. An intergeneric wood-warbler hybrid (*Mniotilta varia* × *Dendroica coronata*) and use of multilocus DNA analyses to diagnose avian hybrid origins. *The Wilson Journal of Ornithology*, 121(2), pp.298-305.

Vogt, L., 2008. Learning from Linnaeus: towards developing the foundation for a general structure concept for morphology. *Zootaxa* 1950:123–152.

White, R.W. and Clausen, A.P., 2002. Rockhopper *Eudyptes chrysocome chrysocome* x macaroni *E. chrysolophus* penguin hybrids apparently breeding in the Falkland Islands. *Marine Ornithology*, 30, pp.40-42.

Winkler, H. and Christie, D.A., 2002. Picidae (Woodpeckers). *Handbook of the Birds of the World* 7:296-555.

Zieliński, P., Iciek, T., Zielińska, M., Szymczak, J., Gajewski, M., Bukaciński, D., Bukacińska, M., Betleja, J., Bednarz, Ł., Loręcki, A. and Kołodziejczyk, P., 2019. Identification of hybrids Mediterranean x Black-headed Gull in Poland. *Dutch Birding*, 41, pp.318-330.

Zusi, R.L., 1993. Patterns of diversity in the avian skull. In: Hanken, J. and B.K. Hall, eds. *The Skull, Volume. 2: Patterns of Structural and Systematic Diversity*. Chicago, IL: University of Chicago Press, pp. 391–437.

# 15. Hybrids, Chromosome Structure, and Speciation within the Horses (Equidae)

HEIKE STEIN-CADENBACH

**Abstract**
With one exception, hybridizations have been observed between all species of the genus *Equus* (horse); thus, all extant species belong to the basic type Equidae, most probably also including the extinct quagga and tarpan. The reasons for the widespread sterility and occasional fertility of equid hybrids can be found in the complex chromosomal variability, which may have happened due to chromosome fusion and chromosome inversion during the microevolutionary diversification of the horses (Equidae) within the framework of the variation potential of this particular basic type. This interpretation is further supported by external similarities and close genetic relationships as well as by hybridization experiments. Various hypotheses regarding the microevolutionary diversification of an ancestral form, resulting in the present-day species, are evaluated. The fact that different character complexes lead to clearly contradicting phylogenetic trees can be interpreted within the framework of a genetically polyvalent ancestral form of the horses.

## 15.1. Introduction

The taxonomy of the Equidae (Table 1) is contradictory and often confusing. Groves and Mazák (1967), for example, distinguish three genera: the genuine horses (*Equus* Linnaeus 1758), zebras (*Hippotigris* H. Smith, 1841) and African / Asian wild asses (*Asinus* Gray 1824), in contrast to the current opinion of only one genus (*Equus*). A taxonomic synopsis of the zebras was produced by Antonius (1951) from exhaustively researched synonymies. It is still widely accepted today, as is shown by comparison with the system of either Volf (1979, p. 599) or Klingel (1987). Of course, Volf does not distinguish between the Damara zebra (*E. quagga antiquorum*) and the phenotypically very similar Chapman's zebra (*E. q. chapmani*). For the plains zebra (*E. quagga* Gmelin, 1788), the synonym Burchell's zebra (Groves 1974) is also used in the literature. It is therefore often unclear whether the species or the subspecies is being referred to. Burchell's zebra (*E. q. burchelli*), according to Antonius (1951), has been extinct since 1910, but, according to Klingel (1987), it survived in small numbers. Meanwhile, newer investigations show that both Burchell's zebra and also the extinct (since 1883) quagga are subspecies of the plains zebra (George and Ryder 1986). The following

Order Perissodactyla (odd-toed ungulates)
    Family Tapiridae (tapirs)
      G *Tapirus* (3 spp.)
      G *Acrocodia* (1 sp.)
    Family Rhinocerotidae (rhinoceroses)
      G *Ceratotherium* (3 spp.)
      G *Diceros* (1 sp.)
      G *Rhinoceros* (2 spp.)
      G *Dicerorhinus* (1 sp.)
    Family Equidae (horses and allies [or equids])
      G *Equus* (horses)
        SG    *Equus* (true horses)
            *E. przewalskii,* Poliakov, 1881 (wild horse)
                *E. p. przewalskii,* Poliakov, 1881 (Przewalski's horse)
                *E. p. gmelini,* Antonius, 1912 (steppe tarpan)†
                *E. p. silvaticus,* Vetulani, 1928 (forest tarpan)†
                *E. p. caballus,* Linné,1758 (domestic horse)
        SG    Asinus (asses)
            *E. asinus,* Linné,1758 (African wild ass)
                *E. a. africanus,* Fitzinger, 1857 (Nubian wild ass)
                *E. a. somalicus,* P. L. Sclater, 1884 (Somali wild ass)
                *E. a. asinus,* Linné,1758 (donkey)
        SG    *Hemionus* (onagers)
            *E. hemionus,* Pallas, 1775 (onager or Asiatic wild ass)
                *E. h. onager,* Boddaerd, 1785 (Persian onager)
                *E. h. kulan,* Groves & Mazak,1967 (kulan or Turkmenian
                onager)
                *E. h. hemionus,* Pallas, 1775 (dziggetai or Mongolian onager)
                *E. h. khur,* Lesson, 1827 (khur or Indian onager)
                *E. h. hemippus,* I. Geoffroy, 1855 (achdari or Syrian onager)
                *E. h. kiang,* Moorcroft, 1841 (kiang or Tibetan onager)
        SG    *Dolichohippus* (Grevy's zebras)
            *E. grevyi,* Oustalet, 1882 (Grevy's zebra)
        SG    *Hippotigris* (true zebras)
            *E. quagga,* Gmelin, 1788 (plains zebra)
                *E. q. boehmi,* Matschie, 1892 (Grant's zebra)
                *E. q. selousi* Pocock, 1897 (Selous' zebra)
                *E. q. chapmani,* Layard, 1865 (Chapman's zebra)
                *E. q. antiquorum,* H. Smith, 1841 (Damara zebra)
                *E. q. burchelli,* Gray, 1824 (Burchell's zebra)
                *E. q. quagga,* Gmelin, 1788 (quagga)†
            *E. zebra,* Linné,1758 (mountain zebra)
                *E. z. hartmannae,* Matschie, 1898 (Hartmann's mountain zebra)
                *E. z. zebra* Linné,1758 (Cape mountain zebra)

Table 1. Systematic arrangement of the horses, modified from Volf (1979). † = extinct, G = genus, SG = subgenus.

is yet another example: Gray (1972) unites the kulan (*E. hemionus kulan*) with the dziggetai (*E. h. hemionus*), while Heptner (after Volf 1979, p. 564) places it within the subspecies of the Persian onager (*E. h. onager*). Klingel (1987), on the other hand, prefers to regard it as a separate species. The species name for the plains zebras (*E. quagga* versus *E.*

*burchelli*) or for the African wild ass (*E. asinus* versus *E. africanus*) is, thus, also chosen differently (Gray 1972; Volf 1979; Klingel 1987).

The examples show that the classification of individuals into biospecies can be challenging. Therefore the question arises whether the unequivocal association of species and subspecies is feasible at a higher taxonomic level. Thus, the present work attempts to apply the basic type concept introduced by Marsh (1976) to the family of the Equidae. Besides crossbreeding data, morphological, cytogenetic, and molecular-biological data will also be taken into account. Hypotheses regarding the diversification of the horses in the course of the evolution of the family are discussed.

## 15.2. External Characteristics

In the following section, some shared, similar and intermediate states of external characteristics in different species, as well as subspecies, are presented briefly (see Figure 1). They indicate a close relationship among the Equidae.

The size of equids is highly variable, ranging from approximately 70 to 150 cm shoulder height. The form of the croup varies with the environment (Groves 1974): asses, with their high and angular croup, (Figure 1c) apparently adapted best to stony desert areas, horses with their wide, rounded croup (Figure 1a, b) to prairies. The plains zebra is described as the most horse-like zebra, the Grevy's zebra as the most donkey-like zebra (Petzsch 1971).

Despite different sounds (horses neigh, asses and onagers bray, Grevy's zebras bellow like deer, plains zebra bark, mountain zebras whistle or neigh with subsequent toot), common traits exist: if no or only little contact to other animals of the same species takes place, the individual species still react to sounds made by animals of the other species (Klingel 1987). The different species of Equidae possess horny calluses of varying size on the inner sides of their front legs (Figure 1c-f). Przewalski's and domestic horses have them on all four legs. There is possibly a tendency towards the loss of this characteristic (Antonius 1951). The horny calluses can for example be missing in Grevy's zebras (Groves and Willoughby 1981). Wild Equidae, in contrast to domesticated species, possess an upright mane (Figure 1a, d-i).

Between the typical horse tail (long flowing hair inserting at the root of the tail, Figure 1b) and the tail of zebras and donkeys (long tail with tassel, Figure 1f-i) there are intermediate forms: in some horse breeds and in Przewalski's horse (Figure 1a) (Schirneker after Lotsy 1922; Groves 1974) the dorsal long hairs insert distal to the base of the tail root. In the kiang (Figure 1e), the longer hairs insert laterally, as far up as the base of the tail (Groves 1974).

Figure 1. Extant species of the genus *Equus* (with their special characteristics).
a) Przewalski's horse: upright mane; dark dorsal streak; tail: dorsal base of the long hair distal to the tail root;
b) domestic horse: wide, rounded off croup; tail: long flowing hair attached directly to the root;
c) domestic donkey: shoulder cross; zebra stripes on the legs; horny callus visible on the right front leg; high angular croup;
d) kulan: dark dorsal streak; long tail with tassel;
e) kiang: dark dorsal streak; shoulder stripe, upright mane; horny callus visible on the left front leg; tail: base of the longer hair on the side up to the root of the tail;
f) Grevy's zebra: narrow stripes; stripes do not run across the belly; upright mane; horny callus visible on the right front leg;
g) Grant's zebra: wide, black stripes on white background, stripes run across the belly; fully developed stripes on legs;
h) Damara zebra: reduced stripes on legs and body, as compared with g); basic body color pale light brown; dark brown and indistinct intermediate stripes; long tail with tassel;
i) Hartmann's zebra: dewlap; stripes do not run across the belly.

The coat coloring of horses and asses shows high variability. As in different horses, the coat of domestic donkeys can be monochrome, variegated or striped. According to van Vleck (1977), almost all variations are possible. The dark dorsal streak (as seen in Figure 1a, c-e, g, h) occurs more or less distinctly in all Equidae, except in a number of domestic breeds of horses. The so-called shoulder cross (a shoulder stripe across the dorsal dark streak) can be present not only in the domestic donkey (Figure 1c), but also in all African and Asiatic wild asses (Figure 1d), as well as in Przewalski's horse (Groves and Mazák 1967; Grzimek 1977, p. 42).

Zebra stripes are widespread among the Equidae. According to Mohr and Volf (1984) and Schirneker (after Lotsy 1922), Przewalski's horses and Somalian wild asses usually have striped legs, which in the tarpan (subspecies of the wild horse, Polish breed) is only occasionally the case. In the tarpans with this pattern, stripes can even run across the back (Goodall 1966). Occasionally, Asiatic wild asses have striped legs (Groves and Mazák 1967), as do the domestic breeds Sorreia pony, Criollo and fjord horse; the latter may have stripes also on the forehead (Goodall 1966, Silver 1978). Zebra-like patterns can also be found on the legs (Figure 1c) or appear on the rear of the domestic donkey (Borwick 1967).

A typical characteristic of the Grevy's zebra is the distinctly narrow stripes (Figure 1f). Plains zebras have fewer and wider stripes (Figure 1g), which in contrast to the two other species of zebras (Figure 1f, i), run across the belly. Mountain zebras have somewaht narrower and more numerous stripes than plains zebras (comp. Figure 1g, i).

Among the zebras, there exists a remarkable variety of stripe patterns. Only basic traits like width, simplicity, or raggedness of the stripes are transmitted, but not the details of the pattern (Klingel 1987). These are formed individually during the ontogeny. The most frequent pattern consists of simple stripes with few ramifications. Besides those, there are variations with multiple ramifications, as well as partly marbled or dotted variations. Occasionally, white or almost white animals or virtually black zebras with white stripes resembling strings of pearls have been observed (Short 1975; Grzimek 1977, 61,67; Klingel 1987). The variety of the stripe patterns is particularly obvious in the plains zebras. Klingel and Antonius describe a variation in stripe colors, background body colors, and leg stripes, which is strikingly graduated along a north-south direction, analogous to the geographical distribution of the subspecies (Figure 2): The Northern forms of the plains zebra (Grant's zebra) have distinct black stripes on a white background (see Figure 1g). In the forms living further south, the background color gradually changes to yellow-brown, the stripes of the legs and body become less distinct (Figure 1h), and lighter intermediate stripes (Figure 1h) are inserted between the main stripes. Thus, Burchell's zebra living in the south of Africa possesses very distinct intermediate stripes, the striping of the upper legs is partially disintegrated, and the striping of the lower legs is missing completely. The lowest degree of striping existed in the extinct

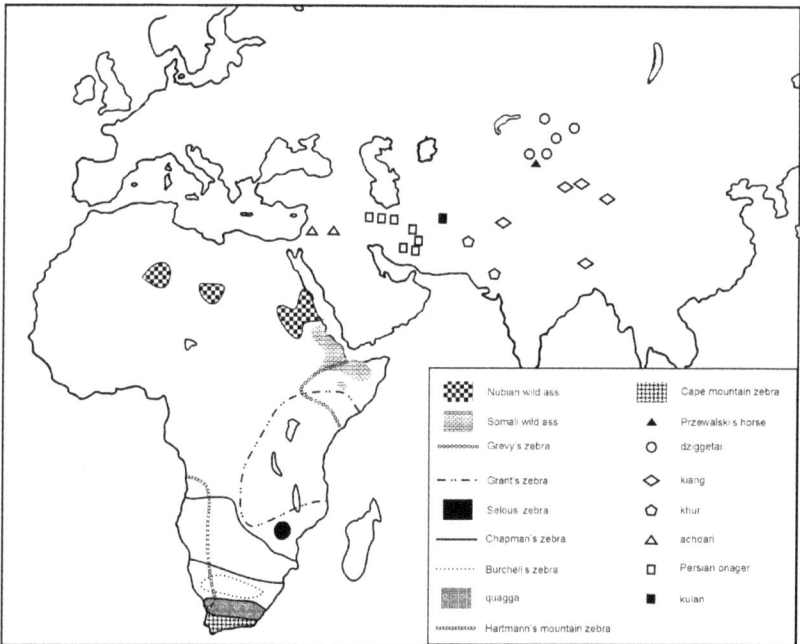

Figure 2. Distribution of the genus *Equus* in historical time (after Antonius 1951, Groves 1974, and Groves & Mazák 1967).

quagga. Its dark brown striping extended only from the head to the shoulders and then disappeared in the pale reddish-brown background. Its legs were beige. There are no distinct boundaries between the colors and patterns of the coat, so that a correct assignment can even fail within one and the same population (Klingel 1987; Antonius 1951).

The examples described above prove the high variability of certain characteristics (like, for example, size, physique, coat color and striping) in the different species of the Equidae. At the same time they raise questions regarding the phylogeny of the Equidae, as well as the causes and mechanisms responsible for the observed variability.

### 15.3. Karyotypes

After the brief presentation of the external features of the Equidae, an overview of the chromosome structure will now be given. With the help of suitable preparatory methods, chromosome charts (karyotypes) with different band patterns can be produced. Such bands allow the identification of homologous chromosomes. At first, the different positions of the centromer attract attention. Their position is used as a systematic criterion (Figure 3) for determining different karyotypes. Maciulis et al. (1984) propose a division into two main groups. Within these groups, the chromosomes are sorted according to their length. With the help of the C-band technique, the constitutive heterochromatin and the position and relative frequency of highly repetitive, non-coding sequences are made visible. Surprisingly, the number of chromosomes varies considerably (Table 2) from species to species, although externally the Equidae are very similar and are able to hybridize.

| two-armed chromosomes | one-armed chromosomes |
|---|---|
| 1. metacentric | acrocentric |
|  ← centromere |  |
| 2. submetacentric | |
|  | |

Figure 3. Different chromosome configurations, based on the position of the centromere.

The tendency towards an increasing number of metacentric autosomes and a decreasing number of acrocentric autosomes, with a simultaneous decrease of the total number of chromosomes, suggests a phylogenetic cause. Although the chromosome number of the domestic horse (2n = 64, 26m + 36a autosomes) and the donkey (2n = 62, 38m + 22a autosomes) are notably different from that of the mountain zebra (2n = 32, 28m + 2a autosomes), the amount of DNA is identical (Table 2). Unfortunately, thus far only an initial examination of the karyotypes of the Equidae has been done (Ryder et al. 1978). Notable results are summarized in the following section.

The chromosomes of the Equidae are characterized both by shared conditions and differences. The karyotype of the donkey possesses few similarities to those of the other Equidae (Benirschke 1969; Ryder et al. 1978; Kopp et al. 1983). Although according to its chromosome number the donkey is closer to the horse than to the onager, these studies point to a closer relationship with the onager (Benirschke and Ryder 1985). A comparison of the banding patterns of the Equidae suggests several pericentric inversions in the chromosomes. For example, the G-banding patterns of the metacentric X-chromosome in Przewalski's horse, the domestic horse, Persian onager, Grevy's zebra, Grant's zebra, and Hartmann's mountain zebra are identical. Only in the donkey, it seems to have changed around the centromere, due to an inversion, because it is no longer located in the center but close to one end, i.e., almost acrocentrically (Ryder et al. 1978). Morphologically, the Persian onager stands between horse and donkey. However, a comparison of the karyotypes shows that there is a higher degree of homology between onager and donkey than between onager and horse or between donkey and horse.

Ryder (1978) detected a chromosomal polymorphism in the onagers. He found a homology between an unpaired metacentric and two non-homologous acrocentric chromosomes in the G-banding karyotype of the kulan (2n = 55). He succeeded in identifying the two acrocentric chromosomes of E. h. onager (2n = 56), which are fused into

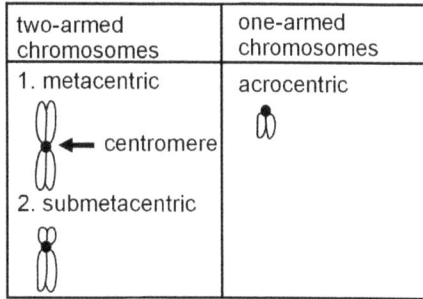

| | diploid set of chromosomes | Autosomes | | | amount of DNA[7] |
|---|---|---|---|---|---|
| | | M | A | NF | |
| E. p. przewalskii | 66 | 24[2] | 40 | 92[2] | |
| E. p. caballus | 64 | 26[2,4] | 36 | 92[2] | .993 (SD±.018)[1] |
| E. a. asinus | 62 | 38[2] | 22 | 102[2] | 1.033 (SD±.014)[5] |
| E. h. onager | 56[2,3] | 44[2] | 10 | 102[2] | |
| E. h. kulan | 55/54[3] | | | | |
| E. h. kiang | 51/52[6] | | | 92[6] | |
| E. grevyi | 46 | 32[2] | 12 | 80[2] | |
| E. q. boehmi | 44 | 34[2] | 8 | 80[2] | |
| E. q. antiquorum | 44 | 34[2] | 8 | 80[2] | |
| E. z. hartmannae | 32 | 28[2] | 2 | 62[2] | .990 (SD±.013)[5] |

Table 2. Chromosome numbers in the Equidae. M: metacentric, A: acrocentric, NF: "nombre fondamental", *i.e.*, number of the chromosome arms. [1]after Atkin *et al.* (1965), [2]after Ryder *et al.* (1978), [3]after Ryder (1978), [4]after Maciulis *et al.* (1984), [5]after Benirschke & Malouf (1967), [6]after Ryder & Chemnick (1990). [7]amount of DNA in relative units, compared to human control cells.

a metacentric chromosome in *E. h. kulan*. In one individual, only 54 chromosomes were found, whereas both parental animals possessed 55 chromosomes. The C-banding karyotype of the kulan showed a loss of constitutive heterochromatin at the centromere of the fused metacentric chromosome in contrast to the acrocentric chromosomes of the Persian onager, but this can occur in the course of such a rearrangement. Ryder and Chemnick (1990) recently discovered a case of Robertson's polymorphism in the kiang. They found individuals with 51 as well as with 52 chromosomes. The fused as well as the two acrocentric elements appeared to be homologous to the element that is part of the Robertson's rearrangement of Perian onager and kulan.

Similarly, a distinctive chromosomal homology exists between the species pairs of Przewalski's horse / domestic horse and Grevy's zebra / plains zebra. In these two cases also, it has to be assumed that a case of Robertson's chromosome fusion occurred (Short *et al.* 1974; Ryder *et al.* 1978). Because the karyotypes of the Grevy's and the plains zebra also show differences in certain chromosome bands, one can assume that further mechanisms of chromosome changes were involved in the speciation process. For example, two acrocentric chromosome pairs of the Grevy's zebra contain both centromeric and telomeric (located at the ends of the chromosomes) C-bands, unlike any of the acrocentric chromosomes of Grant's zebra (the karyotypes of the Damara zebra and of Grant's zebra as the two [examined] subspecies of the plains zebra are virtually identical). Furthermore, the Grevy's zebra / onager hybrids seem to be sterile, in contrast to those of the true horses (comp. Figure 8). The karyotype of Hartmann's mountain zebra is similar to the other two species of zebra only in a few chromosomes. One of its G-banded autosome that is similarly banded in all other Equidae apparently is modified by two inversions. Among the three species of zebras, various banding patterns, which do not exist in other Equidae, closely resemble each other, possibly indicating a closer genetic relationship (Ryder et al. 1978). One individual each of the Somalian wild ass and Grant's zebra was found with a Robertson's fission of a long metacentric chromosome resulting in a chromosome number of $2n = 63$ and $2n = 45$, respectively (Benirschke and Ryder 1985).

The C-banding patterns show that the decreasing number of chromosomes is correlated with a decrease of centromeric heterochromatin and an increase of telomeric and interstitial heterochromatin (Wichman et al. 1991). In Przewalski's horse and the domestic horse for example, one finds constitutive heterochromatin on the position of the centromeres of all chromosome pairs except two. In species with a lower number of chromosomes, fewer centromeric C-bands appear. In such cases, the bands are mostly telomeric and normally located at the ends of the short arms. These telomeric C-bands could have originated due to inversion processes. However, this does not apply in every case as the G-banding pattern, in spite of the changed C-banding patterns, sometimes remains the same (Ryder et al. 1978). The constitutive heterochromatin shows a quantitative heteromorphism in the homologous chromosomes, particularly in the onager (Ryder et al. 1978). Wichman et al. (1991) describe four different classes of repetitive sequences, and on the basis of their distribution in the genome of the different species of Equidae, they were able to confirm their supposition that extensive intragenomic movements of these sequences took place between non-homologous chromosomes and chromosomal fields. The structure of the chromosomes of the Equidae confirms the morphological results of the external appearance and does not contradict the postulate of a common ancestral form (see Discussion).

## 15.4. Molecular Taxonomy

The chromosome differences suggest a number of biochemical differences in the species of the Equidae. The results of electrophoretic separation, for example, of serum albumin and transferrin, show that the degree of variation on a certain gene locus can be higher within the same species (!) than between different species. Esterase tests showed that the degree of variation is higher within the true horses than between different species. A cladogram (Figure 4) based on these results shows a main branch leading to the true horses, with a side branch leading to the onagers, and another main branch leading to the asses, with a side branch to the zebras, from which Hartmann's mountain zebra diverged sooner or later (Kaminski 1979).

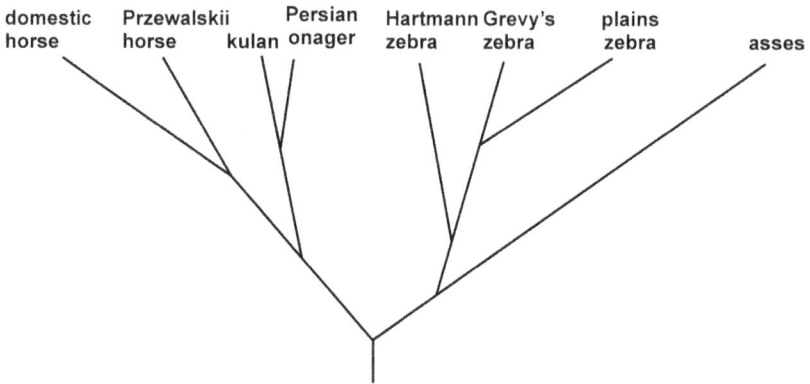

Figure 4. Cladogram based on esterase studies (after Kaminski 1979).

The amino acid sequences of α and β globins, as well as electrophoresis patterns of the serum, showed that asses, onagers, and zebras are much more similar to each other than to the horses. Therefore it would have to be assumed that the true horses have separated from the main line of Equidae before all other species (Kaminski 1979).

George and Ryder (1986) examined the mitochondrial DNA (mtDNA) of different Equidae with the help of restriction analysis (Figure 5). Among 118 gene loci of an mtDNA sequence, 29 were identical in all six species, 14 were highly conserved, i.e., they existed in five out of six species, 44 existed in only one of six species, and 31 were phylogenetically significant. The domestic horse (except the Arabian breed), the Persian onager, and the kulan possess a common gene locus lacking in all the other Equidae (besides that, the Persian onager and kulan show an identical cleavage pattern). George and Ryder mention another restriction endonuclease test, which showed an identical pattern in asses and zebras, whereas the patterns of onagers and true horses differed not only from each other but also from the ass-zebra pattern. Cladistic analysis of the data led to the conclusion that, during the speciation of the Equidae, three main divergences seem to have occurred: 1) the separation of the true horses, 2) the separation of the asses and onagers, and 3) the branching off of the three zebra species, which apparently form a monophyletic group in which Grevy's zebra and the plains zebra separated at a very late stage (Figure 5d, e, f). The phylogenetic positions of E. asinus and E. hemionus remain uncertain because the ancestor of the true horses cannot be distinguished from the ancestors of the asses and onagers. The phylogenetic tree based on protein and chromosome data (Figure 5d) confirms the monophyletic status of the zebras and also shows the above-mentioned three main clades. The systematic position of the quaggas was for a long time obscure. Higuchi et al. (1984), using a skin specimen from a museum, isolated and sequenced a mitochondrial DNA sequence of 229 base pairs that differed in 12 positions from that

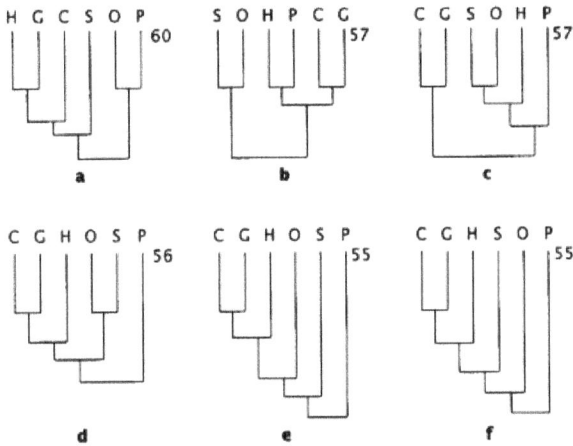

Figure 5. Phylogenetic trees on the basis of the principle of parsimony (after George & Ryder 1986). The number next to the dendrogram is the minimum number of mutations necessary to generate this tree. a, b, c: trees based on morphological criteria: dental (c), cranial (b) and postcranial (a) traits; (d): based on data of protein and chromosome studies; (e) and (f): based on data from mtDNA-restriction-analysis. C-Chapman's zebra (representing the plains zebra), G-Grevy's zebra, H-Hartmann's mountain zebra, O-onager, P-Przewalski's horse, S-Somalian wild ass.

of the mountain zebra and in only one position from that of the plains zebra. An examination of the serum protein likewise supports a much closer association of the quagga with the plains zebra than either the mountian or Grevy's zebra (Lowenstein and Ryder, according to George and Ryder 1986). Groves (after Vogt 1986) showed that the quagga and plains zebras were geographically separated, supporting the quagga as a subspecies of the latter.

## 15.5. Contradictory Phylogenetic Trees

On the basis of different sets of data, George and Ryder (1986) presented the phylogenetic data in the form of dendrograms, based on the principle of parsimony (Figure 5). The dendrogram that contains the fewest number of mutations is considered the most likely one. A comparison of the dendrograms of protein and chromosome studies (Figure 5d), as well as the mtDNA cleavage maps, (Figure 5e, f) shows some agreement as far as the separation of the true horses and the zebras is concerned. Although the cladogram (Figure 4) created by Kaminski (1979) partially agrees with phylogenetic tree f in Figure 5, there is still no general consensus among the different phylogenetic trees. Furthermore, no agreement exists even among the phylogenetic trees a, b, and c in Figure 5, which are based on postcranial, cranial, and dental characteristics, respectively.

|              | E. przewalskii | E. asinus | E. hemionus | E. grevyi | E. quagga | E. zebra |
|--------------|:---:|:---:|:---:|:---:|:---:|:---:|
| E. przewalskii | ● | ● | ● | ● | ● | ● |
| E. asinus      | ● | ● | ● | ● | ● | ● |
| E. hemionus    | ● | ● | ○ |   | ● | ● |
| E. grevyi      | ● | ● |   | ○ | ● | ● |
| E. quagga      | ● | ● | ● | ● | ● | ● |
| E. zebra       | ● | ● | ● | ● | ● | ● |

Figure 6. Crossbreeding matrix of interspecific hybrids. ● = hybridization successful; ○ = hybridization successful according to Gray (1972): achdari × khur, kulan × kiang. No hybrid between onagers and Grevy's zebras has been documented thus far. The points in this matrix do not reflect the direction of the hybridization.

## 15.6. Interspecific Hybrids Among the Equidae

Since interspecific hybridization is the criterion for the classification of species within one and the same basic type, this section gives an overview of interspecific hybrids among the Equidae. In some cases, different species or races of Equidae can be observed coexisting in nature. Normally no permanent contacts develop between them, however similar they might be. Previous descriptions showed that there might have been an intermediate race of mountain zebras in the border region of the Cape mountain zebra and Hartmann's mountain zebra. A hybrid population apparently also existed in the border region between the two wild horses, the Przewalski's horse and the tarpan, which became extinct in 1887. In both cases there was no reproductive isolation (Groves 1974), so both types of wild horses can possibly be regarded as geographical variations (races). Today there are still overlapping distribution areas (see Figure 2) of Przewalski's horses and onagers, of Grevy's zebras and African wild asses, and of Grevy's zebras and plains zebras. In the latter case, mixed herds often develop temporarily. Plains zebras and mountain zebras can also be found in the same areas but do not form mixed herds. This is an interesting phenomenon in the context of isolation mechanisms because in captivity they easily mate, in contrast to the Grevy's zebras (Groves 1974).

Unlike in nature, hybridizations among Equidae under human care are generally possible, as long as no other individuals of the same species are present (Kaminski 1979; Klingel 1987). The results of many years of hybridization experiments which succeed in part only by means of artificial fertilization, are summarized in the form of crossbreeding matrices. Interspecific hybridizations are shown in Figure

| | Pzewalski's horse | domestic horse | Nubian wild ass | Somalian wild ass | donkey | Persian onager | kulan | khur | achdari | kiang | Grevy's zebra | plains zebra | Grant's zebra | Chapmann's zebra | Burchell's zebra | quagga | moutain zebra | Harttmann's zebra | Cape mountain zebra |
|---|---|---|---|---|---|---|---|---|---|---|---|---|---|---|---|---|---|---|---|
| Pzewalski's horse | | 7 | | | | | | | | | 4 | | 3 | | | | 8 | 1 | 1 |
| domestic horse | 3 | | | | 10 | 3 | 11 | 3 | | | 8 | | 1 | | | | 8 | 1 | 1 |
| Nubian wild ass | | | | 3 | 2 | | | | | | | | | | | | | | |
| Somalian wild ass | | | 3 | | 4 | 3 | | | | | | | | 3 | | | 8 | | 1 |
| donkey | 10 | | 2 | | | 3 | | | | 3 | 4 | 6 | 1 | 1 | | | 2 | 1 | 6 |
| Persian onager | 3 | | | 3 | 3 | | | | | | 4 | 1 | | | | | | | |
| kulan | 11 | | 11 | 3 | 3 | | | | 3 | | 3 | | | 3 | 3 | | | | |
| khur | 3 | | | | 3 | | | | | | | | | | | | | | 1 |
| achdari | | | | | 3 | 3 | | | | | | | | | | | | | |
| kiang | 3 | | | 3 | 3 | 3 | | | | | | | | | | | | | |
| Grevy's zebra | 6 | | 1 | | | | | | | | | | 9 | | | | 8 | 1 | |
| plains zebra subsp. unknown | 1 | 4 | | | 3 | | | | | | | | | | | | 4 | | |
| Grant's zebra | 3 | 6 | | | | | | | | | | | | 1 | 3 | | | 3 | |
| Chapman's zebra | 1 | 3 | | | | | | | | | | | 1 | | | | 3 | 1 | 1 |
| Burchell's zebra | 3 | | | | 3 | | | | | | | | | | | | 3 | 1 | 1 |
| quagga | 1 | | | | | | | | | | | | | | | | | | |
| mountain zebra, subsp. unknown | 8 | | 2 | | 4 | | | | | 4 | 4 | | | | | | | | |
| Hartmann's mountain zebra | 1 | | 2 | | | | | | | | | | | | | | | | |
| Cape mountain zebra | 1 | | | | | | | | | | | | | | | | | 1 | |

Figure 7. Crossbreeding matrix among subspecies. The only hybrid that was never documented is the one between Grevy's zebra and onager. A hybrid between the achdari and the khur was claimed, but never proven. According to Short (1967), Benirschke (1969), Short et al. (1974), and Klingel (1987), all Equidae can be hybridized with each other. Literature references for the hybridizations shown in the matrix: 1. Antonius (1951); 2. Benirschke (1967); 3. Gray (1972); 4. Groves (1974); 5. King (1967); 6. King et al. (1966); 7. Koulischer & Frechkop (1966); 8. Mochi & MacClintock (1976); 9. Ryder et al. (1978); 10. Short (1967); 11. Volf (1979, p. 564). If the sex of the parental animals was not stated in the literature, the crossbreeding was included in the matrix reciprocally (in both directions). The maternal parents are listed horizontally, the paternal ones vertically.

6, hybridizations between subspecies in Figure 7. The descendants are all viable but generally sterile (with exception of the hybrids between Przewalski's horse and the domestic horse: Short 1967; Short et al. 1974). The occasional fertility of hybrids is dealt with in the following section. In the literature, no direct crossbreeding between Grevy's zebra and onagers has thus far been mentioned; however, they are indirectly connected because both of them hybridize with mountain zebras or the true horses, for example.

As far as the terminology is concerned, the hybrids between male donkeys and the mares of horses are called mules, whereas the hybrids of reciprocal crosses are called hinnies. Hybrids of zebras and other equids are called zebroids, (zorse, zonkey, etc.). In general, zebroids, as well as mules and hinnies, look like an intermediate of the parental species; the siblings can differ strongly in details such as the length of the ears, the tinge of the background color, or the color of the stripes (Antonius 1951). The characteristic zebra striping appears in varying degrees of intensity either on the whole body or only on the legs (Petzsch 1971). The area between the striping of torso and upper legs shows either a pattern of smaller patches or a lattice pattern. The stripes on the head and the limbs are well defined, the ones on the body less well, and they are much finer, as well as more numerous than on the zebra parent (in Grevy zebroids at least on some body parts) (Groves 1974). This narrow striping is reminiscent of the fine stripes of the Grevy's zebra, even if the zebra parent is not a Grevy's zebra.

The striping of the legs is normally well defined. If both parents possess weak or no leg stripes, it is usually much stronger in the zebroid. In this case, either the horse or the white-legged zebra parent still possesses the respective genetic information, although in a masked state. Antonius mentions the following two examples: A hybrid of the white-legged parents Burchell's zebra and Criollo mare showed a strong striping of the legs. Either the Criollo mare still possessed the genetic information for striped legs, which according to Goodall (1966) now and then occurs within this breed, or the Burchell's stallion possessed this information in masked form. A hybrid between a quagga stallion and a chestnut mare of Arabian descent showed an indistinct striping of the head and body as well as distinctly striped legs. In this case, the maternal chestnut mare apparently possessed the genetic information for a striped leg because the half-siblings from the crossbreeding with an Arabian stallion all showed a very distinct striping of the legs. In a hybrid between a Grevy's zebra stallion and the mare of Hartmann's mountain zebra, indistinct intermediate stripes on the upper legs appeared, present in neither of the parental species (Antonius 1951). It is interesting that this kind of striping is typical of Chapman's zebras and Burchell's zebras, and, thus, of representatives of a third species. Petzsch (1971) thinks that there is a strong phenotypic resemblance between some horse zebroids and the quagga. The monophyletic development of the zebra group was already mentioned.

### 15.7. Sterility of Hybrids

In view of the large number of documented crossbreedings (Figures 6 and 7), the question of the cause of the almost universal sterility of the hybrids arises. The hybrids show an intermediate number of chromosomes of both parental karyotypes (Short 1967). Apart from the fertile hybrid of Przewalski's horse and the domestic horse, only hybrids between representatives of the same subspecies seem to be fertile (Figure 8). Examples are the hybrids between the Nubian wild ass and the domestic donkey as well as various plains zebra hybrids (Benirschke 1967).

In the following section, some histological and genetic data of the male and female hybrids will be presented. The sterility of mules,

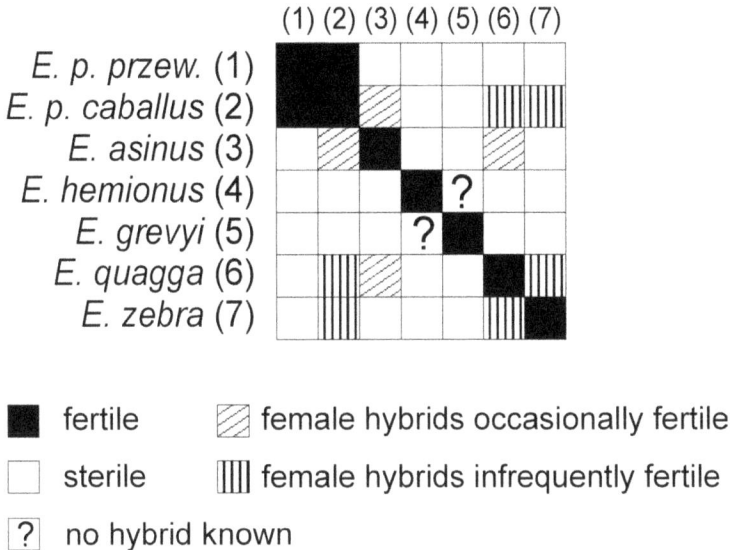

Figure 8. Fertility and sterility of hybrids of the family Equidae; after Gray (1972) and Short (1967).

for example, has been known for at least 2000 years (Taylor and Short 1973). However, what are its causes? The male hybrids possess normal genitals and a normal libido, while there is almost always an absence of sperm cells.

The endocrine function proved to be altogether normal (normal Leydig's cells, normal concentration of testosterone); the exocrine activity of the testicles, however, is reduced. Detailed studies showed that the chromosome sets of horses and asses differ in structure and number. As a result, the combination of homologous chromosomes during the meiotic synapsis may not be possible, and the reproductive cells of the hybrids degenerate (Anderson 1939; Benirschke et al. 1962; Short 1967; Trujillo et al. 1969; Chandley et al. 1974). Examinations of the hybrids between Grevy's zebra and horse, as well as between Grant's zebra and donkey, produced similar results (King et al. 1966; King 1967). The testicles of these two hybrids were morphologically smaller than those of the parental species (King et al. 1966). Incidentally, the seminiferous tubules are hypocellular and appear to be atrophic (Benirschke et al. 1962; King et al. 1966). Histologically, the testicles of mules and hinnies look like those of zebroids (Benirschke 1967).

Zong and Fan (1989) found spermatids in groups of four, the nuclei of which were connected by bridges, as well as some differentiated spermatids, but no spermatozoa (i.e., sperm cells). Even though sperm cells were not observed in the sectioned seminiferous tubules, low numbers of sperm cells were found in the seminal fluid of mules and hinnies, as well as horse zebroids (according to Chandley *et al.* 1974; Zong and Fan 1989). Ohno and Trujillo (after King 1967) suspected that spermatogenesis could proceed to the stage of mature spermatozoa if tetraploid spermatogonia (undifferentiated sperm cells) exist. However, the density of the sperm cells is not high enough to ensure fertility. An examination of the testicle of a hinny by Trujillo *et al.* (1969) showed in different cross-sections of seminiferous tubules various centers of active spermatogenesis with large spermatogonia. Plus, each secondary spermatocyte seemed to contain four instead of two nuclei, which would confirm the above-mentioned assumption of Ohno and Trujillo. The seminal fluid of the hinny was clear and contained only a few immobile sperm cells. In a probe with 1.5 million sperm cells per ml, their morphological appearance was normal; in a later probe with 0.1 million sperm cells per ml, there was a considerable number of abnormal sperm cells with two heads or tails and with a tendency to stick together. (According to Day [after King 1967], a much higher density of sperm cells can be normally found in horses, namely 50-200 million cells per ml of seminal fluid.) The length of the head and tail of the mule sperm cells was not significantly different from that in donkey and horses, which indicates that the sperm cells of hinnies are haploid rather than diploid. The immobility of the sperm cells could be caused by the fact that during the distribution of chromosomes not every sperm cell was equipped with a complete set of haploid chromosomes. The DNA-content in the sperm cells would then have to be variable. Bratanov *et al.* (after Trujillo *et al.* 1969) actually found sperm cells in the seminal fluid of a mule which contained less DNA than the parental gametes. Biochemical differences in the seminal fluids of mules, horses, and donkeys may also play a role in the sterility (Bratanov *et al.* after Trujillo *et al.* 1969; see also below).

Similarly, the endocrine function in the ovaries of female mules seems undisturbed, whereas the exocrine function is reduced, as in the male hybrids. Histological examinations showed that primordial follicles are rare and possess an enlarged cortex that possibly prevents regular ovulation. Follicle cysts, corpora lutea in different developmental stages, corpora albicanta, and hemorrhagic follicles were found sporatically. Mature ova were not discovered, whereas they are easy to find in the ovaries of donkeys and horses (Benirschke and Sullivan 1966; Zong and Fan 1989). As in the male hybrids, the cause of the sterility is found in the first division of meiosis, during which the combination of homologous chromosomes fails after the initial steps. Examinations of hybrid fetuses showed that germ cells migrate to the gonads and undergo mitosis (Taylor and Short 1973). However, they degenerate a few days after birth, when the first stages of meiosis have begun. Some do survive and develop into ova, which apparently induce the development of normal follicle cells. Therefore, the ovary can later be endocrinally active. Also, these ova seem to survive until ovulation (the condition of their chromosomes remains unclear). A development of follicles was observed also in horse zebroids, so that it is altogether possible that complete meiosis can take place in female hybrids. Since the low germ cell density of a male mule is insufficient for fertilization, it is more likely to find a female mule in which a random ovulation and the successful mating with the stallion of a horse or donkey coincide (King et al. 1966). Although supposedly fertile mares of mules were frequently reported, the genetic, serological, or other proof of a genuine hybrid was usually missing (Benirschke et al. 1964; Benirschke and Sullivan 1966).

Ryder et al. (1985) finally succeeded in proving the genuineness of a fertile female mule, using karyotype and blood type analyses. The analysis of the blood type showed that the male descendant of the female mule inherited only the haploid set of maternal chromosomes (and thus the chromosomes of the horse). This would confirm the hypothesis of Anderson (1939), according to whom the paternal set of chromosomes is lost during meiosis as a pole cell. It would also support previous reports that hybrids of fertile female mules and male donkeys were mule-like, whereas hybrids between female mules and horse stallions were horse-like (Anderson 1939; Smith 1930). Chandley (1981) applied Anderson's hypothesis also to hinnies. Zong and Fan (1989) applied phenotype studies to 11 descendants of fertile hinny and mule mares. In only one reciprocal backcross (female mule × horse stallion) could Anderson's hypothesis be confirmed phenotypically as well as genetically. The mule mare had transmitted only her maternal horse-chromosomes to her foal. All other descendants (hinny × horse stallion; mule × donkey stallion; hinny × donkey stallion) were obvious phenotypical hybrids. Anderson (1939) suspects that the high fetal mortality in pregnant mules results from chromosome combinations that are only partially viable, thus originating from such ova lacking the complete maternal set of chromosomes (from the horse parent). According to observations by Zong and Fan, this is not always the case. Zong and Fan examined other, unquestionably fertile, female mules and hinnies, including their offspring. A female descendant's histological data (female mule × horse stallion) shows different follicle stages (as in the mule), but no ova. The seminiferous tubules of male descendants (female hinny × donkey stallion) contained more spermatogonia and primary spermatocytes than those of most mules. Some secondary spermatocytes and spermatids were also found, but no sperm, not even in the seminal fluid. With growing age and after adding PMSG, more spermatids were differentiated and some deformed spermatozoa were found. These changes of spermatogenesis indicate that the chromosomes are probably not the only cause of hybrid sterility (Zong and Fan 1989). Zong and Fan (1989) conclude from their examination of female mules and hinnies that there is a trend towards fertility. The degree of fertility varied greatly from individual to individual. If one could examine even more male hybrids, one might also be able to detect a trend towards fertility.

In view of the general sterility of hybrids in the Equidae, the question arises why the descendants of hybrids between Przewalski's horse and the domestic horse are fertile. Examinations showed that the spermatogenesis including the synapsis proceeded normally. The homology between the parental chromosomes is clearly recognizable. During meiotic synapsis, a trivalent is formed between a metacentric domestic horse chromosome and two homologous acrocentric Przewalski's horse chromosomes (Short et al. 1974). Thus, the fertility of hybrids between Przewalski's horse and the domestic horse can be explained with the close relationship of chromosome number and structure, as well as with the meiotic formation of a trivalent (King 1967).

## 15.8. Discussion

### 15.8.1. The Equidae as Basic Type.
During the Pleistocene, the horse family was distributed across all of America, Asia, Europe and Africa. However, only one extant genus with six species survived. Systematically, the monodactyl (one-toed) horses stand next to the three-toed rhinoceros and the tapirs with four front and three rear toes. These three families of the Perissodactyla are traced back to a common ancestral group in the systematic proximity of the fox-sized *Hyracotherium* (better known as the "dawn horse"), according to classic evolutionary ideas. This viewpoint has found widespread popular acceptance. However, the "dawn horse" has very little in common with the Equidae. Although originally placed amongst the hyrzxes, it displays many similarities to the fossil perissodactyls (Crompton, this volume). The relationship between the three extant perissodactyl families is just as unclear. Wüst (1966, p. 79) for example, states, "that we unite the tapirs and the perissodactyls with the monodactyls, although there are very few similarities. A tiny, conical tail stub, small pig-like eyes of inferior vision, rounded ears, a half-long trunk and a layer of fat have nothing in common with the appearance of extant horses." A critical evaluation of the fossil Equidae is not the object of the present work.

According to Marsh (1976; see also Scherer 1993, this volume) all individuals that are directly or indirectly connected through crossbreeding can be considered to belong to a single basic type. In accordance with this requirement, all Equidae are directly or indirectly connected through hybrids (comp. Figure 6 and 7). The success of the hybridization experiments points to a very close genetic relationship, which is not a surprise given the very similar anatomy of all horses. There are a number of morphological and ethological similarities, such as the dark dorsal streak, the mane, the calluses, voice recognition, as

Stein-Cadenbach - Equidae                                  361

well as characteristics that demonstrate an intermediate position between different species, such as intermediate tail forms. The narrow striping of zebroids that are not offspring of Grevy's zebras indicates both genetic interferences, as well as relationship among the Equidae. The above-mentioned traits, characteristics of hybrids that normally occur only in a third species (like the indistinct intermediate stripes of the hybrid between Grevy's zebra and Hartmann's mountain zebra), and occasional zebra-like or shoulder stripes can be interpreted within the context of the microevolutionary development of a genetically polyvalent ancestral form as a consequence of variable combinations of genes.

The resulting speciation is certainly also influenced by environmental conditions. For example, mountain zebras need to live in the mountains, because their hard hooves would not be worn off sufficiently in the soft ground of the plains and would, thus, grow out and ultimately impede the mobility of the animal. Mountain zebras would then turn into an easy prey for carnivores (Klingel 1987). Newly combined genetic sequences could also lead to the disappearance of a certain trait (for example the calluses on the hind legs) or to the sudden appearance of a known characteristic (e.g., zebra stripes on legs in normally unstriped domestic horses). Furthermore, this concept would also provide an explanation for specific traits, such as the dewlap (Figure 1i), which is typical for the mountain zebra.

The tendency towards a monochrome body color ("stripe degeneration") in the subspecies of plains zebra (see the section on "external characteristics") is used as an argument to interpret the zebra as the most primitive extant equid and the horse as the most highly developed (Krumbiegel 1958; Petzsch 1971). The occasional zebra striping in wild horses, domestic horses, asses, and onagers is subsequently interpreted as the reappearance of an ancestral trait.

In this context, the Grevy's zebra (Figure 1f) is classified as the most primitive form among the zebras due to its "original" characteristics, such as the dense striping and the pure black-and-white coloring (Thenius 1979, p. 544; Volf 1979, p. 544). Antonius (1951) interprets the calluses as unexplained rudiments. If, however, the horse is the highest developed form of the Equidae, then why does it have these calluses on all extremities, in contrast to the other species? The findings about the family Equidae in the newer literature actually do not give any indication that any one equid would have to be classified as more primitive than the others. The facts point to a restructuring in the genetic material in connection with mechanisms like isolation and selection, rather than to increasing complexity. On the basis of the data presented in this work, an original ancestral form of the horses, whose potential variability was considerably higher than that of present-day species, is proposed. All extinct, as well as present-day, representatives of the horse family could

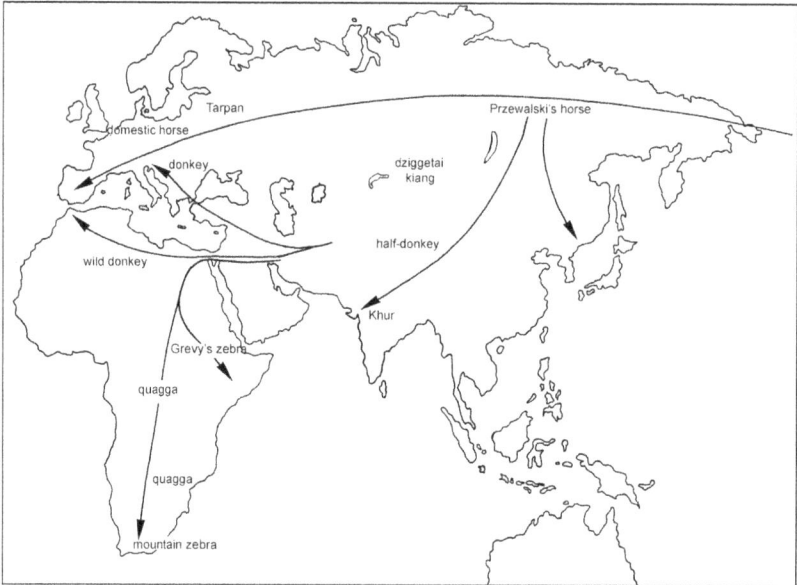

Figure 9. Migration routes of the Equidae from the New World into the Old World. (according to Benirschke 1969 and Thenius 1987).

have developed monophyletically from this ancestral form (radiation) in accordance with the basic type concept. This development cannot be reversed, as the attempt to recreate the tarpan has shown (in the form of the back-bred Polish tarpan) (Groves 1974).

**15.8.2. Chromosome Evolution and Speciation.** Speciation within the Equidae can only be reconstructed hypothetically. Nevertheless, there are some data available (especially on the basis of an analysis of the chromosomes) that allow at least a partial reconstruction of the process of speciation among the horses. Due to their mobility, the family Equidae was capable of dispersing quickly (Figure 9). Thus, on the basis of paleontological results, complex migrations must be assumed in the course of the history of this family, which will not be discussed any further in this work. *Dolichohippus* and *Hippotigris* separated from each other in North America in the Upper Pliocene. It is unknown how *Hippotigris* reached Africa, but it is certain that subsequently it was restricted to Africa during the Pleistocene. There, it divided into the mountain zebra, plain zebra, and the quagga. Whereas fossil quaggas and mountain zebras have never been found outside South Africa, plains zebras were widespread all over Africa, as far north as Algeria (Groves 1974). During the Pleistocene, wild horses populated the plains of Europe and Asia, north of the vast mountain regions— including the Przewalski's horse in Asia and the smaller tarpan in central and eastern Europe (Short 1975). In the course of the distribution of the horses, sympatric speciation through chromosomal changes within the framework of the basic type concept is plausible. When comparing the chromosome numbers of present-day Equidae (Table 2), one notices that the number of chromosomes drops from 66 to 32 from Asia to South Africa (Benirschke 1969; Short et al. 1974). In the course of the speciation of *Equus*, complex chromosome restructurings must therefore have taken place. Apparently, the migration movements are also reflected in the above-mentioned north-to-south "stripe degeneration" of the subgenera of the plains zebra. At the same time, the table shows at least a partial tendency of decreasing acrocentric and increasing metacentric chromosomes. If one interprets this as a line of descent, the new species could have originated through a fusion of acrocentric or a fission of metacentric chromosomes. As described in this work, chromosome fusions and inversions have already been identified as mechanisms (see below). Among other things, an evolution in both directions (fusion and fission) was suggested. However, an evolution through fusion of the wild horse as the most likely forerunner of the present-day Equidae is the preferred version in the literature (Simpson 1977; Benirschke and Malouf 1967; Short et al. 1974). If the Equidae have evolved through chromosome rearrangements, then the same quantity of genetic material would have to exist in every species, however differently arranged. Actually, the domestic horse, the donkey and Hartmann's mountain zebra possess the same DNA-content per diploid cell, compared with human control cells (see Table 2), so that a rearrangement of the genetic information can be assumed, rather than polyploidy or deletion (Benirschke and Malouf 1967; Ryder et al. 1978).

For each of the three pairs of Equidae, Przewalski's horse/domestic horse, Persian onager/kulan, and Grevy's zebra/plains zebra, a fusion of two non-homologous acrocentric chromosomes is highly probable. On the basis of the karyotypes of a hybrid between Przewalski's horse and the domestic horse, it was shown that a metacentric domestic horse chromosome is homologous to two acrocentric Przewalski's horse chromosomes. Ryder et al. (1978) confirmed this through their karyotype studies of the two parental horse species.

In the context of the chromosome polymorphism of onagers mentioned in the section "karyotypes," kulans that possessed the fused chromosome (thus, having a set of 55 chromosomes) produced gametes with 27 and 28 chromosomes. When mating with a typical member of the herd (having the onager complement of 56 chromosomes), their offspring had a 50% chance of possessing 55 or 56 chromosomes. After the assumed original fusion, the polymorphic herd eventually contained a sufficiently high number of descendants with a complement of 55 chromosomes that a pairing between them became likely. Of course, this is assuming that the same mutation did not occur in different individuals more or less simultaneously. With a likelihood of 25%, they produced descendants with 54 chromosomes, which possessed two metacentric instead of four acrocentric chromosomes as in the typical onager. As mentioned earlier, the extreme external similarity of the two subspecies Persian onager and kulan (Klingel 1987) has frequently caused taxonomic difficulties. According to Klingel, a few hundred years ago they most likely possessed a common area of distribution. According to Groves and Mazák (1967), intermediate forms of kulan and onager lived in the border region of Iran and Turkmenistan, since the Elbur Mountains in this area are so flat that they do not present any significant obstacle for a mixed population. Since the kulan and Persian onager today live in separate populations (Klingel 1987), it can be assumed that a kulan karyotype with a diploid set of 54 chromosomes could appear very quickly through further isolation and selection – provided, of course, that the population can be protected from extinction.

In summary, it can be said that chromosome fusion is most likely the mechanism of karyotype modification in the context of the diversification of species and subspecies of the basic type. Besides inversions, further mechanisms that resulted in a change of the karyotype must be assumed. This can be seen in a comparison of Przewalski's horse and Hartmann's mountain zebra. The former possesses 24 metacentric and 40 acrocentric autosomes, whereas the latter has 28 metacentric and 2 acrocentric autosomes (see Table 2). Why such extensive chromosomal restructurings occurred in the short "evolutionary time period" of the Quaternary (from 66 to 32 chromosomes), whereas the karyotype of other families, such as the Camelidae remained unchanged (Benirschke 1969), is unclear. Wichman et al. (1991), through their examinations of the family Equidae, were able to confirm their assumption that repetitive DNA sequences are the driving force of a swift chromosomal evolution. The multiplication and intragenomic movement of these sequences between non-homologous chromosomes are able to cause a number of chromosomal rearrangements within a relatively short time span with low effects on the euchromatic genome. Maybe the heteromorphism in the constitutive heterochromatin of the onager can also be interpreted within the framework of this hypothesis. It has to be assumed that the rearrangement of the genetic material did not proceed orthogenetically (linearly) from Przewalski's horse to the mountain zebra. Migration movements as well as catastrophes, which separated the original ancestral population into different geographical areas, could have resulted in different repeatedly branching lineages. Nothing more can be said regarding the number and type of possible chromosomal intermediate stages or the stability of the karyotypes of hypothetical intermediate forms. According to Short (1975), forms with meiotic trivalents might at least have had a selective disadvantage. Asses, onagers, and mountain zebras seem to have undergone especially complex chromosomal rearrangements (Ryder et al. 1978). The genetic restructuring with its resulting diversity of equids could, through isolation and selection processes, finally have led to reproductive isolation either sympatrically or allopatrically. The manifestation of new combinations or formations of traits is ultimately reflected in the described chromosomal incompatibilities, biochemical differences, different habitats (e.g., asses and onagers in deserts, zebras and horses in the prairies, mountain zebras in the mountains), as well as in different social and sexual behaviors (which have not been further discussed in this work).

**15.8.3. Concluding remarks.** Although the extant Equidae can be classified within a single basic type, many questions regarding their speciation remain unanswered. What was the cause of the complex chromosomal restructurings? Why can such restructurings be observed in some animal groups (e.g., Muridae, Felidae, Bovidae) but not in others (e.g., Camelidae: Benirschke and Malouf 1967; Benirschke 1969)? The exploration of the chromosomal constitution of other animal groups would be an exciting field. Furthermore, an exact karyotype analysis (as in case of the domestic horse, e.g., Maciulis et al. 1984) could also be done for all other Equidae. According to Short (1975), two skins in a museum are all that is left of the original wild form of the tarpans. Therefore, one can only speculate regarding its phylogenetic position. Was the tarpan the connecting link—with 65/64 chromosomes—between Przewalski's horse and the domestic horse? According to Benirschke and Ryder (1985), a specimen of the Polish tarpan (back-bred) indeed possessed 65 chromosomes. Six further specimens had only 64 chromosomes each. Would it be possible—as in the case of the quagga—to gain a sample of mitochondrial DNA from one of the above-mentioned tarpan skins? Also, a complete examination of the DNA content of the different species would be beneficial.

The assumed adaptive radiation of the horses did not necessarily require long time periods. Proceeding from a postulated genetically polyvalent small ancestral population, the formation of different races and species could have happened in relatively short time spans during divergent migrations. An enticing approach for further research in this area is the hypothesis of Wichman et al. (1991) that repetitive sequences are involved as the driving force of the rapid chromosomal evolution (and thus the speciation) of the Equidae. The idea that the characteristics of the hypothetical polyvalent ancestral form have been randomly distributed to the descendants in the various equid branches in the course of the speciation processes would be a simple explanation of why it is not possible to reconstruct a phylogenetic tree that is free of contradictions.

## Acknowledgements

I would like to thank Prof. Dr. Siegfried Scherer for the detailed evaluation of the manuscript and for his encouragement.

# References

Anderson, W.S. 1939. Fertile mare mules. *Journal of Heredity* 30:549-551.

Antonius, O. 1951. Die Tigerpferde. *Monographien der Wildsäugetiere* 11:1-148.

Atkin, N.B, G. Mattinson, W. Becak, and S. Ohno. 1965. The comparative DNA content of 19 species of placental mammals, reptiles, and birds. *Chromosoma* 17:1-10.

Benirschke, K. 1967. Sterility and fertility of interspecific mammalian hybrids. In: Benirschke, K. ed. *Comparative Aspects of Reproductive Failure*. Springer, New York, pp. 218-234.

Benirschke, K. 1969. Cytogenetics in the zoo. *New Scientist* 41:132-133.

Benirschke, K. and N. Malouf. 1967. Chromosome studies of Equidae. *Equus* 1:253-284.

Benirschke, K. and O.A. Ryder. 1985. Genetic aspects of equids with particular reference to their hybrids. *Equine Veterinary Journal Supplement* 3:1-10.

Benirschke, K. and M.M. Sullivan. 1966. Corpora lutea in proven mules. *Fertility and Sterility* 17:24–33.

Benirschke, K., L.E. Brownhill, and M.M. Beath. 1962. Somatic chromosomes of the horse, the donkey and their hybrids, the mule and the hinney. *Journal of Reproduction and Fertility* 4:319-326.

Benirschke, K., R.J. Low, M.M. Sullivan, and R.M. Carter.1964. Chromosome study of an alleged fertile mare mule. *Journal of Heredity* 55:31-38.

Borwick, R. 1967. *Esel*. Ulmer, Stuttgart.

Chandley, A.C., R.C. Jones, H.M. Dott, W.R. Allen, and R.V. Short. 1974. Meiosis in interspecific equine hybrids. *Cytogenetics and Cell Genetics* 13:330-341.

Chandley, A.C. 1981. Does 'affinity' hold the key to fertility in the female mule? *Genetic Resarch* 37:105-109.

George, M., and O.A. Ryder. 1986. Mitochondrial DNA evolution in the genus *Equus*. *Molecular Biology and Evolution* 3:535-546.

Goodall, D.M. 1966. *Pferde der Welt*. Hoffmann, Heidenheim.

Gray, A.P. 1972. *Mammalian Hybrids: A Check-list with Bibliography*. Commonwealth Agricultural Bureaux, Farnham Royal, England.

Groves, C.P. 1974. *Horses, Asses and Zebras in the Wild*. David and Charles, London.

Groves, C.P. and V. Mazák. 1967. On some taxonomic problems of asiatic wild asses; with the description of a new subspecies (Perissodactyla; Equidae). *Zeitschrift Fur Saugetierkunde* 32:321-357.

Groves, C.P. and D.P. Willoughby. 1981. Studies on the taxonomy and phylogeny of the genus Equus, I: Subgeneric classification of the recent species. *Mammalia* 45:321-355.

Grzimek, B. 1977. *Und immer wieder Pferde*. Kindler, Munich.

Higuchi, R., B. Bowman, M. Freiberger, O.A. Ryder, and A.C. Wilson. 1984. DNA sequences from the quagga, an extinct member of the horse family. *Nature* 312:282-284.

Kaminski, M. 1979. Mini review: The biochemical evolution of the horse. *Comparative Biochemistry and Physiology* 63 B:175-178.

King, J.M. 1967. The sterility of two rare equine hybrids. In: Benirschke, K. ed. *Comparative Aspects of Reproductive Failure*. Springer, New York, pp. 235-245.

King, J.M., R.V. Short, D.E. Mutton, and J.L. Hamerton. 1966. The reproductive physiology of male zebra-horse and zebra-donkey hybrids. In: Rowlands, J.W. ed. *Comparative Biology of Reproduction in Mammals*. Academic Press, London, pp. 511-527.

Klingel, H. 1987. Pferde. In: Grzimek, B. ed. *Grzimek Enzyklopädie Säugetiere*. Volume 4. Kindler, Munich, pp. 557-588.

Kopp, E., B. Mayr, and W. Schleger. 1983. Nucleolus organizer regions in the chromosomes of the donkey. *Journal of Heredity* 74:387-388.

Koulischer, L. and S. Frechkop. 1966. Chromosome complement: A fertile hybrid between *Equus przewalskii* and *Equus caballus*. *Science* 151:93-95.

Krumbiegel, I. 1958. Einhufer. *Neue Brehm Bücherei*, Bd. 208. Lutherstadt Wittenberg.

Lotsy, J.P. 1922. Die Aufarbeitung des Kühnschen Kreuzungsmaterials im Institut für Tierzucht der Universität Halle. *Genetica* 4:32-61.

Maciulis, A., T.D. Bunch, J.L. Shupe, and N.C. Leone. 1984. Detailed description and nomenclature of high resolution G-banded horse chromosomes. *Journal of Heredity* 75:265-268.

Marsh, F.L. 1976. *Variation and Fixity in Nature*. Pacific Press Publishing Association, Mountain View, CA.

Mochi, U. and D. MacClintock. 1976. *A Natural History of Zebras*. Scribner, New York.

Mohr, E. and J. Volf. 1984. *Das Urwildpferd: Neue Brehm Bücherei*. Lutherstadt Wittenberg.

Petzsch, H. 1971. *Urania Tierreich, Band 6: Säugetiere*. Harri Deutsch, Frankfurt.

Ryder, O.A. 1978. Chromosomal polymorphism in *Equus hemionus*. *Cytogenetics and Cell Genetics* 21:177-183.

Ryder, O.A and L.G. Chemnick. 1990. Chromosomal and molecular evolution in Asiatic wild asses. *Genetica* 83:67-72.

Ryder, O.A, N.C. Epel, and K. Benirschke. 1978. Chromosome banding studies of the Equidae. *Cytogenetics and Cell Genetics* 20:323-350.

Ryder, O.A, L.G. Chemnick, A.T. Bowling, and K. Benirschke. 1985. Male mule foal qualifies as the offspring of a female mule and jack donkey. *Journal of Heredity* 76:379-381.

Scherer S. 1993. Basic types of life. In: Scherer, S., ed. *Typen des Lebens.* Berlin, pp. 11-30.

Schirneker, B. 1921. *Equus Przewalskii* Poliakoff, seine Geschichte und Einführung in Deutschland unter besonderer Berücksichtigung der Versuche am Tierzucht-Institut in Halle. In: Lotsy, J.P. 1922. Die Aufarbeitung des Kühnschen Kreuzungsmaterials im Institut für Tierzucht der Universität Halle. *Genetica* 4:32-61.

Short, R.V. 1967. Reproduction. *Annual Review of Physiology* 29:373-400.

Short, R.V. 1975. The evolution of the horse. *Journal of Reproduction and Fertility Supplement* 23:1-6.

Short, R.V, A.C. Chandley, R.C. Jones, and W.R. Allen. 1974. Meiosis in interspecific equine hybrids, II: The Przewalski horse/domestic horse hybrid (*Equus przewalskii* × *E. caballus*). *Cytogenetics and Cell Genetics* 13:465–478.

Silver, C. 1978. Pferderassen der Welt. *BLV Bestimmungsbuch* Number 24. Munich.

Simpson, G.G. 1977. *Pferde – die Geschichte der Pferdefamilie in der heutigen Zeit und in 60 Millionen Jahren ihrer Entwicklung.* Parey, Berlin.

Smith, H.H. 1930. A fertile mule from Arizona. *Journal of Heredity* 30:548.

Taylor, M.J. and R.V. Short. 1973. Development of the germ cells in the ovary of the mule and hinney. *Journal of Reproduction and Fertility* 32:441-445.

Thenius, E. 1979. Einhufer oder Pferdeverwandte - Stammesgeschichte. In: Grzimek, B. ed. *Grzimeks Tierleben Enzyklopädie des Tierreichs, Band 3: Säugetiere.* Kindler, Munich, pp. 542-544.

Thenius, E. 1987. Unpaarhufer -Stammesgeschichte. In: Grzimek, B. ed. *Grzimeks Enzyklopädie Säugetiere*, Volume 4. Kindler, Munich, pp. 550-556.

Trujillo, J.M., S. Ohno, J.H. Jardine, and N.B. Atkin. 1969. Spermatogenesis in a male hinny: Histological and cytological studies. *Journal of Heredity* 60:79-84.

van Vleck, L.D. 1977. Breeds in the United States. In: Evans, J.W. ed. *The Horse.* W. H. Freeman and Company, San Francisco, pp. 18-114.

Vogt, H.H. 1986. Quagga—eine Subspecies. *Naturwissenschaftliche Rundschau* 39:126-127.

Volf, J. 1979. Einhufer oder Pferdeverwandte. In: Grzimek, B. ed. *Grzimeks Tierleben. Enzyklopädie des Tierreichs, Band 3: Säugetiere.* Kindler, Munich, pp. 544-546, 564, 599-600.

Wichman, H.A, C.T. Payne, O.A. Ryder, M.J.Hamilton, M. Maltbie, and R.J. Baker. 1991. Genomic distribution of heterochromatic sequences in equids: Implications to rapid chromosomal evolution. *Journal of Heredity* 82:369-377.

Wüst, W. 1966. *Tierkunde, Band 1: Wirbeltiere, Teil I: Säugetiere.* Bayerischer Schulbuch, Munich.

Zong, E and G. Fan. 1989. The variety of sterility and gradual progression to fertility in hybrids of the horse and donkey. *Heredity* 62:393-406.

# 16. Hybridization and Speciation within the Cercopithecinae (Primates, Cercopithecoidea)

### SIGRID HARTWIG-SCHERER

**Abstract**

The Old World monkeys (Cercopithecoidea) include the baboons and baboon-like monkeys, the cercopithecines, (Cercopithecinae) and the colobines, the "leaf-eating monkeys" (Colobinae). Although these two groups can be clearly distinguished from each other, their taxonomic rank has changed several times between the levels of family and subfamily. The classification of the Cercopithecinae is characterized by contradicting divisions into species, genera, and tribes. Some classifications recognize far more than ten genera; others assign all species of the Cercopithecinae to one single genus. Both taxonomic approaches seem justified: "splitting" into several genera reflects the multiple morphological differences, while "lumping" to one genus indicates the close relationship of the species – after all, eight out of nine genera (according to Hill 1966 and Fleagle 1988) are connected through intergeneric hybrids. In order to take into account both aspects, all cercopithecine monkeys are assigned to one basic type, retaining the common division into species and tribes. Thus, the basic type Cercopithecinae includes the two tribes Cercopithecini (the baboon-like monkeys, which are flatter faced) and Papionini (the baboons, which are dog-faced), with a total of 9 genera (*Allenopithecus, Cercopithecus, Erythrocebus, Miopithecus, Cercocebus, Papio, Mandrillus, Theropithecus, Macaca*) and 50-60 species. The high diversity of the cercopithecine monkeys and the high number of their subspecies and emerging new forms may indicate a rather recent radiation of this group of primates. The process of speciation is still in progress, which is indicated by the existence of stable hybridization zones in nature.

## 16.1. Introduction

That monkeys were known to the civilized nations of ancient times is obvious from various Egyptian, Indian, and Eastern Asian temple friezes. Especially the baboons were worshiped as gods, judges, and patrons. Hamadryas baboons *(Papio hamadryas)* and anubis baboons *(P. anubis)* were sacred to the Egyptians. The Egyptian moon deity Thot, who was believed to be the inventor of the art of writing, had a baboon's head, as did Menedes, the goddess of the moon, and Rapi, one of the gods of deceased spirits. The Celebes crested macaque (*Macaca nigra*)

370                              Basic Types of Life

Superfamily Cercopithecoidea with the solitary family Cercopithecidae (both
commonly called the Old World monkeys):

Subfamily Colobinae (colobine or leaf-eating monkeys)

    Subtribe Colobina (colobus monkeys)
      *Colobus* (black-and-white colobus monkeys)
      *Procolobus* (olive colobus monkeys)
      *P. (Pilicolobus)* (red colobus monkeys)

    Subtribe Presbytina (langurs and kin)
      *Presbytis* (leaf monkeys and surilis)
      *Semnopithecus* (gray langurs)
      *S. (Trachypithecus)* (langurs)
      *Pygathrix* (douc langurs)
      *P. (Rhinopithecus)* (snub-nosed monkeys)
      *Nasalis* (proboscis monkeys)
      *N. (Simias)* (simakobou)

Subfamily Cercopithecinae (cercopithecine monkeys)

**Tribe Cercopithecini (baboon-like monkeys)**
    Subtribe Cercopithecina (guenons and kin)
      *Cercopithecus* (guenon monkeys)
      *Erythrocebus* (patas monkeys)
      *Miopithecus* (talapoins)
    Subtribe Allenopithecina (swamp monkeys)
      *Allenopithecus* (Allen's swamp monkeys)
**Tribe Papionini (baboons and kin)**
    Subtribe Macacina (macaques)
      *Macaca* (macaques)
    Subtribe Papionina (baboons)
      *Theropithecus* (gelada baboons)
      *Cercocebus* (mangabeys)
      *Papio* (true baboons)
      *Mandrillus* (mandrills)

Table 1. Classification of the Cercopithecidae modified after Strasser and Delson (1987),
Hill (1966), Thorington and Groves (1970) and Fleagle (1988).

is still worshipped today by the Papua in Bali, and the patas monkeys
(*Erythrocebus patas*) by the West African Ewes tribe.

The baboons and baboon-like monkeys, the cercopithecines,
together with the "leaf-eating monkeys," the colobines, comprise
the Old World monkeys (Cercopithecoidea; Table 1). Based on their
diversity and geographic distribution, the Old World monkeys are the
"most successful" group of primates. They can be found in both Asia and
Africa and include approximately 20 genera, approximately 80 species,
and over 260 subspecies (Napier 1981, 1985).

The two Old World monkey groups are morphologically and
genetically clearly separated from each other. Some of their differences
are directly or indirectly connected with their diet. Whereas the

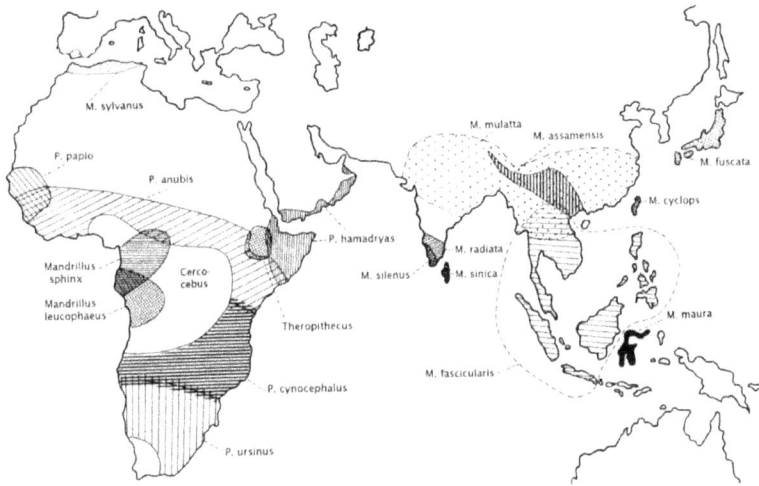

Figure 1. Biogeography of the Papionini. The four species of the genus *Cercocebus* can be found in Central Africa and are not included for the sake of clarity. Likewise, *Macaca hecki*, *M. nigra*, and *M. ochreata* also on Celebes Island are not shown, nor are several *Macaca* species that are widespread in Southeast Asia.

cercopithecine monkeys possess large cheek pouches for short-term storing and predigesting of food, the colobines are characterized by a special bacterial flora in their multi-chambered stomach, with which they digest a diet consisting almost exclusively of leaves. Particular characteristics of the cercopithecine monkeys are a more complex structure of the cerebral sulcus, as well as a longer and narrower face. Colobine monkeys are characterized by a more slender build with longer legs—an adaptation to jumping and swift locomotion. Whether the taxonomic separation of the two groups should take place on the level of families or subfamilies is more or less a question of preference as the history of the classification of the Cercopithecoidea shows. Hill (1966), who prefers the level of families, mentions that since 1875 the classification has changed half a dozen times between the two taxonomic levels.

Today the group is generally divided into two subfamilies (Napier 1981, 1985; Strasser and Delson 1987; Thorington and Groves 1970). However, a return to the older classification can also be observed (Dutrillaux et al. 1988). The classification used by this author (Table 1) follows Strasser and Delson (1987).

Within the Cercopithecinae, too, there have been a large number of different classification attempts. The current classification (Fleagle 1988; Napier 1981, 1985; Thorington and Groves 1970) counts over 50 species in nine genera, divided into two tribes, the Papionini (the baboons) and Cercopithecini (the baboon-like monkeys) (Table 1, 2).

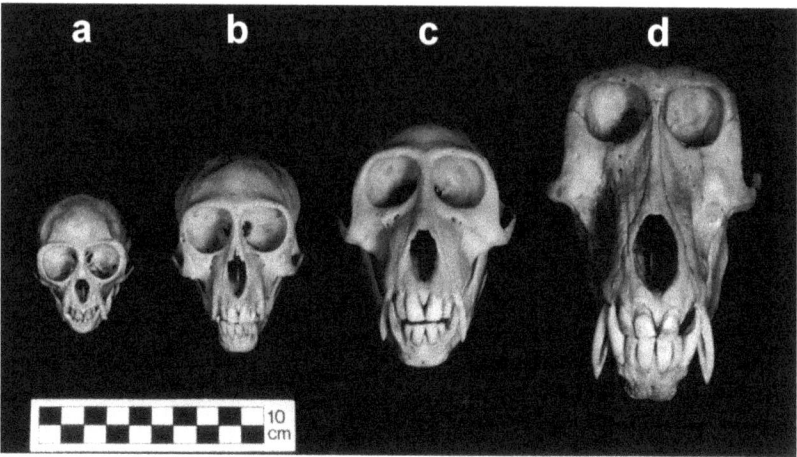

Figure 2. Skulls of adult, male representatives of four species of the Cercopithecinae. The sex of the specimen of *Erythrocebus patas* is unknown. (a) *Miopithecus talapoin*, (b) *Erythrocebus patas*, (c) *Cercocebus torquatus*, (d) *Papio cynocephalus*, collections of the Anthropological Institute and Museum of the University of Zürich.

The cercopithecine monkeys populate wide areas of Africa and Asia (see also Figure 1). In Africa can be found the baboon-like monkey *s. str.* (*Cercopithecus, Miopithecus, Allenopithecus,* and *Erythrocebus*), the mangabeys (*Cercocebus*), the short-tailed baboons (*Mandrillus, Papio*), and the gelada baboons (*Theropithecus*). Southeast Asia, on the other hand, is the domain of the macaques (although one macaque species lives in North Africa and Gibraltar).

The cercopithecine monkeys are a morphologically diverse subfamily (Figure 2). Like the colobine monkeys, they possess additional calluses on their buttocks, enabling them to sit on the branches and use their feet for grasping. On the ground, they run on flat soles. They are usually omnivorous. Their nutrition is versatile and consists of nuts, roots, fruits, leaves, juvenile birds, onions, insects, small animals, and eggs. The larger forms, like mandrills and some macaques, live on the ground, and are, therefore, in greater danger of being attacked by predators. They are characterized by a stricter social order than arboreal forms.

## 16.2. The Cercopithecinae - an Overview

**16.2.1. Macaques.** The genus *Macaca* includes the Barbary macaque, the rhesus monkey, the bonnet macaque and the Celebes crested macaque, to name a few. This genus is distributed over almost all climate zones and fills more niches than any other primate group. Macaques live in trees, on the ground, or on rocks; some species even like the water. They can be found all over Asia, as far north as Japan (no other primates populate such cool regions), and on remote islands such as Bali (Figure 1). The latter is the furthest southeast distributional limit of primates.

| Species # | | | Intergeneric hybrid with species # | Species # | | | Intergeneric hybrid with species # |
|---|---|---|---|---|---|---|---|
| **Genus** | **Cercopithecus** | | | **Genus** | **Macaca** | | |
| 1 | C. aethiops (4) | vervet monkey | 24,26,30, 35,36 | 27 | M. sylvanus | Barbary macaque | 44 |
| 2 | C. pygerythurs (14) | vervet monkey | 35,36,37 | 28 | M arctoides | stump-tailed macaque | |
| 3 | C. sabaeus | green monkey | ?30 | 29 | M. fuscata | Japanese macaque | |
| 4 | C. tantalus (3) | Tantalus monkey | | 30 | M. mulatta | rhesus monkey | 1,?3,52 |
| 5 | C. lhoesti | L'hoest's monkey | | 31 | M. assamensis | Assam macaque | |
| 6 | C. preussi (2) | Preuss's monkey | | 32 | M. cyclopis | Taiwan macaque | |
| 7 | C. solatus | sun-tailed monkey | | 33 | M. silenus | liontail macaque | |
| 8 | C. diana (2) | Diana monkey | | 34 | M. nemestrina | pigtail macaque | 44,48,52, 54 |
| 9 | C. dryas | Dryas monkey | | 35 | M. radiata | bonnet macaque | 1,2 |
| 10 | C. salongo | Salongo monkey | | 36 | M. sinica | toque macaque | 1,2 |
| 11 | C. neglectus | De Brazza's monkey | | 37 | M. fascicularis | long-tailed macaque | 2,44,?49, 53,54 |
| 12 | C. hamlyni | owl-faced monkey | | 38 | M. maura | Moor macaque | 54 |
| 13 | C. ascanius (5) | black-cheeked white-nosed monkey | | 39 | M. ochreata | booted macaque | |
| | | | | 40 | M. nigra | Celebes crested macaque | |
| 14 | C. cephus (2) | moustached monkey | 24 | 41 | M. tonkeana | Tonkean macaque | |
| 15 | C. erythrotis (3) | red-eared monkey | | 42 | M. thibetana | Père David's macaque | |
| 16 | C. erythrogaster | red-bellied monkey | | | | | |
| 17 | C. petaurista (2) | lesser white-nosed monkey | | **Genus** | **Theropithecus** | | |
| | | | | 43 | T. gelada | gelada baboon | 50,21,52 |
| 18 | C. campbelli (2) | Campbell's monkey | | **Genus** | **Cercocebus** | | |
| 19 | C. mona | mona monkey | | 44 | C. torquatus | red-capped mangabey | 23,27,34, 37,53,54 |
| 20 | C. pogonias (3) | crowned guenon | | 45 | C. albigena | gray-cheeked mangabey | |
| 21 | C. wolfi (3) | Wolf's monkey | | 46 | C. aterrimus | black mangabey | |
| 22 | C. nictitans (3) | white-nosed guenon | | 47 | C. galeritus | Tana River mangabey | |
| 23 | C. mitis (8) | blue monkey | 44 | **Genus** | **Papio** | | |
| **Genus** | **Erythrocebus** | | | 48 | P. ursinus | chacma baboon | 34 |
| 24 | E. patas (4) | patas monkey | 1,14 | 49 | P. cynocephalus | yellow baboon | ?37 |
| **Genus** | **Miopithecus** | | | 50 | P. anubis | anubis baboon | 43,53 |
| 25 | M. talapoin | talapoin | | 51 | P. papio | Guinea baboon | 43,54 |
| **Genus** | **Allenopithecus** | | | 52 | P. hamadryas | hamadryas baboon | 30,34,43, 44,53,54 |
| 26 | A. nigroviridis | Allen's swamp monkey | 1 | **Genus** | **Mandrillus** | | |
| | | | | 53 | M. sphinx | mandrill | 37,44,50, 52 |
| | | | | 54 | M. leucophaeus | drill | 34,37,38, 51,52 |

Table 2. The cercopithecine monkeys and their intergeneric hybrids. The numerous intrageneric hybrids are not listed (see Gray 1972 and Lernould 1988). Genera and species after Hill (1966), Lernould (1988) and Fleagle (1988). Numbers in parenthesis indicate the current number of subspecies. Hybrids according to Dunbar & Dunbar (1974), Gray (1972), Lernould (1988), Matsubayabashi (1978) and Markarjian (1974).

The fur color of the macaques is yellow to olive-brown, with a lighter underside and reddish buttock calluses. Older males frequently possess distinct bulges above the eyes. All macaques predominantly eat fruits, but also other plant parts, molluscs, insects, and small vertebrates.

Classification attempts have sought to divide the macaques into subgenera. However, these attempts were abandoned due to a series of conflicting data, for example, hybridization and morphological gradients between individual species.

**16.2.2. Baboons, Mandrills and Gelada Baboons.** The baboons (*Papio, Mandrillus, Theropithecus*) can be found in Africa (south of the Sahara) and in Southern Arabia (Figure 1). They are the largest representatives of the Cercopithecinae and possess a distinctive sexual dimorphism, long skulls with an extended, angular snout, a protruding nose, and small deep-set eyes under powerful bulges. Front and rear extremities are equally long, and the hands and feet are shortened as an adaptation to the terrestrial way of life. The animals seek protection in trees only for the purpose of sleeping, and the formation of groups (protective function) is probably also connected to their terrestrial way of life. The large males have dagger-like cuspids and frequently possess manes or beards. During the mating period, the females display swollen, intensely red buttock calluses. Their food ranges from roots to small mammals. Frequent changes of specific and generic names mark the history of classification of this group (Buettner-Janusch 1966; Dunbar and Dunbar 1974; van Gelder 1977)

**16.2.3. The Mangabeys.** The mangabeys (*Cercocebus*) superficially resemble large guenons because of their arboreal behavior and long tails—the prefix "*cerco*" means tailed. In fact, mangabeys are closer to macaques than to guenons and closer yet to baboons (tribe Papionini): they should be considered "the arboreal baboons." They are slim, good jumpers, and have big cheek pouches and conspicuous bulges above their eyes, which are emphasized by white upper lids. The front and rear extremities are relatively long. They prefer fruits, specifically figs, but also eat invertebrates.

The red-capped mangabey occurs from Cap-Vert, Senegal, to the lower Congo River where its area of distribution overlaps with those of the gray-cheeked mangabey and Tana River mangabey. Its fur coloring is highly variable. The black mangabey lives in the forests south of the Congo River; it shares its territory in part with the Tana River mangabey. The Tana River mangabey and the red-capped mangabey are partly arboreal, partly terrestrial. The strictly arboreal gray-cheeked mangabey and black mangabey are sometimes considered a separate genus (*Lophocebus*).

**16.2.4. The Guenons and kin (baboon-like monkeys).** The Cercopithecini tribe includes the largest number of species of the whole family. Among them are the vervet monkey (*Cercopithecus aethiops*), the patas monkey (*Erythrocebus patas*), and the talapoin. The number of species fluctuates between 15 and 30, with approximately seven subspecies, depending on which specialist counts them: "The guenons are difficult to classify, as testified by the number of classifications, so far proposed and by the number of synonyms for many taxa" (Lernould

1988). Although some authors unite the four genera *Cercopithecus, Erythrocebus, Miopithecus* and *Allenopithecus* as subgenera is the genus *Cercopithecus* (Lernould 1988; Martin and McLarnon 1988), the four genera were retained by this author according to the classification of Strasser and Delson (1987) and Fleagle (1988) (Table 1).

An attempt to unite the numerous species of the genus *Cercopithecus* (Table 2) into superspecies created taxonomic problems and was therefore given up.

The small to medium-sized guenons (*Cercopithecus*) live in forests south of the Sahara. They are characterized by round heads, extremely long tails, and more or less distinctly colorful areas of the face, the fur, and the scrotum. Since they frequently occur sympatrically and are exclusively forest inhabitants, they differ in their preference of forest types as well as tree heights (tiers). In forms that share the same habitat, such as *C. pogonias, C. cephus,* and *C. nictitans,* nutritional niching can be observed (Kingdon 1971). The highly variable face patterns probably serve the purpose of recognition among members of the same species (Kingdon 1971).

## 16.3. Hybridization and Speciation in the Cercopithecinae

A number of interspecific (and some intergeneric) natural hybrids are known. Reciprocal crossbreeding between *Papio hamadryas* and *P. anubis* occurr within a 15-20 km wide overlap zone of both species (Nagel 1973; Phillips-Conroy et al. 1992), as well as between *P. anubis* and *P. cynocephalus* in Kenya (Maples 1972; Samuels and Altmann 1986). Dunbar and Dunbar (1974) report the same for *Theropithecus gelada* and *P. anubis*. Within the genus *Macaca*, too, natural hybrids are more frequent than expected. Polyspecific and even polygeneric associations in nature were described for *Cercopithecus asconius* and *C. mitis* (Struhsaker et al. 1988), for *Macaca fascicularis, M. mulatta* and *M. fuscata*, with hybrids between the first two macaques (Southwick and Southwick 1983), and for *Cercocebus* and *Cercopithecus* (Gartlan and Struhsaker 1972; Lernould 1988). They show that genetic and behavioral barriers are smaller than originally assumed and can definitely be overcome, even in nature.

According to Gautier-Hion (1988), natural hybrids among the Cercopithecini are the rule rather than the exception: "Hybridizations between present wild populations seem the rule near the boundary zone between species... These hybrids are fertile... and they apparently benefit from some advantage over both parent species... Or has this peculiar life style had any effect on guenon speciation through hybridization?"

Moreover, natural hybrid populations display an amazing stability. Maples (1972) observed a high morphological variability in groups of hybrids of *Papio cynocephalus* and *P. anubis*. On the basis of the hypothetical assumption that hybridization causes a lack of genomic stability and thus disadvantages, he wrongly concluded that the hybrid zone could not be very old. Nagel, however, showed that highly variable hybrids in the 20 km wide contact zone between *P. hamadryas* and *P. anubis* must have existed as a balanced system for at least 60 years and probably even longer (Kummer 1968; Nagel 1973). Similarly stable hybrid zones were described for *Macaca mulatta* and *M. fascicularis* in Thailand (Fooden 1964). If high variability can be observed for a relatively long period in relatively stable groups, one can conclude that the advantages of hybridization outweigh its disadvantages.

Several authors have discussed such advantages of hybrid forms over their parent species. An example is the new behavior observed in *Papio* hybrids: Groups including *P. anubis*, *P. hamadas,* as well as their hybrids show three types of reproductive groups: the multiple-males groups typical for *P. anubis*, the harem groups of the *P. hamadryas* and a third form, the formation of groups of pairs, that can be observed neither in the maternal nor in the parental species (Sugawara 1979). Hybrids between *Cercopithecus ascanius* and *C. mitis* show a widened nutritional spectrum and could be an example of nutritional niching and speciation (Struhsaker et al. 1988).

Not only the tendency to hybridize but also the large number of subspecies and "pseudo"-species among the guenons are regarded as examples of the fact "that a group is still in an active stage of speciation" (Lernould 1988). We are, so to speak, witnesses of "evolution in action" (Lernould 1988). The specialists are often not sure whether newly discovered species actually are new species, subspecies, or hybrids: Lernould (1988) describes a new, morphologically distinct form of *Cercopithecus* either as a very rare species, and thus highly endangered, or as a hybrid of two different parental species.

In the context of speciation, the chromosomal differentiation is not necessarily tied to the morphological differentiation. For example, in gibbons chromosomal but not external morphological changes are observed, whereas the opposite is the case in baboons (Godfrey and Mark 1991). This work suggests that morphospeciation (= appearance of diagnostic specific traits) can often be observed not after, but before, biospeciation (= isolation of populations). Baboons possess highly conservative chromosome morphologies but are morphologically quite heterogeneous (Brown et al. 1986). The genus *Cercopithecus*, on the other hand, is characterized by high variability of its chromosome number that is not necessarily reflected in its morphology (see references in Lernould 1988). Godfrey and Marks (1991) emphasize the extent "... to which the fertility of hybrids (and the degree of chromosomal divergence within clades) is uncorrelated with morphological divergence of parent species." The chromosomal evolution of the cercopithecine monkeys is characterized by a successive reduction of the chromosome size with a simultaneous increase of the chromosome number from less than 50 to over 70 (Dutrillaux *et al.* 1988). This increases the number of possible recombinations from crossing over and chromosomal segregation, in turn resulting in increased diversity and, thus, speciation.

Although there are evidently several possibilities, the specific speciation mechanisms are rather unclear. These unsolved questions make the cercopithecine monkeys fascinating subjects for studying speciation (Godfrey and Marks 1991).

## 16.4. The Basic Type of the Cercopithecinae

**16.4.1. Intergeneric Hybrids.** It has long been known that within the cercopithecine monkeys not only interspecific, but also transgeneric, hybrids occur (Table 2). The most important of these hybrids are: *Cercopithecus* × *Erythrocebus* (Matsubayashi *et al.* 1978), *Cercopithecus* × *Macaca* and *Cercocebus* × *Cercopithecus* (Chiarelli 1963), and *Papio* × *Mandrillus*, *Mandrillus* × *Macaca*, *Papio* × *Theropithecus*, and *Allenopithecus* × *Cercopithecus* (Gray 1972).

Hybrids of cercopithecine monkeys show traits which are quite different from those of their parents. The spectrum reaches from the distinctive dominance of one parent through mosaic and intermediate forms to completely new characteristics occurring neither in the paternal nor in the maternal species (Maples 1972; Markarjan *et al.* 1974; Matsubayashi *et al.* 1978; Nagel 1973).

A graphic summary of the currently known intergeneric hybrids within the Ceropithecidae is given in Figure 3. The genera are arranged according to their tribal position. Besides *Miopithecus*, all genera are connected, at least indirectly, through hybrids. Various other indications (see below) support including *Miopithecus* in the same basic type.

It should be emphasized there are no known hybrids between representatives of the cercopithicines and colobines. On the basis of the present crossbreeding data, it is assumed that the Cercopithecinae form a valid basic type, according to the definition given by Scherer (1993, this volume).

**16.4.2. Further Indications of Delimitation.** Thus far, no hybrids of *Miopithecus* have been documented. The fact that *Miopithecus* belongs to the basic type Cercopithecinae can be confirmed with absolute certainty only through successful crossbreeding. The social behavior of *M. talapoin*, which is uncommon for monkeys, could possibly impede a successful hybridization: the sexual interaction is restricted to a short period during the year; plus, the feeding groups consist only of representatives of the same sex. Morphological, serological and chromosomal characteristics point unequivocally to the fact that this species belongs to the Cercopithecinae.

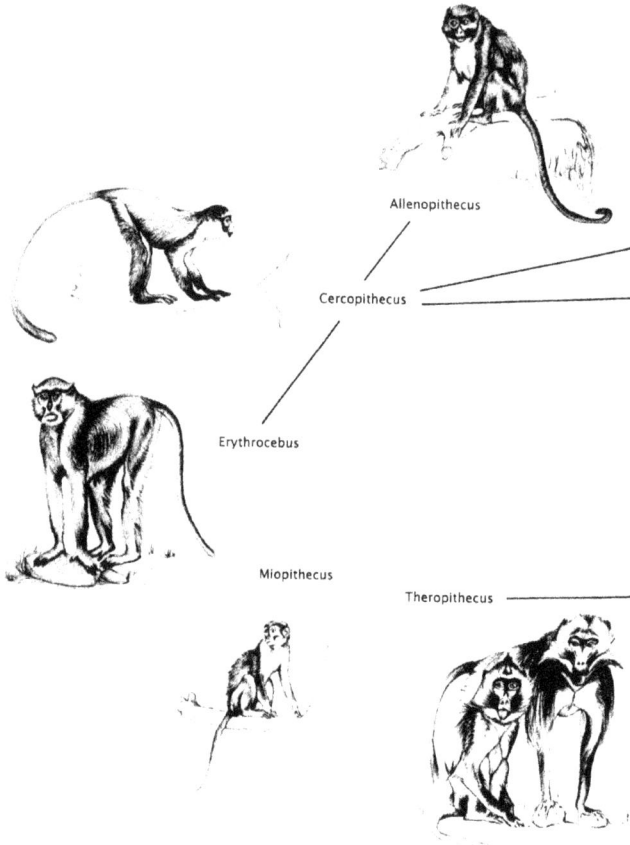

Figure 3. Crossbreeding polygon of the Cercopithecinae. Further information on hybrids can be obtained from table 2. Drawings: Marion Häberle

Shea (1992) showed that *Miopithecus* is a paedomorphic, *i.e.*, arrested or juvenilized, form of *Cercopithecus*, while *Erythrocebus* correspondingly represents the accelerated end of this allometric spectrum. These forms share similar ontogenetic allometries that can explain the profound proportional differences in the adult animal: "… the ontogenetic allometric approach reveals that size differences and allometric factors are the primary determinants of the marked morphological differentiation… as characterizing both the small *talapoin* and large *patas* monkeys" (Shea 1992). *Miopithecus* is repeatedly described as a "neotenous" (more correctly: paedomorphic) form of the Cercopithecini tribe (Strasser and Delson 1987) and is even called a "neotenous" baboon (Napier and Napier 1967). Its relatively large brain and other morphological characteristics support the hypothesis of a phyletic stunting of *Miopithecus*. Besides many other morphological, karyological, and genetic data, these ontogenetic studies

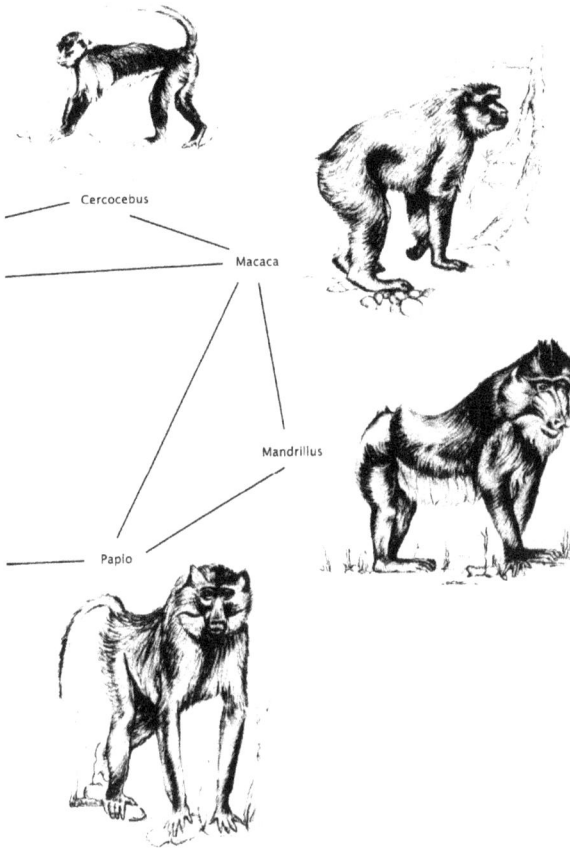

Cercocebus

Macaca

Mandrillus

Papio

of allometry confirm the affiliation of this form with the basic type of the Cercopithecinae.

Many of the observed proportional differences among the species of this basic type could possibly be explained with the help of such common growth processes (*i.e.*, allometry and heterochrony). Such processes can cause considerable morphological divergence with minimal genetic change (Gould 1977; Hartwig-Scherer 1993). These studies, which have gained increasing popularity, are a promising field of research for the delimitation of basic types.

**16.4.3. Conflicts in the Classification and in the Genealogical Relationships.** The classification of the cercopithecine monkeys (*i.e.*, the basic type Cercopithecinae) has been marked by frequent changes in the delimitation of species, genera, and (sub)tribes. One of the reasons for these changes is the network of hybrids. The whole spectrum from "lumping" (all forms united into a single genus) to "splitting" (more than ten genera distinguished) is represented (van Gelder 1977).

Lernould (1988) summarized different suggestions for the taxonomy of the Cercopithecini. He found no consensus whether taxa should be segregated at the level of genera (Napier 1981, 1985), subgenera (Dandelot 1971), or species (Napier and Napier 1967). "If one considers the potential for interbreeding in guenons and their karyological phylogeny, it becomes reasonable to retain only a single genus, *Cercopithecus*, including *Allenopithecus, Miopithecus, Erythrocebus* as subgenera..." (Lernould 1988).

The thought that interspecific hybrids indicate a close taxonomic and evolutionary relationship had already been expressed. Buettner-Janusch (1966) refers to Simpson (1961) when he writes: "If primates which are presumed to be members of separate genera mate and produce living though sterile offspring, they should be considered members of the same genus." Buettner-Janusch therefore united *Theropithecus, Papio,* and *Mandrillus* into a single genus *Papio.* Plus, he discusses the close relationship with the macaques, which also were known to form hybrids with *Papio:* "... the very close relationship of macaques... and baboons, a relationship that should probably be recognized by dropping the generic distinction between these two groups of terrestrial, highly social primates." Van Gelder (1977) goes even further. Based on the distinctive ability to form intergeneric hybrids, he justifies the uniting of all genera into a single genus *Cercopithecus.* According to his viewpoint, *Allenopithecus, Erythrocebus, Miopithecus, Macaca, Papio, Cercocebus, Theropithecus,* and *Mandrillus* are to be understood as synonyms of *Cercopithecus.* This approach corresponds to the basic type proposed by this author, the existence of hybrids being viewed as positive proof for a close taxonomic relationship.

In order to reflect the observed graduated potential for hybridization and use it for detailed taxonomy, Godfrey and Marks (1991) tried to systematize the continuum between complete genetic isolation of two species and the total mixing of two populations. However, using the degree of hybrid fertility (frequent or rare hybridization, sterility, fertility, or restricted fertility) to derive the degree of relationship betwen species and thence a nested classification (Buettner-Janusch 1966; Godfrey and Marks 1991; Markarjan *et al.* 1974) is questionable. It is known that the sterility of hybrids can be traced back to very few alterations of the genome (Godfrey and Marks 1991). On the other hand, fertility was observed over a wide spectrum of morphological, chromosomal, and taxonomic divergences (for example, *Cercopithecus* × *Macaca*).

Morphological and chromosomal characteristics do not result in a completely conflict-free system either. Morphological (Strasser and Delson 1987), morphometric, and allometric comparisons (Martin and

McLarnon 1988; Shea 1992), as well as biochemical and chromosomal characteristics (Dutrillaux 1988), have all been used to design a system for the Cercopithecinae. The separate analyses do not always result in one uniform phylogenetic tree. However, one can postulate a subdivision of the basic type into Papionini and the Cercopithecini. Finer contours are frequently obliterated and cannot be determined unequivocally.

The chromosomal phylogenetic tree is highly complex. The individual genera are difficult to group, particularly within the Cercopithecini: "*Cercopithecus* species have undergone very complex evolution, many chromosome rearrangements being shared by diverse groups of species. This indicates populational or network evolution. All species stem from an initial populational evolution. *Allenopithecus nigroviridis* has remained the most primitive in chromosome terms..." (Dutrillaux 1988). In the opinion of Strasser and Delson (1987), *Allenopithecus* possesses most of the ancestral characteristics of the Cercopithecini. *Allenopithecus* (2n = 48) possesses the lowest chromosome number of the tribe. Together with *Miopithecus* and *Erythrocebus* (2n = 54), *Allenopithecus* comes closest to the chromosome number of the Papionini (2n = 42). Strasser and Delson (1987) regard these three genera as primitive and believe that they separated from the other Cercopithecini early on. Morphological characteristics (Strasser and Delson 1987) place *Miopithecus* between *Erythrocebus* and *Allenopithecus,* and Martin and McLarnon (1988) created a group of the largest and the smallest forms (*Erythrocebus* with *Miopithecus*). In this case, morphological as well as chromosomal characteristics led to the similar assessment of *Allenopithecus* and *Miopithecus* as basal groups of the Cercopithecini.

Within the genus *Cercopithecus*, *C. aethiops* is one of the most variable and generalized species. Due to its accelerated maturation, frequent pregnancies, semi-terrestrial way of life, and high flexibility in group size and territory and food, it possesses the ability to colonize especially quickly, as its wide distribution shows. "If lack of specialization in habitat and dietary requirements gives *C. aethiops* greater ecological maneuverability than their congenerics, their demographic characteristics of rapid maturation, intense allomothering, and generally high fecundity, give them the capacity to adjust group and populational size rapidly" (Fedigan and Fedigan 1988). This guenon might be a good model for an ancestral form, if it were equipped not only with high adaptability but also with a high potency of diversification. However, the ancestral karyotype of the Papionini differs in two translocations, which do not occur in the Cercopithecini (Dutrillaux 1988); Dutrillaux *et al.* 1988). It is therefore unlikely that a baboon or macaque could represent the original form for both tribes. Among the present-day representatives, no species has thus far been detected that could be considered close to the ancestral form at the root of the basic type Cercopithecinae.

## 16.5. Conclusion

Establishment of the basic type "Cercopithecinae" agrees in content with van Gelder's (1977) suggestion to unite all genera into one. However, van Gelder's nomenclature would have the disadvantage of reducing the entire range of genera, from the large, terrestrial baboon (*Mandrillus sphinx*) to the small, arboreal talapoin (*Miopithecus talapoin*), into the one genus *Cercopithecus*. This would mean a considerable loss of information regarding the morphological differences of the forms, which are reflected by division into the various genera. Furthermore, the nomenclatural confusion would be great. Thus, van Gelder's suggestion would probably not be generally accepted. The category of the basic type is therefore placed on a higher level than the genus in order to keep the information content of meaningful lower taxa, such as the tribe and the genus (Scherer 1993, this volume).

Because of the unsolved questions regarding their speciation and fine taxonomy, the cercopithecine monkeys are an extremely attractive group for the study of basic type diversification. The diversity in this basic type provides interesting options for further research: How can different proportions (Shea 1992) be the result of insignificant changes of growth parameters? Which decree of flexibility do the species of this basic type possess, as far as both their morphology (see Figure 2) and their behavior are concerned (Nagel 1973)? And, how fast can the niching into different habitats take place (Lernould 1988, Nagel 1973)?

# References

Brown, C.J., V.G. Dunbar, and D.A. Shafer. 1986. A Comparison of the karyotypes of six species of the genus *Macaca* and a species of the genus *Cercocebus*. *Folia Primatologica* 46:164-172.

Buettner-Janusch, J. 1966. A problem in evolutionary systematics: Nomenclature and classification of baboons, genus *Papio*. *Folia Primatologica* 4:288-308.

Chiarelli, B. 1963. Observations on P.T.C.-tasting and on hybridization in primates. In: Napier, J. and N.A. Barnicot, eds. *Symposium. Zoological Society of London, Volume 10: The Primates*, pp. 89-101.

Dandelot, P. 1971. Suborder Anthropoidea. In: Meester, J. and H.W. Setzer, eds. *The Mammals of Africa: An Identification Manual*, Smithsonian Institution Press, Washington, D.C., pp. 5-45.

Dunbar, R.I.M. and P. Dunbar. 1974. On hybridization between *Theropithecus gelada* and *Papio anubis* in the wild. *Journal of Human Evolution* 3:187-192.

Dutrillaux, B. 1988. Chromosome evolution in primates. *Folia Primatologica* 50:134-135.

Dutrillaux, B., M. Muleris, and J. Coutourier. 1988. Chromosomal evolution of Cercopithecinae. In: Gautier-Hion *et. al.* 1988, pp.150-159.

Fedigan, L. and L.M. Fedigan. 1988. *Cercopithecus aethiops*: A review of field studies. In: Gautier-Hion A, *et. al.* 1988, pp. 389-411.

Fleagle, J.G. 1988. *Primate Adaptation and Evolution*. Academic Press, San Diego.

Fooden, J. 1964. Rhesus and crab-eating macaques: Intergradation in Thailand. *Science* 143:363-365.

Gartlan, J.S. and T.T. Struhsaker. 1972. Polyspecific associations and niche separation of forest anthropoids in Cameroon, West Africa. *Journal of Zoology* 168:221-226.

Gautier-Hion, A. 1988. Polyspecific associations among forest guenons. In: Gautier-Hion *et. al.* 1988, pp. 452-476.

Gautier-Hion, A, F. Bourlière, J.P. Gautier, and J. Kingdon, eds. 1988. *A Primate Radiation: Evolutionary Biology of the African Guenons*, Cambridge University Press, New York.

Godfrey, L. and J. Marks. 1991. The nature and origin of primate species. *American Journal of Physical Anthropology* 34(Supplement 13):34-39.

Gould, S.J. 1977. *Ontogeny and Phylogeny*. Harvard University Press, Cambridge.

Gray, A.P. 1972. *Mammalian Hybrids: A Check-List with Bibliography*. Commonwealth Agricultural Bureaux, Farnham Royal, England.

Hartwig-Scherer, S. 1993. *Allometry in Hominoids: A Comparative Study of Skeletal Growth Trends*. Ph.D. Dissertation, University of Zurich.

Hill, W.C.O. 1966. *Primates: Comparative Anatomy and Taxonomy. VI, Catarrhini: Cercopithecoidea: Cercopithecinae*. Edinburgh University Press, Edinburgh.

Kingdon, J. 1971. *East African Mammals*. Volume 1. Academic Press,

London.

Kummer, H. 1968. Social organization of hamadryas baboons. *Bibliotheca Primatologica*, 6:1-189.

Lernould, J.-M. 1988. Classification and geographical distribution of guenons: A review. In: Gautier-Hion et. al. 1988, pp. 55-78.

Maples, W.R. 1972. Systematic reconsideration and a revision of the nomenclature of Kenya baboons. *American Journal of Physical Anthropology* 36:9-20.

Markarjan, D.S., E.P. Isakov, and G.I. Kondakov. 1974. Intergeneric hybrids of the lower (42-chromosome) monkey species of the Sukhumi Monkey Colony. *Journal of Human Evolution* 3:247-255.

Martin, R.D. and A.M. McLarnon. 1988. Quantitative comparisons of the skull and teeth in guenons. In: Gautier-Hion et. al. 1988, pp. 160-183.

Matsubayashi, K., M. Hirai, T. Watanabe, Y. Ohkura, and K. Nozawa. 1978. A case of patas-vervet hybrid in captivity. *Primates* 19:785-793.

Nagel, U. 1973. A comparison of anubis baboons, hamadryas baboons and their hybrids at a species border in Ethiopia. *Folia Primatologica* 19:104-165.

Napier, J.R. and P.H. Napier. 1967. *A Handbook of Living Primates.* Academic Press, London.

Napier, P.H. 1981. *Catalogue of Primates in the British Museum (Natural History) and Elsewhere in the British Isles, Part II: Family Cercopithecidae, Sub-family Cercopithecinae.* British Museum, London.

Napier, P.H. 1985. *Catalogue of Primates in the British Museum (Natural History) and Elsewhere in the British Isles, Part III: Family Cercopithecidae, Sub-family Colobinae.* British Museum, London.

Phillips-Conroy, J.E., C.J. Jolly, P. Nystrom, and H.A. Hemmalin. 1992. Migration of male hamadryas baboons into anubis groups in the Awash National Park, Ethiopia. *International Journal of Primatology* 13:455-476.

Samuels, A. and J. Altmann. 1986. Immigration of a *Papio anubis* male into a group of *Papio cynocephalus* baboons and evidence for an *anubis-cynocephalus* hybrid zone in Amboseli, Kenya. *International Journal of Primatology* 7:131.

Scherer, S. 1993. Basic Types of Life. In: Scherer, S. ed. *Typen des Lebens*, Pascal, Berlin, pp. 11-30

Shea, B.T. 1992. Ontogenetic scaling of skeletal proportions in the talapoin monkey. *Journal of Human Evolution* 23:283-307.

Simpson, G.G. 1961. *Principles of Animal Taxonomy.* Columbia University Press, New York.

Southwick, C.H. and K.L. Southwick. 1983. Polyspecific groups of macaques on the Kowloon Peninsula, New Territories, Hong Kong. *American Journal of Primatology* 5:17-24.

Strasser, E. and E. Delson. 1987. Cladistic analysis of cercopithecid relationships. *Journal of Human Evolution* 16:81-99.

Struhsaker, T., T.M. Butynski, and J.S. Lwanga. 1988. Hybridization between redtail (*Cercopithecus ascanius schmidti*) and blue (*C. mitis stuhlmanni*) monkeys in the Kibale Forest, Uganda. In: Gautier-Hion et. al. 1988, pp. 477-497.

Sugawara, K. 1979. Sociobiology study of a wild group of hybrid baboons between *Papio anubis* and *Papio hamadryas* in the Awash Valley, Ethiopia. *Primates* 20:21-56.

Thorington, R.W. and C.P. Groves. 1970. An annotated classification of the Cercopithecoidea. In: Napier, J.R. and P.H. Napier, eds. *Old World Monkeys: Evolution, Systematics, and Behavior*, Academic Press, New York, pp. 629-639.

van Gelder, R.G.. 1977. Mammalian hybrids and generic limits. *American Museum Novitates* 2635:1-25.

# Appendix 1. Update To German Edition Concerning Chapters On Aves (Birds)

More recent hybridization data relevant to the chapters featured in this book are as follows.

**A1.1. Upland Game Birds**

Klemm ascribes the Guinea fowl to the status of subfamily, whilst current avian checklists elevate them to the Numididae family level (Dickinson and Remsen 2013; Clements, 2023). Klemm documents the crosses between Guinea fowl and genera within Phasianidae (i.e. Gallus, Pavo), so the promotion of guinea fowl to family rank results in hybridization uniting the Numididae and Phasianidae.

There is also evidence, albeit slender, of hybridization uniting members of Cracidae (guans and curassows) with members of Phasianidae and Numididae. Dean Amadol, an authority on curassows and former curator of birds at the American Museum of Natural History, observed both the parents and hybrid offspring of the following crosses:

*Pipile jaeutinga* (Black-fronted piping-guan) (Cracidae) x *Gallus gallus* (Chicken) (Phasianidae);

*Crax blumenbachii* (Red-billed Curassow) (Cracidae) x *Gallus gallus* (Chicken) (Phasianidae);

*Penelope superciliaris* (Rusty-margined Guan) x *Numida meleagris* (Helmeted Guinea fowl).

In the *Pipile* x *Gallus* cross, the author commented, "The hybrid was very slow and deliberate in its movements like a guan, but its calls were more like those of a chicken."

All of these taxa share numerous features including short, rounded wings, heavy build, a short thick bill with its overhanging upper mandible, stout legs with a short, high-set hallux and commonly one to three spurs on the posterior aspect of the tarsus. Typical calls include clucking and cackling (Martinez 1992). Therefore, it is possible that these families together comprise one basic type, although further evidence is needed for the links with Cracidae. Klemm states that no hybridization has been reported within the megapodes (brush-turkeys). Hybridization data cited by McCarthy (2006) suggests that nine species hybridize out of the 22 species of brush-turkeys (41 %), although these are limited to inter-specific crosses, which do not link the 7 genera together. However, hybrids between megapodes and other game birds still remain unknown. Although similar in morphology to other game birds, the megapodes are distinctive in laying eggs in a mound of earth/vegetation, which may lead to reproductive isolation from other game birds. Alternatively, a lack of hybridization may result from developmental peculiarities.

### A1.2. Diurnal Birds of Prey

In the account of the Accipitridae (hawks and eagles, kites, OW vultures), Zimbelmann (Chapter 12) no hybrids among the eagles were known, but more recent records link up to 6 species of eagles together. For instance, *Clanga clanga* hybridizes with *Clanga pomarina* (Greater Spotted eagle x Lesser Spotted Eagle) (Maciorowski et al. 2015). However, there remains a lack of documentary evidence of hybrids between eagles and other accipitrids. Taken together, this data gives further confirmation of Zimmermann's proposed basic type status to the true eagles. There is also a field report of hybridisation between *Milvus migrans* and *Buteo buteo* (Black kite x Common buzzard) (Corso and Gildi 1998). This report makes an extensive morphological analysis of the observations, and provides photographic evidence of the hybrids and detailed reasoning for excluding other species parentage. The implication of this report is that Milvus could be added to the hawks and buzzards basic type assemblage.

### A1.3. Anatidae

Analysis of more recent hybridization data (Tyler Chapter 14) yet further strengthens the designation of the Anatidae as a basic type. Scherer (Chapter 10) reported over 400 interspecific hybrids: McCarthy (2006) cited up to 532 such hybrids, yet there are still no known hybrids between the Anatidae and members of any neighbouring taxa.

### A1.4. Finch-like Assemblage

Fehrer (Chapter 13) concludes that two basic types of finches can be recognised, each united by a wealth of hybridization data: the Cardueline and Estrildid finches. Within the Fringillidae, Fehrer states that there is also some evidence of hybridization between the Fringillinae and Carduelinae sub-families, such as via the cross *Carduelis chloris* (greenfinch) x *Fringilla coelebs* (chaffinch). This is further validated by contemporary breeders (e.g. Staffordshire British Bird Mule & Hybrid Club, at http://www.birdinfo.co.uk/sites/Mules_Hybrids/greenfinch_crosses.html), who provide photographic evidence and are familiar with this cross. There is also a greenfinch x chaffinch hybrid specimen located in the Te Papa Tongarewa museum of New Zealand (https://collections. tepapa.govt.nz/object/634755). Photographic evidence can also be found in Carr 1980.

The systematics of finch-like birds has been in a state of flux for over a century. For instance, the number of species within Fringillidae has varied from 137 to 993 (Payevsky 2015). Membership within the

Thraupidae (tanagers), typified by their magnificently colourful plumage, is also in a state of great flux. The bill is typically stout and finch-like but similar bill morphology is found in many emberizine finches. As a result of DNA-DNA hybridization and DNA-sequence data, numerous genera until recently located within Emberizidae, Cardinalidae and Coerebidae have been allied to tanagers, whilst strangely some classic tanagers have been removed from the so-called tanager family and placed elsewhere (reviewed by Hilty 2011).

McCarthy (2006) and Gray (1958) cite over 70 records of hybridization data which potentially could link up to nine finch-like families into a super-family assemblage: the Cardinalidae, Emberizidae, Estrildidae, Fringillidae, Icteridae, Passeridae, Thraupidae, Passerellidae and Ploceidae. These consist of zoologists' and breeders' reports in antiquity, with documentary evidence to substantiate the crosses being limited. Relatively more recently are the reports and observations by Helmut Sick (Sick et al 1993), distinguished ornithologist and former Curator of Birds at the National Museum in Rio de Janeiro (Vuilleumier 1998). Sick records the following crosses, which would unite the Fringillidae, Icteridae and Thraupidae[1]:

*Chrysomus ruficapillus* [ICT] x *Sicalis flaveola* [THR];
*Paroaria coronata* [THR] x *Molothrus bonariensis* [ICT];
*Volatinia jacarina* [THR] x *Serinus domesticus* (canary) [FRI];
*Coryphospingus pileatus* [THR] x *Serinus donesticus* [FRI].

However, even this evidence needs further confirmation, which is lacking in contemporary breeders' records. This could be because the cross is indeed not possible. Alternatively, there may be no record of the crosses being attempted, or that the crosses are challenging to breed, requiring the greatest of aviculturist skill and experience (Carr 1980).

## References

Carr, V.A.V., 1980. Cage bird hybrids. Saiga publications, Surrey, England.

Clements, J. F., Rasmussen, P. C., Iliff, S. M. J., Fredericks, T. A., Gerbracht, J. A., Lepage, D., Spencer, A., Billerman, S. M., Sullivan, B. L. and Wood, C. L., 2023. The eBird/Clements checklist of Birds of the World: v2023. Downloaded from https://www.birds.cornell.edu/clementschecklist/download/

Corso, A., and Gildi, R., 1998. Hybrids between Black Kite and Common Buzzard in Italy 1996. Dutch Birding 20:226-233.

Dickinson, E.C. and Remsen, J.V., eds., 2013. Howard and Moore Complete Checklist of the Birds of the World, 4th edition, Vol. 1. Aves Press, Eastbourne, UK.

Gray, A. 1958. Bird Hybrids. Commonwealth Agricultural Bureaux, Slough, England.

---

1    Abbreviations: FRI, Fringillidae; ICT, Icteridae; THR, Thraupidae.

Hilty, S. 2011. Family Thraupidae (Tanagers). Handbook of the Birds of the World 16:201-217.

Martinez, I., 1992. Numididae. In Del Hoyo J, Del Hoyo J, Elliott A, Sargatal J. eds. Handbook of the Birds of the World 3. Lynx Edicions. Barcelona.

McCarthy, E., 2006. Handbook of avian hybrids of the world. Oxford University Press, England.

Maciorowski, G., Mirski, P. and Väli, Ü., 2015. Hybridisation dynamics between the Greater Spotted Eagles Aquila clanga and Lesser Spotted Eagles Aquila pomarina in the Biebrza River Valley (NE Poland). Acta ornithologica, 50(1), pp.33-41.

Payevsky, V.A., 2015. Taxonomy of true finches (Fringillidae, Passeriformes): A review of problems. Biology Bulletin, 42, pp.713-723.

Sick, H., Barruel, P. and O'Neill, J.P., 1993. Birds in Brazil: a natural history (pp. xviii+-703). Princeton: Princeton University Press.

Vuilleumier, F., 1998. In Memoriam: Helmut Sick, 1910-1991. The Auk 115(2):470-472.

# Appendix 2: List Of Further Avian Families In Which At Least 20% Of The Taxon Hybridize Between Species

This list is in addition to groups already described in Sections 2 and 3 of Tyler: Chapter 14. Within each of the taxa below, inter-specific hybridization links a number of species together into "core groups". With data from supplementary criteria or further hybridization data, these may reveal further BT candidates in the future.

The list also compares number of species hybridizing (3rd column) with numbers of species hybridizing if species merged primarily due to hybridization are re-split (4th column). This can be relevant to discerning prospective BTs which might otherwise be overlooked. If merging primarily due to hybridization is taken into account, several families (which formerly had under 20 % hybridization) now have well over 20 % inter-specific hybridization throughout the family (e.g Neosittidae, Pardalotidae).

Data analysed from hybridization citations in McCarthy, E.M., 2006. Handbook of avian hybrids of the world. Oxford University Press, England.

| Taxon | Common name | No. of spp. hybridizing [% of total no. of spp] | No. of spp. hybridizing, if spp. merged are split | No. of genera linked by hybridization [total no. of genera] |
|---|---|---|---|---|
| Acanthizidae | Thornbills | 8 [13 %] | 18 [26 %] | 0 [14] |
| Alcidae | Auks, Puffins | 5 [21 %] | 11 [38 %] | 2 [11] |
| Charadriidae | Plovers | 13 [20 %] | 16 [24 %] | 2 [10] |
| Cinclosomatidae | Quail-thrushes | 3 [60 %] | 3 [60 %] | SG |
| Coraciidae | Rollers | 4 [30 %] | 9 [56 %] | 0 [2] |
| Corvidae | Crows, Ravens | 32 [27 %] | 43 [34 %] | 2 [24] |
| Dicruridae | Drongos | 8 [36 %] | 11 [44 %] | 0 [2] |
| Diomedeidae | Albatrosses | 3 [23 %] | 7 [41 %] | 0 [4] |

| Icteridae | NW Blackbirds | 29 [30 %] | 42 [40 %] | 8 [26] |
|---|---|---|---|---|
| Lybiidae | African Barbets | 5 [12 %] | 12 [25 %] | 0 [6] |
| Maluridae | Fairy Wrens | 4 [14 %] | 9 [28 %] | 0 [5] |
| Malaconitidae | Bushshrikes | 6 [12 %] | 15 [26 %] | 0 [10] |
| Mimidae | Mockingbirds | 8 [24 %] | 8 [24 %] | 2 [12] |
| Motacillidae | Pipits, Wagtails | 11 [17 %] | 23 [31 %] | 0 [5] |
| Musophagidae | Turacos | 11 [48 %] | 11 [48 %] | 2 [6] |
| Neosittidae | Sittelas | 0 | 5 [63 %] | SG |
| Odontophoridae | NW Quails | 9 [28%] | 11 [32 %] | 0 [9] |
| Oriolidae | Orioles, Figbirds | 5 [17 %] | 7 [23 %] | 0 [2] |
| Pachycephalidae | Whistlers | 5 [12 %] | 11 [25 %] | 0 [6] |
| Pardalotidae | Pardalotids | 0 | 7 [47 %] | SG |
| Passeridae | OW Sparrows | 11 [28 %] | 15 [34 %] | 0 [11] |
| Phalacrocoraci-dae | Cormorants, Shags | 11 [31 %] | 12 [32 %] | SG |
| Philepittidae | Asities | 2 [50 %] | 2 [50 %] | 0 [2] |
| Phoeniculidae | Wood-hoopoes | 5 [63 %] | 5 [63 %] | 0 [2] |
| Ploceidae | Weavers, Bish-ops | 26 [24 %] | 40 [33 %] | 2 [22] |
| Procellaridae | Fulmars | 17 [23 %] | 21 [28 %] | 0 [14] |
| Promeropidae | Sugarbirds | 2 [100 %] | 2 [100 %] | SG |
| Pycnonotidae | Bulbuls | 19 [16 %] | 28 [23 %] | 0 [22] |
| Scolopacidae | Sandpipers | 25 [27 %] | 25 [27 %] | 3 [23] |
| Sittidae | Nuthatches | 4 [16 %] | 8 [28 %] | 0 [2] |
| Thraupidae | Tanagers | 38 [19 %] | 42 [20 %] | 6 [50] |
| Turdidae | Thrushes | 38 [23 %] | 40 [24 %] | 0 [24] |
| Vireonidae | Vireos, Greenlets | 9 [17 %] | 11 [21 %] | 0 [4] |

Abbreviations: SG = single genus taxon; spp = species; NW, New World; OW, Old World

# Appendix 3. Prospective Further Basic Types Of Plants And Animals And Update To German Edition Concerning Chapters On Plants

The following Table provides preliminary data, compiled by Herfried Kutzelnigg, to indicate prospective basic type (BT) status for a further 18 plant taxa and 20 animal taxa, in addition to the BTs featured in this book. The Chapters on plants featured in the book were mainly revised by the authors. However, Kutzelnigg also includes an update summary on these Chapters (indicated with an asterisk).

Abbreviations: SG = single genus taxon; spp = species; NW, New World; OW, Old World

**Plants**

| Name of Basic Type | Genera + spp. | I-gs | Ref. | Notes |
|---|---|---|---|---|
| BRYOPHYTA [Mosses] | | | | Since the diploid phase in mosses is limited to the sporophyte, hybrids cannot be easily identified. At least, intergeneric hybrids from 7 moss families have been described. |
| Fam. Funariaceae | 15: 200-350 | 4 | 1 | Probable BT, but further data needed to confirm this. |

| PTERIDOPHYTA [Ferns] | | | | |
|---|---|---|---|---|
| Fam. Aspleniaceae* [Spleenworts] | 1-9: 670 | 0-7 | 13 | Number of genera varies, depending on the author, between 1 and 9. This changes no. of i-g hybrids, but not BT situation. There are numerous hybrids within the genus Asplenium, e.g. 32 in Central Europe alone. |
| Fam. Cystopteridaceae [Bladder ferns] | 3:32 | 1 | 12 | More recent taxonomy recognises only 3 genera, of which the two larger ones have formed a natural hybrid despite a separation of 60 million years. There are several interspecific hybrids in Cystopteris. |
| SPERMATOPHYTA [flowering plants] | | | | |
| Fam. Acoraceae [Sweet flag family] | 1: 2-4 | – | 12 | The family is isolated and contains only the one genus Acorus. The order Acorales includes only this family. |
| Aizoaceae: Ruschioideae: Tribe Ruschieae* [Living stones] | 101: 1563 | 37 | 16 | Family with 4 sub-families. Sub-family Ruschioideae is considered monophyletic. The 3 other tribes are not clearly separated from Ruschieae, so the BT could be at the sub-family, rather than tribe level. |

| Fam. Amaryllidaceae Sub-fam Amaryllidoideae [Amaryllis] | 65: 900 | 18 | 12 | Family with 3 sub-families. The Amarylloideae are a clearly defined group. 6 of the 13 tribes are connected by hybridization. |
|---|---|---|---|---|
| Fam. Amborellaceae [Amborella] | 1:.1 | | 12 | Very isolated family with only 1 species in its own order. Considered a basal group of the Angiospermae. |
| Apocynaceae: Ceropegieae: Sub-tribe Stapeliinae [Stapelia etc]] | 30: 330 | 33 | 12 | The BT possibly includes the entire tribe Ceropegieae (40: 440). Further study needed |
| Asphodelaceae: Asphodeloideae Tribe Aloeae [Aloes] | 5-8: 550 | 24 | 12 | Clearly defined group All genera (at least according to their former boundaries) are directly or indirectly connected through crosses. |
| Brassicaceae: Tribe Brassiceae [Brassicas] | 46-50: 230 | 33 | 20 | It remains to be examined whether other tribes or perhaps even entire family belong to the BT. |
| Bromeliaceae: Sub-fam. Bromelioideae [Bromeliads] | 29: 425-750 | 37 | 18 | |
| Bromeliaceae: Sub-fam. Pitcairnioideae | 16: 500 | 7 | 18 | The family is now divided into 5 subfamilies, of which the Pitcairnioideae are interconnected with the Hechtioideae and Puyoideae. Whether the remaining genera belong to the same BT remains to be examined. |

| | | | | |
|---|---|---|---|---|
| Bromeliaceae: Sub-fam. Tillandsioideae | 6: 800 | 6 | 18 | |
| Cactaceae: Sub-fam. Cactoideae [Cacti] | 92: 1250 | 110 | 12 | 8 of the 10 tribes are interconnected. There are numerous trigeneric hybrids. The family includes 2 subfamilies. Opuntioideae are likely to represent a separate BT. |
| Crassulaceae: Sub-fam. Sempervivoideae [Houseleeks] | 28: 975 | 31 | 12 | |
| Fam. Dipsacaceae [Teasel] | 7: 160-290 | 1 | 12 | Clearly defined group. |
| Fam. Grossulariaceae [Gooseberry] | 1-2: 150 | 0-1 | 12 | Clearly defined group. Approx. 23 known interspecific crosses. The genus Grossularia, now included in Ribes, crosses with Ribes. |
| Orchidaceae [Orchids]: Orchidoideae Tribe Cymbidieae | 150: 3800 | 286 | 12 | All 11 subtribes are connected by a total of 50 crosses: the most species-rich BT. |
| Orchidaceae: Orchidoideae Tribe Vandeae | 207: 2340 | 241 | 12 | 4 of the 5 subtribes are connected by a total of 27 crosses. |
| Orchidaceae: Orchidoideae Tribe Orchideae | 64: 2185 | 52 | 12 | All representatives of the sub-tribe Orchidinae (54: 1865) belong to a BT. Too little crossbreeding data is known for the other 4 subtribes (together only 10: 320). |

| | | | | |
|---|---|---|---|---|
| Poaceae: Pooideae Tribe Triticeae* [Wheat, barley, rye] | 18-36: 330 | 68-103 | 6 | The number of intergeneric hybrids depends on how broadly the genera are defined. The BT possibly also includes the Bromeae and thus the supertribe Tricodae (40: 500). |
| Rosaceae: Sub-fam. Maloideae* [Pome fruit plants e.g. apples, pears] | 24: 950 | 28 | 12 | According to more recent systematics, the Maloideae are classified as subtribe Malinae within the tribe Maleae of the now broad subfamily Amygdaloideae. |
| Rosaceae: Rosoideae Tribe Geeae* [Avens etc] | 3-9: 42 | 1-6 | 7 | The Geeae are now considered expanded by Fallugia - as part of the monophyletic Tribe Colurieae. Thus Fallugia also belongs to this BT. |
| Fam. Salicaceae (excl. Flacourtiaceae) [Willow, poplar] | 2-3: 530 | 1-2 | 12 | In the genus Salix (willow), large numbers of hybrids have been known for a long time. There are also hybrids with the genus Chosenia. New are hybrids with the neighboring genus Populus (poplar), which could be obtained through embryo culture. |

## Animals

| Name of Basic Type | Genera and Species | i-gs | Refs | Notes |
|---|---|---|---|---|
| MAMMALIA [mammals] | | | | |
| Bovidae: Sub- fam. Caprinae [Goats, sheep] | 9-12: 35 | 5 | 11 | |

| Fam. Camelidae [Camels, llama] | 2-3: 6 | 3 | 8 | All 2 or 3 genera are linked, including Camelus and Lama. |
|---|---|---|---|---|
| Fam. Canidae [Dogs, foxes, wolves] | 15: 33 | 4 | 4 | Recent crosses and the results of molecular studies that crosses connect widely separated subgroups confirm the basic type. |
| Fam. Felidae [Cats, lion, tiger] | 10-15: 38 | 11 | 5 | The two tribes of big cats and small cats are linked through crossbreeding. |
| Fam. Ursidae [Bears] | 5: 8 | 5 | 17 | 2 of the 3 tribes are interconnected. Only the giant panda Ailuropoda has not yet been cross-bred. |
| Fam. Balaenopteridae [Rorquals and humpback whale] | 2: 9 | 1 | 12 | Hybridization interconnects humpback and blue whales.Often classified as belonging to different tribes. |
| Fam. Phocoenidae [Porpoises] | 3: 6 | 1 | 12 | 2 of the 3 genera are linked by hybridization |
| Fam. Delphinidae [Dolphins] | 16: 40 | 8 | 12 | Species that are assigned to different subfamilies are also linked by hybridization |
| Fam. Hominidae [Humans] | 1: 1 | – | 9 | |
| Fam. Hylobatidae [Gibbons] | 4: 15 | 3 | 12 | All 4 genera are interconnected by hybridization |
| Fam. Elephantidae [Elephants] | 2: 3 | 1 | 10 | Hybridization interconnects both African and Indian elephant genera |

| | | | | |
|---|---|---|---|---|
| Orycteropodidae [Aardvark] | 1: 1 | – | 11 | The family is anatomically clearly defined. The aardvark is the only extant species of the order. |
| PISCES [fish] Because of external fertilization, problems that might otherwise prevent spontaneous hybridization are eliminated. The number of hybrids observed so far is correspondingly high. For example, in the carp fish there are 276 intergeneric hybrids, approaching numbers in the orchid group, Cymbidieae (278). | | | | |
| Cyprinidae [Carp] | 220: 2400 | 276 | 12 | All 12 subfamilies are directly or indirectly linked by crossing. |
| Gasterosteidae [Sticklebacks] | 5: 16 | 3 | 12 | |
| Lepisosteidae [Gars] | 2: 7 | 1 | 12 | |
| Centrarchidae [Sunfish] | 8: 27 | 17 | 12 | See also Cavanaugh & Sternberg 2004. |
| Cichlidae [Cichlids] | 220: 1000 | 22 | 12 | Strong radiation in the recent past. See Scherer 1998 and Junker & Scherer 2014. |
| Petromyzontidae [Lampreys] | 8: 39 | 4 | 12 | Very isolated order |
| Salmonidae [Salmon, trout] | 11: 190 | 16 | 12 | Only family of the order. The three subfamilies are interconnected |
| REPTILIA [reptiles] | | | | |
| Cheloniidae [Sea turtles] | 5: 7 | 4 | 12 | Plus 5 extinct genera |
| Geoemydidae [Old World pond turtles] | 25: 70 | 16 | 12 | The two subfamilies are linked by crossing through several hybrids. See Brophy et al. 2006. |

# References

Adler M (1993) Merkmalsausbildung und Hybridisierung bei Funariaceen (Bryophyta, Musci). p. 67-70 in Scherer S (Hrsg.) Typen des Lebens. Berlin: Pascal.

Brophy TR, Frair W & Clark D (2006) A review of interspecific hybridization in the order Testudines. p. 17 in Sanders R (ed.) Proc. Fifth Int. Conf. BSG Group. Occas. Papers BSG Study Group.

Cavanaugh DP & Sternberg RV (2004) Analysis of morphological groupings using ANOPA, a pattern recognition and multivariate statistical method: a case study involving centrarchid fishes. J. Biological Systems 12, 137-67.

Crompton N (1993) A review of selected features of the family Canidae with reference to its fundamental taxonomic status. p. 217-224 in Scherer S (Hrsg.) Typen des Lebens: Berlin: Pascal.

Crompton N & Winkler N (2006) Die Katzenartigen – ein klar abgegrenzter Grundtyp. Stud. Integrale J. 13, 68-72.

Junker R (1993a) Der Grundtyp der Weizenartigen (Poaceae, Tribus Triticeae). p. 75-93 in Scherer S (Hrsg.) Typen des Lebens. Berlin: Pascal.

Junker R (1993b) Die Gattungen Geum (Nelkenwurz), Coluria und Waldsteinia (Rosaceae, Tribus Geeae). p. 95-111 in Scherer S (Hrsg.) Typen des Lebens. Berlin: Pascal.

Junker R (2000) Mischling aus Dromedar und Lama – die Cameliden: ein Grundtyp. Stud. Integrale J. 7 (1), 41-42.

Junker R & Scherer S (Hrsg.) (2014) Evolution. Ein kritisches Lehrbuch. 7. Aufl. Gießen: Weyel.

Klöckner P (2004) Elefantenevolution in Bewegung. Morphologie und Moleküle einmal mehr im Konflikt. Stud. Integrale J. 11 (1), 36-38

Klöckner Unpublished.

Kutzelnigg Unpublished.

Kutzelnigg H (1993a) Verwandtschaftliche Beziehungen zwischen den Gattungen und Arten der Kernobstgewächse (Rosaceae, Unterfamilie Maloideae). p. 113-127 in Scherer S (Hrsg.) Typen des Lebens. Berlin: Pascal.

Kutzelnigg H (1993b) Die Streifenfarngewächse (Filicatae, Aspleniaceae) im Grundtypmodell. p. 71-74 in Scherer S (Hrsg.) Typen des Lebens. Berlin: Pascal.

Kutzelnigg H (2003) Neues zur Systematik und Evolution der Kernobstgewächse (Rosaceae: Maloideae). Stud. Integrale J. 10 ()12), 32-34.

Kutzelnigg H (2010) Die „Lebenden Steine" und ihre Verwandten (Aizoaceae: Ruschieae). Ein neuer Grundtyp und ein weiteres Beispiel für besonders schnelle Evolution. Stud. Integrale J. 16 (2), 100-104.

Mohr unpublished.

Neuhaus K (1995) Die Familie der Bromeliaceen – ein oder mehrere Grundtypen? Stud. Int. J. 2, 15-19.

Scherer S (1998) Abnehmender Sexappeal von männlichen Buntbarschen durch Umweltverschmutzung des Viktoriasees. Stud. Integrale J. 5, 85-86.

Sickinger & Kutzelnigg Unpublished.